Wildlife Conservation in a Changing Climate

Wildlife

Conservation

in a

Changing

Climate EDITED BY

JEDEDIAH F. BRODIE,

ERIC POST,

AND

DANIEL F. DOAK

The
University
of Chicago
Press
Chicago
and
London

Jedediah F. Brodie is assistant professor of conservation ecology at the University of British Columbia. Eric Post is professor of biology at the Pennsylvania State University. Daniel F. Doak is professor in the Environmental Studies Program at the University of Colorado Boulder.

The University of Chicago Press, Chicago 60637
The University of Chicago Press, Ltd., London
© 2013 by The University of Chicago
All rights reserved. Published 2013.
Printed in the United States of America

22 21 20 19 18 17 16 15 14   2 3 4 5

ISBN-13: 978-0-226-07462-7 (cloth)
ISBN-13: 978-0-226-07463-4 (paper)
ISBN-13: 978-0-226-07464-1 (electronic)
ISBN-10: 0-226-07462-5 (cloth)
ISBN-10: 0-226-07463-3 (paper)
ISBN-10: 0-226-07464-1 (electronic)

Library of Congress Cataloging-in-Publication Data

Wildlife conservation in a changing climate / edited by Jedediah F. Brodie, Eric Post, and Daniel F. Doak.
    pages ; cm
    Includes bibliographical references and index.
    ISBN 978-0-226-07462-7 (cloth : alkaline paper) — ISBN 0-226-07462-5 (cloth : alkaline paper) — ISBN 978-0-226-07463-4 (paperback : alkaline paper) — ISBN 0-226-07463-3 (paperback : alkaline paper) — ISBN 978-0-226-07464-1 (e-book) — ISBN 0-226-07464-1 (e-book)
    1. Wildlife conservation.   2. Climate changes — Environmental aspects.   3. Bioclimatically.
    4. Wildlife management.   I. Brodie, Jedediah F. (Jedediah Farrell)   II. Post, Eric S. (Eric Stephen)   III. Doak, Daniel F., 1961–
    QL82.WS53 2012
    333.95'4—dc23
                                    2012017787

♾ This paper meets the requirements of ANSI/NISO Z39.48-1992 (Permanence of Paper).

Contents

Color plates follow page 152.

Climate Change and Wildlife Conservation

Jedediah F. Brodie, Eric Post,
and Daniel F. Doak

Climate change is one of the paramount conservation issues of our time, with the potential to devastate biodiversity as we know it as well as disrupt human societies. While earth's climate has altered dramatically in the past, at least two factors make today's changes different. First is the unprecedented rate at which these changes are occurring. Even the incredibly rapid warming at the Paleocene-Eocene boundary (56 million years ago) is estimated to have been an order of magnitude slower than the global heating that is going on right now and which may continue, and even accelerate, over at least the next century. Second is the fact that climatic changes today are operating on ecosystems that are already heavily impacted by humans. Biodiversity increased in response to Paleocene-Eocene warming (Jaramillo et al. 2010) in part because organisms were able to undergo huge and unimpeded shifts in their distribution in response to changing physical conditions. Such shifts will be impossible in most areas today due to habitat loss and fragmentation on continental scales (Brodie et al. 2012a). Moreover, the resilience of many ecosystems to climate change may be weakened by the host of other anthropogenic threats that these systems face, such as overexploitation, invasive species, and pollution (Hansen et al. 2010). Indeed, the average risk of extinction by 2100, across taxa and regions, was recently estimated at 11% (Maclean and Wilson 2011).

Climate change has captured the attention of the public and scientists. Figure 1.1 shows the rapid rise of climate change literature published over the last three decades. Yet our understanding of how climate change affects biodiversity remains quite superficial (McMahon et al. 2011). For example, we have abundant evidence showing that climate change can lead to shifts in the timing of life-history events (e.g., earlier migration, advanced breeding dates, etc.), but relatively little knowledge about whether or how such phenological changes actually affect population dynamics or extinction risk (Miller-Rushing et al. 2010). A huge body of literature on the ecological impacts of climate change has focused on shifting species distributions using various forms of bioclimatic envelope models, which

1

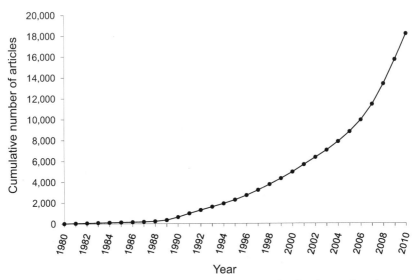

Figure 1.1. Climate change research from 1980 to 2011, showing the number of articles returned from a search using the terms TI = (climate change* OR global warming) in the Science Citation Index of ISI Web of Science.[SM]

assume that species ranges are governed solely or primarily by climate rather than by ecological interactions or historical factors. While there is evidence that many species are already shifting (or at least expanding) their ranges upward in latitude or elevation, there is also striking variation between species in these responses, calling into question how robust the predictions of bioclimatic models can actually be (e.g., Root et al. 2003, Parmesan 2006, Tingley and Beissinger 2009). Furthermore, most bioclimatic envelope approaches do not have evidence of actual shifts, but rather use current distributional data to infer future range changes. This strategy relies on numerous assumptions, many of which are likely to be biologically untenable. As such, these models may afford us little in the way of accurate predictions of net climate change impacts on natural communities (Schmitz et al. 2003, McMahon et al. 2011).

The purpose of this book is to delve deeper into the impact of climate change on populations and communities. We focus on terrestrial vertebrates across a range of ecosystems. While these species are clearly only a small subset of the world's biodiversity, they are the foci of many current and anticipated conservation plans, and also represent a suitably limited range of problems and considerations for a single book. Chapters herein assess details of climate change ecology such as demographic implications for individual populations, evolutionary responses, impacts on movement patterns, and alterations of species interactions. We also ask a critical question about the ecological impact of climate change: What can we do about it? The contributors to the book present a number of actual and potential actions by which conservationists and managers can counter or ameliorate the on-the-ground impacts of climate change. We focus specifically on the theme of resilience. While there is little that local wildlife managers can do about atmospheric CO_2 or rising global temperatures, they do have the ability to influence the vulnerability of populations to climatic stressors. As mentioned above, many of the familiar conservation threats—exotic species, exploitation, pollution—act in synergy with climate change by reducing the resilience of native wildlife populations. Conversely, certain management actions can increase resilience; these tactics need to be identified, tested, and deployed on a large scale over the coming decades (also see Mawdsley et al. 2009).

Assessing the population-level effects of climate change has received too little attention. It is no longer sufficient to predict that species may shift their range boundaries due to climate change or alter their life histories in ways that may or may not be important. Instead, we need to explore the demographic details of wildlife responses to changing abiotic conditions, and thereby provide a mechanistic understanding of how survival, fecundity, and population dynamics are affected by climate change. Understanding how climate change affects the demography of individual populations affords us a much more accurate, precise, and nuanced understanding of how climate change potentially alters demographic rates and extinction risk (Botkin et al. 2007, Post et al. 2009, Doak and Morris 2010).

Finally, a demographic approach can provide useful solutions for ameliorating the impacts of climate change, generally by addressing concurrent stressors. A demographic focus is particularly useful if there are enough data to understand population dynamics across an entire region or throughout the whole distribution of a species. In these cases we can often identify "hot spots" where biological responses to climate change are particularly intense. One such analysis, of the dynamics of red deer (*Cervus elaphus*) and caribou (*Rangifer tarandus*) across the Northern Hemisphere, showed that different populations responded to temperature anomalies very idiosyncratically (Post et al. 2009). Thus, in contrast to the simplistic expectations of a bioclimatic envelope approach, we believe that there may be no way to predict the magnitude or even the direction of response to warming simply on the basis of a population's latitude or its position within the geographical range of the species.

We also emphasize the importance of understanding species interactions in the context of climate change. Emerging evidence shows that altered species interactions may be more important to community structure than are direct changes in abiotic conditions. For example, the direct ecophysiological effects of declining snowpack on the recruitment of aspen (*Populus tremuloides*) in Yellowstone are outweighed by the fact that reduced snowpack increases the impact of herbivory by elk (*Cervus elaphus*; Brodie et al. 2012b). Likewise, plant community responses to climate change may depend much more on altered species interactions than on changed abiotic conditions (Suttle et al. 2007, Post and Pedersen 2008, Clark et al. 2011).

Despite the vast literature on the ecological impacts of climate change, the field as a whole has only recently begun to focus on the question of what we can do about it, in terms of integrating climate change impact directly into conservation plans (Dawson et al. 2011, Poiani et al. 2011). Strategies to address climate change are often described as falling into two categories: direct mitigation (reducing atmospheric carbon) or adaptation (ameliorating or adjusting to the effects). Some conservationists are opposed to climate change adaptation on the grounds that it may detract from the more important mitigation efforts (e.g., Orr 2009). We agree that our foremost strategy should be mitigation; we clearly need to immediately and drastically reduce humanity's carbon emissions. But we reject the notion that adaptation and mitigation are mutually exclusive. Though conservation resources are indeed finite and often scarce, they are seldom perfectly transferable; funds for adaptation and mitigation will likely come from separate sources. In other words, there is little reason to believe that a dollar not spent on adaptation will be available for mitigation, or vice versa. Climate change adaptation will become ever more critical over the coming decades. The time lag between emissions and atmospheric response ensures that our past discharges have not yet caught up with us; even if we were to stop emitting fossil fuels tomorrow, climatic warming would continue. It is therefore critical that we develop on-the-ground means of alleviating and adapting to the impact of climate change on species and ecosystems.

Many conservationists are also worried that concern about climate change has distracted attention and diverted resources away from more urgent conservation problems, to the detriment of overall biodiversity protection. A NewsFocus article in the prestigious international journal *Science* on efforts to conserve highly endangered giant freshwater fish in the Mekong River system of Southeast Asia reported, "Whether such measures will succeed in the long run will depend largely on community acceptance–and funding. . . . The Mekong Wetlands Biodiversity

Conservation and Sustainable Use Programme kicked in $50,000 for receivers and tags [research equipment] . . . before the program was killed last year and its budget redirected for climate change research" (Stone 2007: 1688). This is but one example of a serious problem, where single-minded approaches to complex conservation issues may indeed create direct conflicts in resource use, to the detriment of overall conservation efforts. Yet we argue that climate change adaptation and biodiversity protection should go hand in hand (also see Hannah 2010, Hansen et al. 2010). Throughout this book, our contributing authors make the point that many of the ways in which we can alleviate climate change impacts are through the same tried-and-true strategies that conservation biologists have used for decades. Habitat connectivity, adjusting harvest levels, controlling invasive species, reintroducing top carnivores: these are all well-known conservation goals, but are also among our best tools for ameliorating the impacts of climate change.

As a positive example of human responses to climate change, witness the growing effort to reduce emissions from deforestation and degradation (REDD). Tropical deforestation and degradation are responsible for an estimated 20% of global carbon emissions to the atmosphere (Myers 2007). In response, REDD protocols are being implemented by the United Nations Framework Convention on Climate Change. The conservation community has been ineffective at halting tropical deforestation to date; REDD could be our last chance. But it also stands to provide a huge gain in terms of wildlife habitat. This is especially true if, as some have proposed, REDD protocols were extended to not just protect trees but also seed-dispersing animals (Brodie and Gibbs 2009) and other strong interactors in forest communities (cf. Soulé et al. 2003, Soulé et al. 2005).

Indeed, it is a recurring theme throughout this book that addressing synergistic anthropogenic threats can facilitate climate change adaptation. In the best case, the threat of climate change will breathe new life into the promotion of specific, effective management methods that we already know work to stabilize or restore threatened populations. Many conservation scientists have noted that our influence with the public declined dramatically in recent years, until climate change grabbed society's attention. We now have a unique opportunity to reinvigorate the massive, landscape-level conservation efforts needed to sustain wildlife populations in today's changing world (Beier and Brost 2010, Hannah 2010).

However, conservation of some populations affected by climate change will require completely new approaches or novel uses of existing tools. Some examples include assisted colonization of species and the explicit incorporation of climate change effects into harvest models. Several of the chapters in this book assess such novel conservation strategies and how or whether they will work as effective responses to climate change. We stress that, as with all conservation strategies, climate change amelioration efforts should be very clear about their goals, and should explicitly account for and report uncertainty (Millar et al. 2007, Prato 2009).

Structure of the Book

This book is organized into three parts. The chapters in part 1 deal with specific issues in the assessment and prediction of climate change impacts on wildlife. Part I opens with an overview by Carney et al. of recent and future climatic changes, with an eye toward those changes that could particularly influence wildlife. Among the important points to emerge from this chapter are a discussion of climate-ecosystem feedback loops, and also the point that the ways in which humans respond to climatic changes may be as important for wildlife conservation as the actual physical changes themselves. This idea has not yet been widely discussed in the scientific literature (Brodie et al. 2012a), but is a hot topic in certain conservation and policy circles.

Next, Austin et al. discuss the potential for wildlife populations to respond to climate change via evolution and phenotypic plasticity. While many chapters later in the book address human impacts on wildlife resilience to climate change, this one looks at the resilience inherent in the populations themselves. The potential for evolutionary or plastic responses to climate change is ignored all too often in coarse-scale assessments of extinction risk in the face of climate change (Sgro et al. 2011). The next two chapters present novel modeling approaches for assessing climate change impacts on populations. Matthews et al. explicitly incorporate climate-induced changes in life-history strategies into structured population models. Matrix models are a powerful tool for conservation assessments (Morris and Doak 2002), and developing our understanding of how to use them in the context of climate change is an important advance. Fordham et al. then explore a critical impact of climate change: the facilitation of invasion by exotic species. They do so by developing a new quantitative framework that uses dynamic, spatially explicit metapopulation models coupling climatic and demographic processes. Where data are available to build such models, they will prove a powerful tool for assessing climate change impacts on biological invasions and wildlife in general.

The final two chapters in part 1 provide ways to predict climate influences on wildlife. Paull and Johnson look at another important climate change impact, enhanced spread of diseases, and discuss a series of promising new tools for predicting these effects, including biophysical modeling, experimental manipulations of coupled direct and indirect climatic effects, and several technological innovations. Finally, Young et al. recognize that resources for conservation are scarce and that not all species can be individually assessed for their responses to climate change. Using cross-taxon analyses, they develop a system to predict taxa or life-history strategies that are consistently affected by climate change in similar ways. Their "climate change vulnerability index" provides managers with a useful tool for rapidly assessing susceptibility to climate change for vertebrates within a defined geographical area.

Part 2 looks at several case studies of climate change impacts on wildlife conservation across a variety of species and ecosystems. First, Owen-Smith and Ogutu assess the impacts of altered rainfall in savannah ecosystems of Africa—communities that harbor some of the densest and most diverse concentrations of large mammals on the planet. Many ungulates in these systems migrate seasonally to track water and forage; shifts in their migration routes to track altered precipitation patterns could be very difficult due to fences on park boundaries and intense human land use outside of protected areas. However, there is hope that if climate shifts are correctly predicted, these climate change impacts could be ameliorated through proactive expansion of parks, extension of conservation efforts beyond protected area boundaries, or the restoration of ecosystem heterogeneity within parks.

Le Galliard et al. then discuss the impacts of warming on a very different guild, squamate reptiles in Europe. The authors have assembled a massive dataset from across the continent; initial declines are revealed for southern populations of the common lizard (*Zootoca vivipara*) and many additional future impacts are predicted. Preserving and restoring habitat connectivity (no small task in the highly altered landscapes of Europe), along with some fairly simple habitat manipulations, could help ameliorate some of these impacts. Next, Zack and Liebezeit assess the impacts of multiple physical and landscape changes on arctic-breeding shorebirds. Current and predicted changes in temperature and precipitation are more severe in the Arctic than at lower latitudes, with important potential consequences for wildlife, especially in conjunction with the all-too-frequent expansion of energy development and other human infrastructures. The authors promote a worldwide identification and prioritization of high-latitude wetland habitats, and strong improvements to the legal protection of critical sites.

Next, Manne addresses the impacts of sea level rise and a variety of concomitant anthropogenic stressors on wildlife species endemic to oceanic islands. The chapter outlines the major threats to herpetofauna, seabirds, and mammals that cannot simply shift their distributions to track climatic changes. The chapter also provides some important recommendations for ways in which some of these threats can be ameliorated.

Moving away from guild- or group-level case studies, the final two chapters in part 2 analyze the details of climate change impacts on individual species. The plight of the American pika, as discussed by Ray et al., is especially worrisome. Pikas are strongly and directly affected by warming temperatures, and their ability to shift distribution upslope to track their thermal niche is precluded, in much of the US Great Basin, by their literally "running out of mountain." While some synergistic threats, particularly grazing, also affect pika populations, it is not clear to what extent; therefore it may or may not be possible to ameliorate the net impacts on pikas by addressing concomitant stressors such as grazing. Next, Tews et al. assess the influence of increasingly frequent extreme weather events on arctic caribou. These stochastic events can cause very high mortality in caribou herds. As with pikas, there may be no clear way to relieve this increasing, climate-driven stress on population health. The Peary caribou herd is hunted by local people, but hunt levels are already low enough that any further reductions in harvest, to try to offset climate change impacts, may not be effective as a management method. These two chapters drive home the point that the threats to some organisms may simply be insurmountable; it is important that we be able to recognize these situations, as well as to determine in which instances this realization is due to prolonged inaction on the part of conservationists and managers. Such an understanding will help motivate monitoring and adaptive management of currently noncritically endangered species.

Finally, part 3 presents chapters that more directly explore management options to promote resilience in wildlife populations and thus reduce the ecological impacts of climate change. Boyce et al. show how harvest models can explicitly incorporate climate change effects. Many wildlife species of conservation concern are both harvested by humans and potentially affected by climate change. Shifts in harvest levels could thus become a powerful management tool for climate change adaptation if they are made in response to changing physical conditions so as to help offset the net impacts on the harvested population. Another primary conservation strategy for addressing climate change impacts on wildlife is the protection and restoration of habitat connectivity. This issue has been discussed extensively in the literature, but often at very coarse scales. Here, Cross et al. provide some of the first in-depth assessment of exactly what it means to maintain habitat connectivity in a changing climate, and how this strategy can be effectively employed in a variety of settings. Next, Wilmers et al. discuss a newer climate change adaptation strategy, the restoration of large carnivores. While carnivores are widely discussed in conservation biology, their ability to ameliorate climate change impacts by stabilizing herbivore population dynamics is only beginning to be explored. Carnivores could possibly even help mitigate CO_2 levels under some circumstances via cascading impacts on plant biomass.

The next chapter addresses a conservation strategy unique to climate change adaptation: the issue of assisted colonization. This is a topic of intense debate, which could become even more contentious in the future. Popescu and Hunter's discussion balances the pros and cons of this strategy from a neutral standpoint. They provide a set of guidelines for designing an assisted colonization project incorporating both ecological and socioeconomic factors, and specify conditions under which this strategy might or might not work.

Finally, Blay and Dombeck move our discussion of climate change ecology firmly into the policy realm. They focus on forest protection in the United tates as a tool for both carbon mitigation and wildlife protection. They present a series of policy recommendations at different scales, from local planning, prioritization, and management options up to a national-level forest carbon offset program. The recommendations provide the template for a coordinated response to climate change in the United States, which could then be exported to other countries as well.

Clearly a book like this is intended for a wide audience. With the broad backgrounds and affiliations of our various contributors, we hope to reach students and professors of wildlife ecology and conservation biology as well as wildlife managers and conservation professionals. Too often there is a gap between academia and wildlife management; we strive to do our small part to bridge that gap. Finally, we hope to reach politicians, policy makers, and conservationists in the nongovernmental community. Our ability to effectively address the impacts of climate change on wildlife and biodiversity depends on influencing policy at multiple levels, from national and international strategies to small-scale tactics employed locally. Finally, it is our hope that the book as a whole will serve as a seed crystal for an evolving discussion among scientists, managers, policy makers, and the public about the multifaceted ways in which we can address climate change so as to effectively conserve wildlife over the coming century.

LITERATURE CITED

Beier, P. and B. Brost. 2010. Use of land facets to plan for climate change: conserving the arenas, not the actors. *Conservation Biology* 24:701–10.

Botkin, D. B., H. Saxe, M. B. Araujo, R. Betts, R. H. W. Bradshaw, T. Cedhagen, P. Chesson, T. P. Dawson, J. R. Etterson, D. P. Faith, S. Ferrier, A. Guisan, A. S. Hansen, D. W. Hilbert, C. Loehle, C. Margules, M. New, M. J. Sobel, and D. R. B. Stockwell. 2007. Forecasting the effects of global warming on biodiversity. *Bioscience* 57:227–36.

Brodie, J. F. and H. K. Gibbs. 2009. Bushmeat hunting as climate threat. *Science* 326:364–65.

Brodie, J. F., E. Post, and W. F. Laurance. 2012a. Climate change and tropical biodiversity: A new focus. *Trends in Ecology and Evolution* 27:145–50.

Brodie, J. F., E. Post, F. Watson, and J. Berger. 2012b. Climate intensification of herbivore impacts on tree recruitment. *Proceedings of the Royal Society B* 279:1366–70.

Clark, J. S., D. M. Bell, M. H. Hersh, and L. Nichols. 2011. Climate change vulnerability of forest biodiversity: Climate and competition tracking of demographic rates. *Global Change Biology* 17:1834–49.

Dawson, T. P., S. T. Jackson, J. I. House, I. C. Prentice, and G. M. Mace. 2011. Beyond predictions: Biodiversity conservation in a changing climate. *Science* 332:53–58.

Doak, D. F. and W. F. Morris. 2010. Demographic compensation and tipping points in climate-induced range shifts. *Nature* 467:959–62.

Hannah, L. 2010. A global conservation system for climate-change adaptation. *Conservation Biology* 24:70–77.

Hansen, L., J. Hoffman, C. Drews, and E. Mielbrecht. 2010. Designing climate-smart conservation: Guidance and case studies. *Conservation Biology* 24:63–69.

Jaramillo, C., D. Ochoa, L. Contreras, M. Pagani, H. Carvajal-Ortiz, L. M. Pratt, S. Krishnan, A. Cardona, M. Romero, L. Quiroz, G. Rodriguez, M. J. Rueda, F. de la Parra, S. Moron, W. Green, G. Bayona, C. Montes, O. Quintero, R. Ramirez, G. Mora, S. Schouten, H. Bermudez, R. Navarrete, F. Parra, M. Alvaran, J. Osorno, J. L. Crowley, V. Valencia, and J. Vervoort. 2010. Effects of rapid global warming at the Paleocene-Eocene boundary on Neotropical vegetation. *Science* 330:957–61.

Maclean, I. M. D., and R. J. Wilson. 2011. Recent ecological responses to climate change support predictions of high extinction risk. *Proceedings of the National Academy of Sciences, USA* 108:12337–42.

Mawdsley, J. R., R. O'Malley, and D. S. Ojima. 2009. A review of climate-change adaptation strategies for wildlife management and biodiversity conservation. *Conservation Biology* 23:1080–89.

McMahon, S. M., S. P. Harrison, W. S. Armbruster, P. J. Bartlein, C. M. Beale, M. E. Edwards, J. Kattge, G. Midgley, X. Morin, and I. C. Prentice. 2011. Improving assessment and modelling of climate change impacts on global terrestrial biodiversity. *Trends in Ecology and Evolution* 26:249–59.

Millar, C. I., N. L. Stephenson, and S. L. Stephens. 2007. Climate change and forests of the future: managing in the face of uncertainty. *Ecological Applications* 17:2145–51.

Miller-Rushing, A. J., T. T. Hoye, D. W. Inouye, and E. Post. 2010. The effects of phenological mismatches on demography. *Philosophical Transactions of the Royal Society B* 365:3177–86.

Morris, W. F., and D. F. Doak. 2002. *Quantitative Conservation Biology*. Sunderland: Sinauer Associates.

Myers, E. C. 2007. Policies to reduce emissions from deforestation and degradation (REDD) in tropical forests. Resources for the Future DP 07–50, Washington.

Orr, D. W. 2009. Baggage: The case for climate mitigation. *Conservation Biology* 23:790–93.

Parmesan, C. 2006. Ecological and evolutionary responses to recent climate change. *Annual Review of Ecology Evolution and Systematics* 37:637–69.

Poiani, K. A., R. L. Goldman, J. Hobson, J. M. Hoekstra, and K. S. Nelson. 2011. Redesigning biodiversity conservation projects for climate change: examples from the field. *Biodiversity and Conservation* 20:185–201.

Post, E., J. F. Brodie, M. Hebblewhite, A. D. Anders, J. A. K. Maier, and C. C. Wilmers. 2009. Global population dynamics and hot spots of response to climate change. *Bioscience* 59:489–97.

Post, E. and C. Pedersen. 2008. Opposing plant community responses to warming with and without herbivores. *Proceedings of the National Academy of Sciences, USA* 105:12353–58.

Prato, T. 2009. Evaluating and managing wildlife impacts of climate change under uncertainty. *Ecological Modelling* 220:923–30.

Root, T. L., J. T. Price, K. R. Hall, S. H. Schneider, C. Rosenzweig, and J. A. Pounds. 2003. Fingerprints of global warming on wild animals and plants. *Nature* 421:57–60.

Schmitz, O. J., E. Post, C. E. Burns, and K. M. Johnston. 2003. Ecosystem responses to global climate change: moving beyond color mapping. *Bioscience* 53:1199–1205.

Sgro, C. M., A. J. Lowe, and A. A. Hoffmann. 2011. Building evolutionary resilience for conserving biodiversity under climate change. *Evolutionary Applications* 4:326–37.

Soulé, M., J. Estes, J. Berger, and C. Martinez del Rio. 2003. Ecological effectiveness: Conservation goals for interactive species. *Conservation Biology* 17:1238–50.

Soulé, M. E., J. A. Estes, B. Miller, and D. L. Honnold. 2005. Strongly interacting species: Conservation policy, management, and ethics. *Bioscience* 55:168–76.

Stone, R. 2007. The last of the leviathans. *Science* 316:1684–88.

Suttle, K. B., M. A. Thomsen, and M. E. Power. 2007. Species interactions reverse grassland responses to changing climate. *Science* 315:640–642.

Tingley, M. W. and S. R. Beissinger. 2009. Detecting range shifts from historical species occurrences: New perspectives on old data. *Trends in Ecology and Evolution* 24:625–33.

PART I

Assessing and Predicting Climate Change Impacts on Wildlife

2

Recent and Future Climatic Change and Its Potential Implications for Species and Ecosystems

Karen M. Carney, Brian Lazar, Charles Rodgers, Diana R. Lane, Russell Jones, Scott Morlando, and Allison E. Ebbets

Until recently, conservation biologists have focused primarily on threats to biodiversity from land use change, overexploitation of natural resources, pollution, and illegal trade. However, climate change and its impacts have become widely recognized as a grave threat to biodiversity worldwide (Millennium Ecosystem Assessment 2005). Later chapters of this book focus on specific strategies and methods for achieving conservation goals in the face of climate change. This chapter provides a brief overview of current scientific understanding about key aspects of future climate change and how they might affect species dynamics and ecosystem processes of particular concern to conservation biologists.

This chapter is composed of nine sections. The first section provides a very brief, general overview of the development of climate change projections, including major sources of uncertainty in climate change modeling. The next four sections each discuss recent observations and future projections of key aspects of future climate change (i.e., temperature change, precipitation change, snow and ice loss, and sea level rise). The sixth section describes the potential implications of climate change for some important ecosystem and species dynamics. The seventh section describes how ecosystem responses to climate change can either accelerate climate change or slow it down through positive or negative feedbacks. The eighth section discusses briefly how human actions taken to mitigate or adapt to climate change can threaten biodiversity. The last section concludes with a brief discussion of how conservation biologists and wildlife managers can adjust their actions to take into account these impending changes.

Projecting Future Climate Change

It is widely accepted that human-induced climate change is driven primarily through increasing atmospheric concentrations of greenhouse gases (GHGs) such as CO_2 and methane. Because GHGs absorb infrared radiation emitted by the earth's surface, temperatures generally rise with increasing GHG concentrations. The amount of climate change that occurs in the future will depend directly on global GHG emissions trajectories as well as equilibrium climate sensitivity, or the amount of warming that result from a given concentration of GHGs in the atmosphere.

As it is virtually impossible to know exactly what future GHG emissions levels will be, scientists associated with the Intergovernmental Panel on Climate Change (IPCC) have developed a range of different emissions scenarios, based on assumptions about population growth, economic development, technological change, and reliance on fossil fuels. The scenarios result in quite divergent GHG emissions levels in the future, from relatively high (akin to continuing on our current trajectory) to relatively low (see table 2.1). Interestingly, climate assessments may be moving away from scenarios driven by socioeconomic storylines, and toward a range of plausible emissions scenarios that could result from any number of economic, technological, demographic, policy and institutional futures (Moss et al. 2010.) These new types of scenarios should enable researchers and policy makers to understand the various economic, demographic, and policy futures that can result in a given concentration pathway.

To determine how much warming will occur in the future, scientists input future GHG emissions levels into general circulation models (GCMs), which are numerical models that represent processes in the atmosphere, ocean, cryosphere (frozen water and ground), and land surface that shape the climate. The Intergovernmental Panel on Climate Change's (IPCC) Fourth Assessment Report (AR4) uses different combinations of scenarios and GCMs to capture a wide range of possible climate change outcomes. The IPCC often integrates results across either all or most of 23 different GCMs to report "multiple ensemble means," which represent the average projection across the different models.

Even though highly sophisticated models are used to project future climate change, the projections have numerous sources of uncertainty, and here we discuss three major ones: climate driver uncertainty, climate system uncertainty, and downscaling uncertainty. We briefly describe each type of uncertainty below; for a fuller discussion, see Barsugli et al. (2009).

1. *Driver uncertainty*. It is not possible to accurately predict future GHG emissions trajectories or other anthropogenic drivers of climate change, and this introduces an important source of uncertainty into climate projections. While the IPCC scenarios discussed above have been used to help delineate some potential future outcomes, future GHG emissions remain highly uncertain.

Table 2.1. IPCC special report on emissions scenarios

Scenario family	Description
A1	This story line and scenario family describes a future world of very rapid economic growth, global population that peaks in mid-century and declines thereafter, and the rapid introduction of new and more efficient technologies. Major underlying themes are convergence among regions, capacity building, and increased cultural and social interactions, with a substantial reduction in regional differences in per capita income. The A1 scenario family develops into three groups that describe alternative directions of technological change in the energy system. The three A1 groups are distinguished by their technological emphases: fossil intensive (A1FI), non-fossil energy sources (A1T), and a balance across all sources (A1B).
A2	This story line and scenario family describes a very heterogeneous world. The underlying theme is self-reliance and preservation of local identities. Fertility patterns across regions converge very slowly, resulting in continuously increasing global population. Economic development is primarily regionally oriented, and per capita economic growth and technological change are more fragmented and slower than in other story lines.
B1	This story line and scenario family describes a convergent world with the same global population as in A1 and A2, which peaks in mid-century and declines thereafter, as in the A1 storyline, but with rapid changes in economic structure toward a service and information economy with reductions in material intensity and the introduction of clean and resource-efficient technologies. The emphasis is on global solutions to economic, social, and environmental sustainability, including improved equity, but without additional climate initiatives.
B2	This story line and scenario family describes a world in which the emphasis is on local solutions to economic, social, and environmental sustainability. It is a world with continuously increasing global population at a rate lower than A2, intermediate levels of economic development, and less rapid and more diverse technological change than in the B1 and A1 story lines. While the scenario is also oriented toward environmental protection and social equity, it focuses on local and regional levels.

Source: Nakićenovic et al. 2000, p. 4–5.

In addition, other climate drivers may materialize, such as the occurrence of volcanic eruptions, which can reduce the influx of radiation, or changes in solar output, though these types of effects are well understood and are included in historical climate records.

2. *Climate system uncertainty.* This refers to the uncertainty in
   (a) how atmospheric GHG concentrations through the carbon

cycle will change, (b) the climate response to changes in atmospheric GHG concentrations (e.g., how much average global temperatures will increase), and (c) how much natural variability there is in the climate system.

3. *Downscaling uncertainty.* GCMs work at relatively coarse scales, so when examining the impact of climate change on any given ecosystem, species, or natural resource, it is necessary to downscale to a spatial and temporal resolution that will be of use to natural resource managers. This is generally done either statistically or dynamically. Statistical downscaling involves the use of empirically observed relationships between large-scale and local climatic conditions. Statistical associations are used to produce local climate conditions from GCM outputs. A key potential limitation of statistical downscaling is that local climatic processes cannot feed back to affect larger-scale climatic conditions. This may not be a critical issue in regions where large-scale climatic patterns tend to drive local climate conditions. In addition, the relationship between large-scale climate features and local climate is assumed to be constant. In dynamical downscaling, global climate model output is used to drive a process-based regional climate model (RCM), which covers a limited geographic area. RCMs are able to simulate local mechanistic climate processes affected by local features, such as topography or nearby water bodies. These kinds of details are absent from GCMs, and thus potentially produce more accuracy on regional scales. However, a key constraint is that developing and applying regional climate models requires a considerable investment of time and resources.

Temperature Change

In this section we first discuss recent observations of increasing global average temperature, most of which have very likely been driven by anthropogenic GHG emissions since the beginning of the industrial revolution (IPCC 2007a). We then discuss future temperature projections based on different GHG emissions scenarios and GCM simulations.

Over the last 100 years, global mean surface temperatures have risen by about 0.74°C, with the rate of temperature increase nearly doubling over the last 50 years (Trenberth et al. 2007). Warming over land has been greater than over the oceans, increasing at nearly twice the rate after 1979 (0.27°C per decade after 1979 versus 0.13°C per decade before 1979; Trenberth et al. 2007). The amount of observed warming varies substantially geographically. For example, while most of the globe has warmed since 1979, some areas, specifically the middle latitudes of the Southern Hemisphere oceans, have actually cooled slightly (Smith and Reynolds, 2005, Trenberth et al. 2007). While urban heat island effects are present, they tend to be localized and only have a minor effect on long-term trends in temperature (Peterson et al. 1999).

Between approximately 2010 and 2030, temperatures are projected to rise by about 0.5°C, regardless of the emissions scenario considered. After this time, projected temperatures diverge according to the emissions scenarios. An ensemble of GCM projections shows that by 2100, mean global temperatures are projected to increase by a total of 3.1°C over 1980–99 temperatures under the high-emissions scenario (A2), 2.7°C under the medium-emissions scenario (A1B), and 1.8°C under the low-emissions scenario (B1) (Meehl et al. 2007; also see figure 2.1).

However, projected temperature increases are dependent on climate sensitivity, which varies across GCMs. A multitude of studies have been dedicated to better elucidating likely ranges for climate sensitivity, and results vary (Knutti and Heger 2008, Hegerl et al. 2006, Räisänen 2005, Wigley et al. 2005). Overall, the studies suggest that equilibrium climate sensitivity, the amount of warming expected from a doubling of atmospheric CO_2 concentrations, will likely fall between 2°C and 4.5°C (Meehl

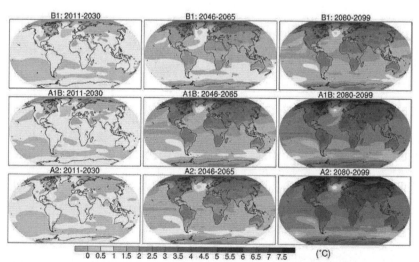

Figure 2.1. Multi-model mean of annual mean surface warming (surface air temperature change, °C) for the scenarios B1 (*top*), A1B (*middle*) and A2 (*bottom*), and three time periods, 2011 to 2030 (*left*), 2046 to 2065 (*middle*) and 2080 to 2099 (*right*). Source: IPCC 2007a.

et al. 2007, Meinhausen 2006). Thus, the warming observed may be higher or lower than the means cited above, depending on how sensitive the climate actually is to GHGs. Consistent with recent observations, mean warming over land is expected to exceed mean warming over water, with the greatest warming expected to occur in the northern latitudes, and the least in the southern ocean (Meehl et al. 2007). In addition to global mean average temperature increases, it is also very likely that heat waves will be more intense, frequent, and longer in duration in the future.

Substantial intraregional variation is also expected across all continents. For example, in North America the western, southern, and eastern continental edges are projected to experience annual mean warming of 2°C to 3°C by 2100, while northern regions could experience an increase in annual mean temperature of more than 5°C. Winter temperatures in northern Alaska and Canada may increase by as much at 10°C over the same time period. In eastern, central, and western North America away from the continental edges, temperature changes and seasonal differences are expected to be much more modest (Christensen et al., 2007).

Changes in Precipitation

While future changes in the direction and magnitude of temperatures are projected with relatively high confidence (IPCC 2007a), projected changes in global and regional hydrology (i.e., rainfall, runoff) are less certain. Generally, the hydrologic cycle is expected to intensify with warming because air can hold more water vapor at higher temperatures. For each degree Celsius of warming, atmospheric water-holding capacity is expected to increase by about 7%, so globally-averaged annual precipitation should also increase (Trenberth et al. 2003). In addition, since water vapor is itself a GHG, increases in atmospheric humidity may act as a positive feedback, contributing to further warming of the atmosphere. Physical theory and climate models also predict an increase in the frequency and magnitude of extreme or heavy precipitation events, due to increases in both atmospheric water and the energy available to drive convective processes (Trenberth et al. 2003). Below, we discuss what has recently been observed regarding precipitation changes as well as what climate models project for the future.

Recent Observations

Historical climate data generally support the prediction of increasing precipitation with increasing temperature. Annual precipitation over land areas has generally increased over the 20th century at mid- and upper latitudes (30° N–85° N) and in the deep tropics of the Southern Hemisphere. However, decreases in precipitation have been observed from 10°S to 30°N, most significantly since the 1960s (Bates et al. 2008). Spatial patterns of change in precipitation tend to be less coherent than patterns of temperature change. While many factors potentially contribute to observed changes in precipitation, recent anthropogenic climate change is estimated to have played a dominant role (Zhang et al. 2007).

Climate change is expected to affect not only the overall amount of annual precipitation, but also the distribution of that precipitation across the year. For example, rainfall events classified as "heavy," "very heavy," or "extreme" have increased, in both absolute and relative terms (Alexander et al. 2006). Furthermore, Groisman et al. (2005) found that while total annual precipitation volumes over the United States have increased by only 1.2% per decade from 1970 to 1999, the share of annual precipitation associated with extreme events increased by 14% per decade over the same time period.

Potential Future Precipitation Change

GCM projections prepared for the IPCC generally project a continuation of recently observed changes in precipitation patterns. Specifically, annual precipitation is projected to generally increase at higher latitudes and in the tropics, and to generally decrease in the subtropics and lower to middle- latitudes (Bates et al. 2008). Plate 1 shows projected changes in Northern Hemisphere winter and summer precipitation for 2080 to 2099, relative to those from 1980 to 1999. Certain regions, such as the southwestern United States, northeastern Brazil, southern Africa, and the Mediterranean basin, are projected to experience annual average decreases in precipitation. Other regions, including northern North America and Asia, will likely become wetter. Across 17 different GCMs, global average precipitation is projected to increase between +1.8% and +5.4% in a "low" emissions scenario (B1), and between +1.7% and +7.5% in a "high" emissions scenario (A2), over the next century. In general, historically wet regions are projected to become wetter, and dry regions are projected to become drier. This suggests that the magnitude of change in precipitation increases with higher temperatures. These models are also in agreement with historical data in that they project a shift toward more intense and extreme precipitation events, even in regions projected to experience a decline in total annual precipitation (Sun et al. 2007). Through the combined changes in frequency of rainy days and tendency for rainfall to occur in high-intensity events, both flood and drought risks are likely to increase by the mid- to late 21st century.

Changes in Runoff and Discharge

Projected changes in the water resources of major river basins reflect, to a great extent, projected changes in precipitation. Plate 2 shows projected late-21st-century runoff relative to that of the late 20th century. A major source of uncertainty in runoff projections, in addition to the uncertainties associated with GCMs, is the mismatch in scale between the global atmospheric circulation patterns produced by GCMs and the regional- and local-scale characteristics that influence catchment water balances. Such regional and local characteristics include soils, topography, vegetation, and land use, which are not well represented in GCMs (Laoiciga et al. 1996). So the patterns in plate 2 should be interpreted as indicative only on the subcontinental scale. In addition, future regional patterns of runoff will reflect long-term changes in patterns of human settlement, demographics, economic activity, and technological change. Arnell (2004) notes that by the mid-21st century, differences in the demographic and economic assumptions underlying the SRES scenarios may have a greater impact on regional water scarcity than differences in the climate scenarios (Bates et al. 2008).

Declines in Snow and Ice Cover

Snow and ice cover on earth is highly vulnerable to projected future increases in temperature. The rate and extent of snow and ice loss will have a critical impact on the nature and dynamics of a wide range of ecosystems. Below, we discuss trajectories in losses of arctic summer sea ice, snowpack, glaciers, and ice sheets, and discuss the ecological implications of these changes.

Losses of Summer Sea Ice in the Arctic

Arctic sea ice influences regional and global climate by modulating ocean circulation patterns and surface energy balance (Meehl et al. 2007). Changes in sea ice and resultant changes in arctic ocean temperatures will affect habitat for species ranging from bacteria and algae to polar bears and whales. Loss of sea ice also has important implications for indigenous populations, shipping routes, and energy exploration (UNEP 2007).

Arctic sea ice covers an area of approximately 15 million km^2 in winter and 7 million km^2 in summer. The September sea ice minimum has been decreasing at an average rate of about 8.6% per decade since 1978 (Serreze et al. 2007). Climate models project increases in global average temperature throughout the rest of the century, as well as a continuation and possible acceleration of these observed trends in sea ice extent (Meehl et al. 2007).

Warming is projected to be more pronounced at high latitudes, due to feedback mechanisms related to the high albedo of snow and ice cover (Meehl et al. 2007). Put simply, reductions in highly reflective snow and ice allow oceans and land to absorb more of the sun's energy, thus leading to more warming. While projections for annual arctic sea ice extent range from nearly no change to significant and rapid reductions (Zhang and Walsh 2006), summer sea ice area is projected to decline at a relatively rapid rate, indicating a movement from permanent towards seasonal sea ice cover in the Arctic (Gordon and O'Farrel 1997). In fact, recent modeling suggests that by the end of the 21st century, the Arctic could be ice-free in summer (UNEP 2007, Boé et al. 2009).

Snowpack Declines

Seasonal snowpacks supply water for roughly one-sixth of the world's population (UNEP 2007). They also play a critical role in maintaining stream flow (Mote et al. 2005, Barnett et al. 2005), providing dry-season water sources for widlife and regulating salinity in near-shore estuaries and bays (Knowles and Cayan, 2002). Snowpacks have decreased in the latter part of the 20th century; snow coverage has declined, snowlines have moved up in elevation, and more precipitation has come as rain than as snow (Lettenmaier et al. 2008, Bates et al. 2008, Lemke et al. 2007, UNEP 2007; Cayan et al. 2006; see plate 3).

Reductions in snowpack are strongly correlated with increases in air temperature, and snowpacks are thus projected to continue to decline substantially throughout the 21st century. In fact, a reduction of 60% to 80% in snow-water equivalent is projected in many mid-latitude regions (ESG 2007, UNEP 2007). The onset of snow accumulation is projected to begin later, the start of snowpack melt to begin earlier, and the fractional snow coverage to decrease during the snow season (Stewart et al. 2005, UNEP 2007, Bates et al. 2008).

Snowmelt serves as the sole source of water during the dry summer months in many locations (Saunders et al. 2008). Many of these areas could face shortfalls in water storage capacity due to increased winter runoff and finite reservoir capacity, potentially resulting in the loss of large volumes of available freshwater downstream or to the oceans (Lettenmaier et al. 2008, UNEP 2007, Barnett 2005). Earlier snowpack melt, which is already being observed (Mote et al. 2005, Lemke et al. 2007), is exacerbating summer water shortages in Colorado and California, with serious implications for aquatic ecosystems, fisheries, agriculture, and residential water use (Snyder et al. 2002, Field et al. 2007; Saunders et al. 2008).

Glacial Melt

Glaciers provide water to a large portion of the world's population and to important natural ecosystems, particularly during dry seasons when precipitation is limited (Bates et al. 2008, Barnett et al. 2005). Glaciers worldwide have shrunk in the last century as a direct result of climate change, with the greatest reductions occurring since 1980 (Lemke et al. 2007, UNEP 2007). The strongest mass losses per unit area have occurred in Patagonia, Alaska, the northwest United States, and southwest Canada (Larsen et al. 2007, Lemke et al. 2007, Dyurgerov and Meier 2005).

The current trend of rapid glacial reduction is projected to continue or accelerate throughout the 21st century (Pfeffer et al. 2008; Meier et al. 2007); it is possible that many glaciers will lose 60% of their volume by 2050 (Schneeberger et al. 2003) and that many mountain ranges will experience complete deglaciation by 2100 (Bates et al. 2008, UNEP 2007, Bradley et al. 2004). Glacier dynamics and melting could contribute an additional 0.1 to 0.55 m of sea level rise by 2100 (Pfeffer et al. 2008, Meier et al. 2007, UNEP 2007). The number of glacial lakes considered vulnerable to catastrophic flooding and associated debris flows is expected to grow as glaciers recede and fill the vacated depressions with meltwater (UNEP 2002, Ghimire et al. 2005, IPCC 2007b).

Ice Sheet Melt

The ice sheets of Greenland and Antarctica hold approximately half of earth's fresh water and 99% of its freshwater ice (the equivalent of 64 m of sea level rise; UNEP 2007), and therefore any changes to the ice sheets have profound implications for sea level rise, earth's surface energy balance, and ocean circulation. Dynamic losses and melting of the Antarctic and Greenland ice sheets have very likely contributed to observed sea level rise since at least 1993 (Meier et al 2007, Lemke et al. 2007), and could contribute an additional 0.3 to 1.2 m of sea level rise by 2100 (Pfeffer et al. 2008).

Annual loss of the Greenland ice sheet has approximately doubled since 1990, with increased rates of loss since 2005 (Thomas et al. 2006). Changes to the Antarctic ice sheet are less certain, but losses have likely occurred during this same time period (Rignot 2008, Lemke et al. 2007, UNEP 2007). Uncertainty in the physics and modeling of the ice sheets make well-constrained projections of their responses to climate change difficult (Pfeffer et al. 2008, Shepard and Wingham 2007). It is possible that the thinning and loss of ice shelves along both major ice sheets will continue throughout the 21st century, which could allow the outflowing glaciers to drain the interior of the ice sheets at increasing rates.

Sea Level Rise

The term "sea level" generally refers to the mean level of the ocean's tides measured over a period of 20 years (CCSP 2009). When discussing sea level rise (SLR), it is important to differentiate between two different types: eustatic (or global) sea level rise, and relative sea level rise. Eustatic SLR refers to the average increase in the mean sea level over the world's oceans, while relative sea level rise refers to the sea level relative to the elevation of the land surface. Eustatic SLR is due primarily to thermal expansion of the oceans and melting of land-based ice sheets and glaciers. Relative SLR is the combination of eustatic and local vertical land movement caused by subsidence or rising of the land surface because of natural factors (e.g., crustal rebound from the last glaciation) or human factors (e.g., groundwater withdrawals). While most estimates of SLR are made in reference to the eustatic rate, it is important to realize that the relative SLR in any specific location may be much higher or lower than the global average, due to the variety of local factors that may be in play.

Over the 20th century the total global rate of SLR has been estimated at 1.7 mm/year (CCSP 2009, IPCC 2007a) for a total rise of 17 cm (6.7 inches). However, this rate appears to be increasing. In its latest assessment, the Intergovernmental Panel on Climate Change (IPCC) has estimated that from 1961 to 2003, the average rate of SLR was approximately 1.8 mm/yr, while from 1993 to 2003 that rate increased to approximately 3.1 mm/yr (IPCC 2007a).

Based on climate model projections, the IPCC has estimated a global increase in sea level between 18 and 59 cm (7–23 inches) by the end of the century (IPCC 2007a), depending on the emissions scenario considered (see figure 2.2). However, it is important to note that these estimates do not address the potential for massive changes in flows from the Greenland and Antarctic ice sheets. More recent studies that consider such impacts have projected between 0.2 to 2 m of SLR by 2100 (Pfeffer et al. 2008). The wide range in estimates results from an incomplete understanding of ice sheet dynamics. As the science of ice sheet modeling improves, these estimates may be further refined and could potentially increase.

Potential Impact of Climate Change on Ecosystem and Species Dynamics

In this section we provide examples of key ecosystem or species dynamics that may be affected by recent and future changes in climate. This section is meant not to be comprehensive, but rather to highlight some key issues that wildlife and natural resource managers may face as cli-

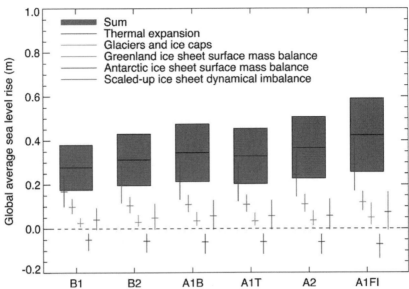

Figure 2.2. Projections and uncertainties (5% to 95% ranges) of global average sea level rise and its components for 2090–99 (relative to 1980–99) for the six IPCC SRES scenarios. The projected sea level rise assumes that the part of the present-day ice sheet mass imbalance that is due to recent ice flow acceleration will persist unchanged. It does not include the contribution of scaled-up ice sheet discharge, which is an alternative possibility. It is also possible that the present imbalance might be transient, in which case the projected sea level rise would be reduced by 0.02 m. It must be emphasized that the likelihood of any of these three alternatives cannot be assessed. Source: IPCC 2007a.

mate change continues to intensify. Specifically, we discuss the potential impacts of climate change on drought severity and extent, inland and coastal wetland extent and functioning, and snow-driven changes in dry season length. Note that we do not address the impact of climate change on species range shifts and phenology, as this issue is covered well in other chapters of this book.

Drought Severity and Extent

The severity and extent of drought are expected to increase under climate change, particularly in historically dry regions. Even without a change in precipitation, regions may experience droughtlike conditions because higher evaporative demand from higher temperatures may lead to increased drought stress (Kling et al. 2003). Drought can have a wide variety of impacts on species and ecosystems, including direct mortality of water-stressed animals and plants, decreased resource availability, and shifts in vegetation type, among others (Breshears et al. 2005, Previtali et al. 2009). For example, in the southwestern United States, drought in the 1950s led to a major long-term retraction in Ponderosa pine (Allen and Breshears 1998), and recent drought (2002–3) has led to widespread mortality of pinyon pine (Breshears et al. 2005).

Changes in Coastal and Interior Wetlands

While coastal wetlands have historically been able to keep pace with sea level rise by moving vertically through the trapping of sediments and organic matter (a process called accretion), the increased rate of future sea level rise will likely cause either degradation or loss of coastal wetland habitat in many areas. Wetland plant die-off can result from increased stress due to salinity increases, or from submergence. If sea level rise is accompanied by decreases in precipitation and runoff, wetland stress and die-off may be further accelerated (Day et al. 2005). Humans may exacerbate these problems through their efforts to adapt to climate change. Shoreline armoring structures (e.g., seawalls) will preclude the inland movement of most tidal wetlands and will reduce the exchange of nutrients and organisms from watersheds to estuaries (Jones and Strange, in press). Armoring structures may also increase coastal erosion rates, as the energy from waves is redirected after hitting these structures.

Interior freshwater wetlands are likely to be affected by changes in precipitation and temperature under climate change (Burkett and Kusler 2000). Decreases in precipitation or changes in precipitation timing, especially in combination with increased evaporation from higher temperatures, can result in a shrinking of wetland area, an earlier disappearance of seasonal wetlands, a shift from wetland to dry land, or a change in wetland type (Burkett and Kusler 2000, Kling et al. 2003). For example, increased temperatures of just a few degrees Celsius in the prairie pothole region may have significant effects on the size, location, and productivity of intermittent prairie pothole wetlands. Increased temperatures cause increased evaporation, a shortened time period of "wetting" for these intermittent wetlands, and an increase in the frequency and duration of drought (Johnson et al. 2005). Conditions favorable for intermittent wetlands are predicted to shift to the east into a region dominated by more intensive agriculture (cropland instead of rangeland), where productive wetland habitats are less likely to develop (Johnson et al. 2005).

Permafrost wetlands and alpine wetlands are at risk of loss primarily from increased temperatures, while peatlands are at risk from changes in groundwater level that may be caused by alterations in precipitation and recharge (Burkett and Kusler 2000). Changes in precipitation intensity (i.e, larger and more intense storm events) may increase erosion and contaminant mobilization, resulting in degraded water quality for the water that feeds wetlands (Poff et al. 2002). Alternatively, longer periods of low flows and higher water temperatures could exacerbate water pollution and can lead to algal blooms or other problems (Poff et al. 2002, Kundzewicz et al. 2007).

Longer Dry Sseasons due to Losses of Snow

Snow reflects a large portion of incoming solar radiation, and a reduction in the duration of snow cover therefore will allow for more pronounced soil warming during the warm spring and summer months (Armstrong and Brun 2008). Snow also insulates the soil from cold air temperature during the fall and early winter months, and prevents soil moisture from evaporating. The combined effects of later snow accumulation and earlier snowmelt results in a longer growing season for vegetation with the potential for prolonged dry seasons and droughts. This observed trend is projected to continue and become more pronounced throughout the 21st century (Lettenmaier et al. 2008, Bates et al. 2008, Lemke et al. 2007, Stewart et al. 2005, Mote et al. 2005).

A longer snow-free season that decreases soil moisture can exacerbate the impact of drought caused by reductions in precipitation. Periodic or prolonged droughts have the potential to alter fire behavior, including changing fire frequency and fire size (McKenzie et al. 2004). For example, Westerling et al. (2006) evaluated the relative influences of climate and management on the incidence and severity of fires in the western United States. They examined wildfires in the period between 1970 and 2003 and found that increases in wildfires have been concentrated between 1,680 m and 2,690 m in elevation, and are strongly associated with spring snowmelt timing. Early spring snowmelt leads to reductions in surface and soil water later in the season, effectively prolonging the dry conditions that are conducive to wildfire. Warmer temperatures also increase potential evapotranspiration, further drying soil and vegetation.

Drought and heat stress can also make forests more vulnerable to insect outbreaks (Colorado Division of Forestry 2004). For example, in recent years, pine forests in the Rocky Mountains of the western United States and British Columbia have experienced widespread outbreaks of mountain pine beetles that can kill or damage trees (Hamann and Wang 2006, Kurz et al. 2008). While these outbreaks have generally been attributed to warmer temperatures that are more favorable to the pine beetle, reduced vigor is known to be a critical determinant of the severity and extent of pine beetle outbreaks (Raffa et al. 2008).

Ecosystem-Related Climate Feedback

Ecosystems are not only affected by climate but also help to shape it. One important way they do this is through affecting where carbon or other nutrients reside. For example, ecosystems can affect climate by storing carbon, removing it from the atmosphere (which is termed sequestration) or releasing it into the atmosphere as methane or carbon dioxide (Nilsson and Schopfhauser 1995, Foley et al. 1998, Lal 2004, Lehmann 2007, Reay et al. 2007). Terrestrial ecosystems compose a vast and important pool in the global carbon budget, storing up to four times more carbon than the atmosphere (Lal 2004), primarily in plants and soils. In fact, the destruction or degradation of tropical forests contributed approximately 12% of the net flux of carbon into the atmosphere during the 1990s (van der Werf 2009). The critical role that ecosystems, particularly tropical forests, play in regulating climate has led to widespread international interest in funding forest conservation as a way to mitigate climate change.

While terrestrial ecosystems, and forests in particular, are often viewed as a potential tool for slowing the pace of climate change, the responses of natural ecosystems can also accelerate climate change through positive feedbacks. As noted earlier, increases in pest outbreaks, such as those being observed in western North America, may lead to large-scale losses of forest carbon to the atmosphere (Kurtz et al. 2008). In the Amazon, the reduced precipitation projected by climate models may lead to a large-scale die-back of the tropical forests in the region (Cox et al. 2004). This would lead to a loss of 70 Gt of carbon, thus providing a positive feedback to climate change (Cox et al. 2004). On a global scale, increasing temperatures may stimulate the decomposition of soil organic matter to such an extent that the terrestrial biosphere moves from a net sink to a net source of carbon (Heimann, 2008).

The effects of climate change on permafrost could also be critical to the rate and extent of future climate change. Permafrost is a repository for vast stores of carbon, as biological material has been frozen into the soil column rather than degraded. Thawing permafrost could therefore result in significant release of CO_2 and methane into the atmosphere. Recent studies suggest that the release of carbon may be significant enough to create a positive feedback in climate change (Walter et al. 2007a, 2007b; Zhuang et al. 2006; Khvorostyanov et al. 2008). However, some researchers argue that increasing temperatures could actually reduce the rate of permafrost degradation. As the active layer increases in thickness, its insulating capacity increases as well, which may reduce permafrost degradation during the summer months. Changes in precipitation may impact permafrost through changes in snow depth, cover, and timing, which may act as either positive or negative feedbacks, depending on the dynamics of precipitation change (Zhang 2005, Osterkamp 2007). While there is potential for permafrost thaw to provide a positive feedback, it is not yet clear how important this ecological effect would be relative to others (Anisimov 2007, Delisle 2007, CCSP 2008).

Human Adaptation and Mitigation Measures:
Additional Threats to Biodiversity

While the direct impacts of climate change are a major concern for conservation biologists and natural resource managers, the indirect effects of climate change may be equally important. Specifically, the strategies and technologies that people use to slow climate change or adapt to its effects will have potentially major implications for biodiversity conservation efforts. In essence, climate change mitigation and adaptation efforts may alter or magnify the non–climate change drivers that already serve to threaten wildlife. As managers develop strategies to adapt to climate change, it will be critical to continually assess the strategies by which human respond to climate change in addition to the physical changes wrought by climatic change itself.

The nature of the threats posed by climate change mitigation efforts will be highly specific to the site or ecosystem in question. Here we list a few examples to highlight issues worthy of consideration. One obvious avenue through which biodiversity could be threatened is the expansion of biofuels. In fact, tropical forests are being converted at alarming rates to make way for palm oil plantations, with devastating effects on biodiversity (Fitzherbert et al. 2008, Danielson et al. 2009). Another potential threat is the planting of forests for carbon sequestration. While this seems like a potential "win" for conservation biologists, if natural forests or savannahs capture and store less carbon than fast-growing non-native species, natural ecosystems may be replaced by monotypic stands of those non-native species (Huston and Marland 2003). Geologic sequestration of CO_2 may pose a threat to ecosystems if the CO_2 is not properly contained. Large-scale leaks of CO_2 could cause plant mortality or acidify nearby fresh- and salt-water systems (IPCC 2005). Even seemingly environmentally benign energy generation approaches, such as wind, solar, and tidal power, have the potential to have serious environmental impacts if done at industrial scales.

Threats posed by adaptation measures will also vary with the specific site being considered. One potential adaptation-related threat has been mentioned briefly above: construction of coastal barriers to protect coastal communities from sea level rise. Construction of these barriers can destroy coastal habitat, and sea walls can prevent the movement of materials and organisms among marine and freshwater ecosystems. Another possible threat could be posed by human migration (IOM 2008). People may move away from coastal and arid areas, which could lead to increased development in other areas, some of which may not be currently experiencing this kind of growth and development. Finally, the type and location of agricultural activities may shift in response to climate change (FAO 2007). Changes in water availability and temperature may drive farmers to locations with climates previously less desirable for agriculture. Alternatively, farmers may stay where they are but grow new crops. New management techniques, such as the use of greater amounts of fertilizer, novel pesticides, or irrigation, may accompany such a change and introduce or intensify threats to biodiversity.

Conclusion

Given all the potential changes to the earth's system due to future climate change, it is clear that conservation biologists and wildlife managers need to start adjusting their strategies and management plans to ensure that gains made now are sustained in the future. Specifically, considering likely climatic changes may alter the priority of tasks, change the focus on species or ecosystems, or alter the areas in which management interventions take place. For example, if sea level rise over the next few decades is likely to inundate coastal wetland areas currently slotted for restoration or vegetation enhancement, funds allocated to that task will likely be better used elsewhere. Likewise, increases in the temperatures of alpine ecosystems may lead to the local extinction of certain plants or animals; these species might no longer be the ones upon which conservation planning in the area is based. Finally, human responses to climate change, through climate mitigation or adaptation-related activities, may introduce new threats to ecological systems. Long-term conservation or management plans should take such changing threats into account.

LITERATURE CITED

Alexander, L. V., X. Zhang, T. C. Peterson, J. Caeser, B. Gleason, A. M. G. Klein Tank, M. Haylock, D. Collins, B. Trewin, F. Rahimzadeh, A. Tagipour, K. Rupa Kumar, J. Revadekar, G. Griffiths, L. Vincent, D. B. Stephenson, J. Burn, E. Aguilar, M. Brunet, M. Taylor, M. New, P. Zhai, M. Rusticucci, and J. L. Vazquez-Aguirre. 2006. Global observed changes in daily climate extremes of temperature and precipitation. *Journal of Geophysical Research* 111:D05109.

Allen, C. D., and D. D. Breshears. 1998. Drought-induced shift of a forest/woodland ecotone: rapid landscape response to climate variation. *Proceedings of the National Academy of Sciences of the United States of America* 95:14839–42.

Anisimov, O. 2007. Potential feedback of thawing permafrost to the global climate system through methane emission. *Environmental Research Letters* 2: 045016.

Armstrong, R. L., and E. Brun, eds. 2008. *Snow and Climate: Physical Processes, Surface Energy Exchange and Modeling*. Cambridge and New York: Cambridge University Press.

Arnell, N. W. 2004. Climate change and global water resources: SRES emissions and socioeconomic scenarios. *Global Environmental Change* 14:31–52.

Barnett, T. P., J. C. Adam, and D. P. Lettenmaier. 2005. Potential impacts of warming climate on water availability in snow-dominated regions. *Nature* 438: 303–309.

Barsugli, J. C. Anderson, J. B. Smith, and J. M. Vogel. 2009. Options for improving climate modeling to assist water utility planning for climate change. Water Utility Climate Alliance, http://www.amwa.net/galleries/climate-change/WUCA_models_whitepaper_120909.pdf. Accessed September 29, 2010.

Bates, B. C., Z. W. Kundzewicz, S. Wu, and J. P. Palutikof (eds). 2008. Climate Change and Water. Technical paper of the Intergovernmental Panel on Climate Change, IPCC Secretariat, Geneva.

Boé, J., A. Hall and X. Qu. 2009. September sea-ice cover in the Arctic Ocean projected to vanish by 2100. *Nature Geoscience*: 1–3.

Bradley, R. S., F. T. Keimig and H. F. Diaz. 2004. Projected temperature changes along the American cordillera and the planned GCOS network. *Geophysical Research Letters* 31:L16210.

Breshears, D. D., N. S. Cobb, P. M. Rich, K. P. Price, C. D. Allen, R. G. Balice, W. H. Romme, J. H. Kastens, M. L. Floyd, J. Belnap, J. J. Anderson, O. B. Myers, and C. W. Meyer. 2005. Regional vegetation die-off in response to global-change-type drought. *Proceedings of the National Academy of Sciences USA* 102:15144–48.

Burkett, V., and J. Kusler. 2000. Climate change: Potential impacts and interactions in wetlands of the United States. *Journal of the American Water Resources Association* 36:313.

Cayan, D., A. L. Luers, M. Hanemann, G. Franco, and B. Croes. 2006. Scenarios of climate change in California: An overview. California Climate Change Center, Sacramento.

CCSP. 2008. Abrupt climate change. A report by the U.S. Climate Change Science Program and the Subcommittee on Global Change Research. Clark, P. U., A. J. Weaver, E. Brook, E. R. Cook, T. L. Delworth, and K. Steffen. US Geological Survey, Reston, VA.

CCSP. 2009. Coastal sensitivity to sea-level rise: A focus on the mid-Atlantic region. A report by the US Climate Change Science Program and the Subcommittee on Global Change Research. J. G. Titus, K. E. Anderson, D. R. Cahoon, D. B. Gesch, S. K. Gill, B. T. Gutierrez, E. R. Thieler, and S. J. Williams. US Environmental Protection Agency, Washington.

Christensen, J. H., B. Hewitson, A. Busuioc, A. Chen, X. Gao, I. Held, R. Jones, R. K. Kolli, W. T. Kwon, R. Laprise, V. Magaña Rueda, L. Mearns, C. G. Menéndez, J. Räisänen, A. Rinke, A. Sarr, and P. Whetton. 2007. Regional climate projections. In *Climate Change 2007: The Physical Science Basis*. Contribution of Working Group I to the Fourth Assessment Report of the Intergovernmental Panel on Climate Change.

Colorado Division of Forestry. 2004. Report on the health of Colorado's forests. 2004. Colorado Department of Natural Resources, Division of Forestry, Denver.

Cox, P. M., R. A. Betts, M. Collins, P. P. Harris, C. Huntingford, and C. D. Jones. 2004 Amazonian forest die-back under climate-carbon cycle projections for the 21st century. *Theoretical and Applied Climatology* 78:137–56.

Danielson, F., H. Beukema, N. D. Burgess, F. Parish, C. A. Brühl, P. F. Donald, D. Murdiyarso, B. Phalan, L. Reijnders, M. Struebig, and E. B. Fitzherbert. 2008. Biofuels plantations on forested lands: double jeopardy for biodiversity and climate. *Conservation Biology* 23:348–58.

Day, J.W., J. Barras, E. Clairain, J. Johnston, D. Justic, G. P. Kemp, J. Ko, R. Lane, W. J. Mitsch, G. Steyer, P. Templet, and A. Yanez-Arancibia. 2005. Implications of global climate change and energy cost and availability for the restoration of the Mississippi delta. *Ecological Engineering* 24:253–65.

Delisle, G. 2007. Near-surface permafrost degradation: How severe during the 21st century? *Geophysical Research Letters* 34:10.1029/2007GL029323.

Dyurgerov, M., and M. F. Meier. 2005. Glaciers and the changing Earth system: A 2004 snapshot. Occasional Paper 58, Institute of Arctic and Alpine Research, University of Colorado, Boulder.

ESG. 2007. WCRP CMIP3 Multi-Model Dataset. Earth System Grid, https://esg.llnl .gov:8443/. Accessed February 2007.

FAO (Food and Agriculture Organization of the United Nations), 2007. Adaptation to climate change in agriculture, forestry, and fisheries: Perspective, framework, and priorities. Interdepartmental Working Group on Climate Change, Rome, Italy.

Field, C.B., L. D. Mortsch, M. Brklacich, D. L. Forbes, P. Kovacs, J. A. Patz, S. W. Running, and M. J. Scott. 2007. North America. In *Climate Change 2007: Impacts, Adaptation and Vulnerability. Contribution of Working Group II to the Fourth Assessment Report of the Intergovernmental Panel on Climate Change*, edited by M. L. Parry, O. F. Canziani, J. P. Palutikof, P. J. van der Linden, and C. E. Hanson, 617–52. Cambridge: Cambridge University Press.

Fitzherbert, E. B., M. J. Struebig, A. Morel, F. Danielsen, C. A. Bruhl, P. F. Donald, and B. Phalan. 2008. How will oil palm expansion affect biodiversity? *Trends in Ecology and Evolution* 23:538–45.

Foley, J. A., S. Levis, I. C. Prentice, D. Pollard, and S. L. Thompson. 1998. Coupling dynamic models of climate and vegetation. *Global Change Biology* 4:561–79.

Ghimire, M. Mirza, M. M. Q. and Q. K. Ahmad, eds. 2005. Climate change and glacier lake outburst floods and the associated vulnerability in Nepal and Bhutan. In *Climate Change and Water Resources in South Asia*, 137–54. Leiden: A.A. Balkema.

Gordon, H. B., and S. P. O'Farrell. 1997. Transient climate change in the CSIRO coupled model with dynamic sea ice. *Monthly Weather Review* 125:875–907.

Groisman, P.Y., R. W. Knight, D. R. Easterling, T. R. Karl, G. C. Hegerle and V. N. Razuvaev. 2005. Trends in intense precipitation in the climate record. *Journal of Climate* 18:1326–50.

Hamann, A. and T. Wang. 2006. Potential effects of climate change on ecosystem and tree species distribution in British Columbia. *Ecology* 87:2773–86.

Hegerl, G. C., T. J. Crowley, W. T. Hyde, and D. J. Frame. 2006. Climate sensitivity constrained by temperature reconstructions over the past seven centuries. *Nature* 440:1029–32.

Heimann, M. and M. Reichstein. 2008. Terrestrial ecosystem carbon dynamics and climate feedbacks. *Nature* 45:289–92.

Huston, M.A. and G. Marland. 2003. Carbon management and biodiversity. *Journal of Environmental Management* 67:77–86.

IOM (International Organization for Migration). 2008. *Migration and Climate Change*. IOM Migration Research Series, no. 31. Geneva.

IPCC. 2005. IPCC special report on carbon dioxide capture and storage. Prepared by Working Group III of the Intergovernmental Panel on Climate Change. B. Metz, O. Davidson, H. C. de Coninck, M. Loos, and L. A. Meyer, eds. Cambridge and New York: Cambridge University Press.

IPCC, 2007a. Summary for policymakers. In *Climate Change 2007: The Physical Science Basis. Contribution of Working Group I to the Fourth Assessment Report of the Intergovernmental Panel on Climate Change*, edited by S. Solomon, D. Qin, M. Manning, Z. Chen, M. Marquis, K. B. Averyt, M.Tignor, and H. L. Miller. Cambridge: Cambridge University Press.

IPCC. 2007b. *Climate Change 2007: Impacts, adaptation and vulnerability. Contribution of Working Group II to the Fourth Assessment Report of the Intergovernmental Panel on Climate Change*. M. L. Parry, O. F. Canziani, J. P. Palutikof, P. J. van der Linden, and C. E. Hanson, eds. Cambridge: Cambridge University Press.

Johnson, W. C., B. V. Millett, T. Gilmanov, R. A. Voldseth, G. R. Guntenspergen, and D. E. Naugle. 2005. Vulnerability of northern prairie wetlands to climate change. *BioScience* 55:863–72.

Jones, R., and L. Strange. 2009. An analytical tool for evaluating the impacts of sea level rise response strategies. *Management of Environmental Quality* 20:383–407.

Khvorostyanov, D. V., P. Ciais, G. Krinner, S. A. Zimov, C. H. Corradi, and G. Guggenberger. 2008. Vulnerability of permafrost carbon to global warming. Part II: Sensitivity of permafrost carbon stock to global warming. *Tellus* 60B:265–275.

Kling et al. 2003. *Confronting Climate Change in the Great Lakes Region: Impacts on Our Communities and Ecosystems*. Executive summary updated 2005. Union of Concerned Scientists.

Knowles, N. and D. R. Cayan. 2002. Potential effects of global warming on the Sacramento/San Joaquin watershed and the San Francisco estuary. *Geophysical Research Letters* 29:10.1029/2001GL014339.

Knutti, R. and G. Hegerl. 2008. The Equilibrium Sensitivity of the Earth's Temperature to Radiation Changes. *Nature GeoScience* 1:735–43.

Kundzewicz, Z.W., L. J. Mata, N. W. Arnell, P. Döll, P. Kabat, B. Jiménez, K. A. Miller, T. Oki, Z. Sen and I. A. Shiklomanov. 2007. Freshwater resources and their management. In *Climate Change 2007: Impacts, Adaptation and Vulnerability. Contribution of Working Group II to the Fourth Assessment Report of the Intergovernmental Panel on Climate Change*, edited by M. L. Parry, O. F. Canziani, J. P. Palutikof, P. J. van der Linden and C. E. Hanson, 173–210. Cambridge: Cambridge University Press.

Kurz, W. A., C. C. Dymond, G. Stinson, G. J. Rampley, E. T. Neilson, A. L. Carroll, T. Ebata, and L. Safranyik. 2008. Mountain pine beetle and forest carbon feedback to climate change. *Nature* 24:987–90.

Lal, R. 2004. Soil carbon sequestration impacts on global climate change and food security. *Science* 304:1623–27.

Larsen, C. F., R. J. Motyka, A. A. Arendt, K. A. Echelmeyer and P. E. Geissler.

2007. Glacier changes in southeast Alaska and northwest British Columbia and contribution to sea level rise. *Journal of Geophysical Research* 112:F01007, doi:10.1029/2006JF000586.

Lehmann, J. 2007. A handful of carbon. *Nature* 447:143–144.

Lemke, P., J. Ren, R. B. Alley, I. Allison, J. Carrasco, G. Flato, Y. Fujii, G. Kaser, P. Mote, R. H. Thomas and T. Zhang. 2007. Observations: Changes in snow, ice and frozen Ground. In *Climate Change 2007: The Physical Science Basis. Contribution of Working Group I to the Fourth Assessment Report of the Intergovernmental Panel on Climate Change*, edited by S. Solomon, D. Qin, M. Manning, Z. Chen, M. Marquis, K. B. Averyt, M. Tignor, and H. L. Miller. Cambridge: Cambridge University Press.

Lettenmaier, D., D. Major, L. Poff, and S. Running. 2008. Water resources. In *The Effects of Climate Change on Agriculture, Land Resources, Water Resources, and Biodiversity*. Report by the US Climate Change Science Program and the subcommittee on Global Change Research, Washington.

Loaiciga, H. A., J. B. Valdes, R. Vogel, J. Garvey, and H. Schwarz. 1996. Global warming and the hydrologic cycle. *Journal of Hydrology* 174:83–127.

McKenzie, D., Z. Gedalof, D. Peterson, and P. Mote. 2004. Climatic change, wildfire, and conservation. *Conservation Biology* 18:890–902.

Meehl, G. A., T. F. Stocker, W. D. Collins, P. Friedlingstein, A. T. Gaye, J. M. Gregory, A. Kitoh, R. Knutti, J. M. Murphy, A. Noda, S. C. B. Raper, I. G. Watterson, A. J. Weaver and Z. C. Zhao. 2007. Global climate projections. In *Climate Change 2007: The Physical Science Basis. Contribution of Working Group I to the Fourth Assessment Report of the Intergovernmental Panel on Climate Change*, edited by S. Solomon, D. Qin, M. Manning, Z. Chen, M. Marquis, K. B. Averyt, M. Tignor and H. L. Miller. Cambridge: Cambridge University Press.

Meier, M. F, M. B. Dyurgerov, U. K. Rick, S. O'Neel, W. T. Pfeffer, R. S. Anderson, S. P. Anderson, and A. F. Glazovsky. 2007. Glaciers dominate eustatic sea-level rise in the 21st century. *Science* 317:1064–67.

Meinshausen, M. 2006. What does a 2°C target mean for greenhouse gas concentrations? A brief analysis based on multi-gas emission pathways and several climate sensitivity uncertainty estimates. In *Avoiding Dangerous Climate Change*, edited by H. J. Schellnhuber et al., 265–79. New York: Cambridge University Press.

Millennium Ecosystem Assessment. 2005. *Ecosystems and Human Well-Being: Biodiversity Synthesis*. Washington: World Resources Institute.

Milly, P. C. D., K. A. Dunne and A. V. Vecchia. 2005. Global pattern of trends in streamflow and water availability in a changing climate. *Nature* 438:347–50.

Moss, R. H., J. A. Edmonds, K. A. Hibbard, M.R. Manning, S. K. Rose, D. P van Vuuren, T. R. Carter, S. Emori, M. Kainuma, T. Kram, G. A. Meehl, J. F. B. Mitchell, N. Nakicenovic, K. Riahi, S. J. Smith, R. J. Stouffer, A. M. Thomson, J. P. Weyant, and T. J. Wilbanks. 2010. The next generation of scenarios fro climate change research and assessment. *Nature* 463:747–56.

Mote, P. W., A. F. Hamlet, M. P. Clark, and D. P. Lettenmaier. 2005. Declining snowpack in western North America. *Bulletin of the American Meteorological Society* 86:39–49.

Nakićenovic, N., J. Alcamo, G. Davis, B. de Vries, J. Fenhann, S. Gaffin, K. Gregory, A. Grubler, T.Y. Jung, T. Kram, E.L. La Rovere, L. Michaelis, S. Mori, T. Morita,

W. Pepper, H. Pitcher, L. Price, K. Raihi, A. Roehrl, H.-H. Rogner, A. Sankovski, M. Schlesinger, P. Shukla, S. Smith, R. Swart, S. van Rooijen, N. Victor, and Z. Dadi. 2000. *Emissions Scenarios. A Special Report of Working Group III of the Intergovernmental Panel on Climate Change.* , New York: Cambridge University Press.

Nilsson S. and W. Schopfhauser. 1995. The carbon-sequestration potential of a global afforestation program. *Climatic Change* 30:267–93.

Osterkamp, T.E. 2007. Characteristics of the recent warming of permafrost in Alaska. *Journal of Geophysical Research* 112:1–10.

Peterson, T.C., et al.1999. Global rural temperature trends. *Geophysical Research Letters* 26:329–32.

Pfeffer W.T., J. T. Harper and S. O'Neel. 2008. Kinematic constraints on glacier contributions to 21st-century sea-level rise. *Science* 321:1340–43.

Poff, N. L., M. Brinson and W. J. Day. 2002. Aquatic ecosystems and global climate change: Potential impacts on inland freshwater and coastal wetland ecosystems in the United States. Prepared for the Pew Center on Global Climate Change.

Previtali, M.A., M. Lima, P. L. Meserve, D. A. Kelt, and J. R. Gutierrez. 2009. Population dynamics of two sympatric rodents in a variable environment: Rainfall, resource availability, and predation. *Ecology* 90:1996–2006.

Raffa, K. F., B. H. Aukema, B. J. Bentz, A. L. Carroll, J. A. Hicke, M. G. Turner, and W. H. Romme. 2008. Cross-scale drivers of natural disturbances prone to anthropogenic amplification: The dynamics of bark beetle eruptions. *Bioscience* 58:501–17.

Räisänen, J., 2005. Probability distributions of CO_2-induced global warming as inferred directly from multimodel ensemble simulations. *Geophysica* 41:19–30.

Reay, D., C. Sabine, P. Smith, and G. Hymus. 2007. Climate change 2007: Spring-time for sinks. *Nature* 446:727–28.

Rignot, E., J. L. Bamber, M. R. van den Broeke, C. Davis, Y. Li, W. J. van de Berg, and E. van Meijgaard. 2008. Recent Antarctic ice mass loss from radar interferometry and regional climate modelling. *Nature Geoscience* 1:106–10.

Saunders, S., C. Montgomery, T. Easley, and T. Spencer. 2008. *Hotter and Drier: The West's Changed Climate*. Rocky Mountain Climate Organization and the Natural Resources Defense Council. New York: NRDC Publications Department.

Schneeberger, C., H. Blatter, A. Abe-Ouchi, and M. Wild. 2003. Modeling changes in the mass balance of glaciers of the northern hemisphere for a transient $2\times CO_2$ scenario. *Journal of Hydrology* 282:145–63.

Serreze, M. C., M. M. Holland, and J. Stroeve. 2007. Perspectives on the Arctic's shrinking sea-ice cover. *Science* 315:1533–36.

Shepherd, A. and D. Wingham. 2007. Recent sea-level contributions of the Antarctic and Greenland ice sheets. *Science* 315:1529–32.

Smith, T. M., and R. W. Reynolds. 2005. A global merged land and sea surface temperature reconstruction based on historical observations (1880–1997). *Journal of Climate* 18:2021–36.

Snyder, M. A., J. L. Bell, and L. C. Sloan. 2002. Climate responses to a doubling of atmospheric carbon dioxide for a climatically vulnerable region. *Geophysical Research Letters* 29:1514.

Stewart, I. T., D. R. Cayan, and M. D. Dettinger. 2005. Changes toward earlier streamflow timing across western North America. *Journal of Climate* 18:1136–55.

Sun, Y., S. Solomon, A. Dai, and R. W. Portmann. 2007. How often will it rain? *Journal of Climate* 20:4801–18.

Thomas, R., E. Frederick, W. Krabill, S. Manizade, C. and Martin. 2006. Progressive increase in ice loss from Greenland. *Geophysical Research Letters* 33:L10503.

Trenberth, K. E., D. Aiguo, R. M. Rasmussen and D. B. Parsons. 2003. The changing face of precipitation. *Bulletin of the American Meteorological Society* (September): 1205–17.

Trenberth, K. E., P. D. Jones, P. Ambenje, R. Bojariu, D. Easterling, A. Klein Tank, D. Parker, F. Rahimzadeh, J. A. Renwick, M. Rusticucci, B. Soden, and P. Zhai. 2007. Observations: Surface and Atmospheric Climate Change. In *Climate Change 2007: The Physical Science Basis. Contribution of Working Group I to the Fourth Assessment Report of the Intergovernmental Panel on Climate Change*, edited by S. Solomon, S., D. Qin, M. Manning, Z. Chen, M. Marquis, K. B. Averyt, M. Tignor and H. L. Miller. Cambridge: Cambridge University Press.

UNEP. 2002. United Nations Environmental Programme (UNEP) News Release 2002/20. http://www.unep.org/Documents.Multilingual/Default.asp?Document ID=245&ArticleID=3042. Accessed January 7, 2007.

UNEP. 2007. Global outlook for ice and snow. United Nations Environment Programme, Nairobi, Kenya.

Van der Werf. 2009. CO_2 emissions from forest loss. *Nature Geoscience* 2:737–38.

Walter, K. M., M. E. Edwards, G. Grosse, S. A. Zimov and F. S. Chapin III. 2007b. Thermokarst lakes as a source of atmospheric CH_4 during the last deglaciation. *Science* 318:633–36.

Walter, K. M., L. C. Smith and F. S. Chapin III. 2007a. Methane bubbling from northern lakes: Present and future contributions to the global methane budget. *Philosophical Transactions of the Royal Society of London A* 365:1657–76.

Westerling, A. L., H. G. Hidalgo, D. R. Cayan, and T. W. Swetnam. 2006. Warming and earlier spring increase western U.S. forest wildfire activity. *Science* 313: 940–43.

Wigley, T. M. L., C. M. Ammann, B. D. Santer and S. C. B. Raper. 2005. Effect of climate sensitivity on the response to volcanic forcing. *Journal of Geophysical Research* 110:D09107.

Zhang, T. 2005. Influence of the seasonal snow cover on the ground thermal regime: An overview. *Reviews of Geophysics* 43:RG4002.

Zhang, X., F. W. Zwiers, G. C. Hegerl, F. H. Lambert, N. P. Gillett, S. Solomon, P. A. Stott and T. Nozawa. 2007. Detection of human influence on twentieth-century precipitation trends. *Nature* 448:461–465.

Zhang, X. D., and J. E. Walsh. 2006. Toward a seasonally ice-covered Arctic Ocean: Scenarios from the IPCC AR4 model simulations. *Journal of Climate* 19:1730–47.

Zhuang, Q., J. M. Melillo, M. C. Sarofim, D. W. Kicklighter, A. D. McGuire, B. S. Felzer, A. Sokolov, R. G. Prinn, P. A. Steudler, and S. Hu. 2006. CO_2 and CH_4 exchanges between land ecosystems and the atmosphere in northern high latitudes over the 21st century. *Geophysical Research Letters* 33:L17403.

3

Natural Selection and Phenotypic Plasticity in Wildlife Adaptation to Climate Change

James D. Austin, Christine W. Miller, and Robert J. Fletcher Jr.

Climate-change models predict that during the course of the 21st century the resilience of many species is likely to be outpaced by an unprecedented combination of climate change and other global alterations (especially land-use change and overexploitation). By 2100, ecosystems will be exposed to atmospheric CO_2 levels and global temperatures substantially higher than in the past 650,000 years (Solomon et al. 2007). As a consequence, sea level rise, increased heterogeneity in weather, and associated habitat change will alter the biodiversity and functioning of most ecosystems. This understanding has led to an increase in research on the ability of wildlife populations to cope with the effects of climate change and avoid extinction (Brodie et al., this volume).

Despite increasing theoretical and empirical research on how and at what pace wildlife can adapt, we still have only limited understanding of what to expect in terms of the impacts of climate change on species and ecosystems (Heino et al. 2009). The term "adaptation" can be used broadly in wildlife management and conservation. The Intergovernmental Panel on Climate Change (IPCC) defines adaptation as an "adjustment in natural or human systems in response to actual or expected climatic stimuli or their effects, which moderates harm or exploits beneficial opportunities." The US Fish and Wildlife Service and some state wildlife agencies are also developing wildlife adaptation plans in response to climate change. These different meanings and uses of adaptation require careful consideration as biologists and practitioners need be able to understand the intent of a particular action. That is, "incorporating adaptation into wildlife responses to climate change" means different things to different parties. Here we use the term "adaptation" from an evolutionary perspective (see table 3.1 for relevant terms).

Most wildlife populations are able to accommodate "normal" levels of environmental variability experienced within a lifetime. However, the predicted increase of climate-induced environmental variance may lead to variability too great for individuals to survive or reproduce successfully, so that the population begins to decline. The likely result will be

Table 3.1. Glossary of terms used in this chapter

Adaptation (evolutionary). The process of genetic change, the creation of new phenotypes from that change, and the corresponding increased fitness of individuals possessing the new genetic trait.

Canalization. The ability of an organism to produce a consistent phenotype regardless of environmental conditions. This ability often requires a developmental system with the ability to resist or buffer environmental influences on phenotypes (Stearns 1989).

Climate envelope models. A type of species distribution model in which envelopes of suitability are generated from climate data and information on known occurrence of a species. BIOCLIM is a common climate envelope model.

Coadapted gene complex. The interaction of alleles from different genes to produce viable or well-adapted phenotypes. At the intraspecific level, coadapted gene complexes are more likely to be common in species where gene flow between populations is low because high gene flow would disrupt adaptive complexes from forming.

Community genetics. A type of genetics that emphasizes evolutionary genetic processes that occur among interacting populations in communities, and which permits the evaluation of ecosystem consequences of species interactions (Antonovics 1992).

Evolution. The change in a form of phenotype that occurs over generations. Frequently a distinction is made between macroevolution and microevolution.

Fitness. In the evolutionary sense, the average greater reproductive success of an individual that can be attributed to a particular genotype.

Gene flow. The movement and incorporation (through reproduction) of alleles from conspecific populations.

Genetic drift. The alteration of populations' allele frequencies through sampling error (chance) rather than by natural selection, mutation, or immigration. In small populations genetic drift is a major factor, due to high variance in reproductive success among individuals from generation to generation.

Genetic effective population size (N_e). The number of breeding individuals in an *idealized population* that show the same amount of inbreeding or loss of diversity as the census population under consideration. The idealized population is a mathematically convenient one, having constant size through time, an even sex ratio, and equal reproductive success among individuals, among other assumptions. If a real population were to have or approach the ideal, then the N_e would be similar to the census size (N). Most natural populations have Ne far lower than N because of differential reproductive success, fluctuating population size, skewed sex ratios, and other ecological factors.

Heritability. The proportion of a phenotype's total variation that is attributable to the average effect of genes in a particular environment.

Natural selection. The process by which the relative frequency of a particular genotype changes from generation to generation because of differential fitness of phenotypes controlled by genes in question.

Ontogeny. The growth and development of an individual's anatomy from cell to maturity.

(Continued)

Table 3.1. (*Continued*)

Phenology. The timing of a recurring phenomenon, The phenomenon may be a response to one or more environmental factors such as photoperiod, climate, or drought.

Phenotypic integration. The correlation among certain traits, some of which may develop under natural selection. Phenotypic plasticity can evolve not only because of adaptive value, but also because of correlations with other traits (Pigliucci and Preston 2004).

Phenotypic plasticity. The ability of an organism to respond to its environment with a change in form, state, movement, or rate of activity. The broadest definition of phenotypic plasticity includes responses that are reversible and irreversible, adaptive and nonadaptive, active and passive, and continuously and discontinuously variable (West-Eberhard 2003).

Selection. As used in this chapter, natural selection: the differential fitness of genotypes that results in the change in relative frequencies of alleles over generations.

Standing genetic variation. The presence of neutral or slightly deleterious alleles found in a population, as opposed to the presence of allelic variants that appear by new mutation events.

---

to change patterns of population growth and abundance (Bell and Collins 2008). As a consequence, wildlife populations will respond to biotic and abiotic changes associated with increased climate variability in three fundamental ways: (1) by going extinct; (2) by expressing different behavioral or physical phenotypes under altered environmental conditions, including behavioral changes and increased dispersal to track suitable environments (phenotypic plasticity); or (3) by adaptive evolutionary responses to selection imposed by climate change (Holt 1990).

Numerous examples of distributional and phenological shifts with a changing climate have been documented for wildlife populations (e.g., Parmesan and Yohe 2003, Davis et al. 2005, Primack et al. 2009). Yet with continued habitat alteration, many species will not have the opportunity to track favorable conditions (Groom and Schumaker 1993). In these instances populations will need to adapt by evolving, responding plastically (other than dispersal), or both. For instance, although the timing of laying eggs is to some degree genetically determined, environmental cues that an individual experiences may signal the most appropriate time to lay eggs. Both evolution and phenotypic plasticity can be considered forms of adaptation in a general sense, though they have different implications for wildlife population persistence and management.

Not all species, or even populations within species, will adapt similarly to climatic (or other) variation (Young et al., this volume). Identifying and prioritizing conservation strategies will be improved with an understanding of the potential for plastic and evolutionary responses of populations. For example, recent climate-envelope models predicting large-scale extinctions (e.g., Lawler et al. 2009) often contain a high level of uncertainty. Such uncertainty will likely be reduced with a better understanding of the role of adaptation (e.g., Morin and Thuiller 2009). Finally, recent practices in climate-change management (e.g., "assisted migrations" and captive breeding) often ignore the implications of local adaptation in determining the long-term success of these management options (Popescu and Hunter, this volume).

Here we describe key evolutionary components of populations and species that are relevant to evolutionary adaptation in changing climates. We then highlight the concept of phenotypic plasticity and its relevance to climate change, as well as how the interplay of phenotypic plasticity and evolutionary adaptation may facilitate or hinder the probability of species persistence. We identify current gaps in the application of concepts of plasticity and evolution to climate-change–related biodiversity problems. Finally, we discuss some strategies for conservation and management under climate change.

Evolutionary Responses

Natural selection depends on several factors, including the rate at which new genetic material can be replenished through either *mutation* or *gene flow*, the amount of standing genetic variation in a population, the strength and consistency of selection on a particular trait, and how long selection occurs relative to the generation time and age structure of the organism in question. Given that these factors generally are nonexclusive, predicting and managing evolutionary change will be context-specific.

Adaptive evolution cannot occur without genetic-based trait variation and differential fitness among phenotypes. Mutations are the ultimate source of new variation, but their occurrence is exceedingly rare. Because of this, most adaptation in response to projected change will rely on *standing genetic variation*, as has been demonstrated from decades of research on quantitative genetics (Hill 1982, Roff 2007). Distinguishing between new mutations and standing variation is important because the process of adaptation is expected to be quite different between them. The persistence of new mutations depends on the magnitude of beneficial effect of that mutation and the effective population size (N_e). Since most new mutations are neutral or deleterious and wildlife populations are typically small (particularly those of conservation concern), many new mutations are expected to be nonadaptive or lost through drift. In contrast, standing genetic variation reflects the cumulative effect of recurrent mutations and genetic drift acting to maintain some mutation-generated neutral variation. Given the relationship between mutation, drift, and standing genetic variation, the latter is expected to be the product of many generations of accumulation of new allelic variants. Thus, the amount of standing variation may help (if high) or hinder (if low) adaptation. Under environmental change, existing neutral variation may become advantageous, and thus allow for adaptation (Barrett and Schluter 2007).

Even when adaptive traits are predicted to respond positively to climate change, the underlying genetic architecture can constrain adaptation. Etterson and Shaw (2001) found that the multivariate adaptive response of legume plants (*Chamaecrista fasciculate*) was much slower than when adaptively important traits (fecundity, reproductive state, leaf size, and number) were considered individually. This is because negative interactions among important traits can constrain overall adaptation. Although such genetic constraints can be overcome with time, in many cases this will be too slow to avoid extinction (Gomulkiewicz and Houle 2009).

Gene flow is for many species an important means of gaining or maintaining genetic variation. As a source of new variation, a small proportion of immigrants can offset the detrimental effects of population isolation by increasing N_e (Mills and Allendorf 1996). N_e is important because it, not census size (N), predicts the rate of inbreeding and the rate of fixation or loss of neutral and (future) adaptively important alleles. However, gene flow can sometimes have negative effects on evolutionary processes when spatial patterns of natural selection result in local evolutionary adaptation. In these situations, different populations may vary in a particular trait due to local environments favoring different variations of the trait. Gene flow in these instances can inhibit local adaptation by continually introducing locally nonadapted alleles (Hendry 2004).

Environmental selection is often strong (Hereford et al. 2004) and may increase in intensity under climate change in many habitats. Selection can also frequently change direction, meaning that the selective pressures faced by parents might be very different from those faced by their offspring due to temporal or spatial variation. Such fluctuating selection pressures can result in imprecisely adapted populations (Bell and Collins 2008). From a purely natural selection perspective, environments that constantly change will keep populations "on their toes" with respect to how well they are adapted. Importantly, rapid and constant directional change (e.g., increasingly warmer mean temperatures, drier conditions) may outpace the potential rate of adaptation for organisms with longer generation times, and high environmental variability can further impede directional selection.

Phenotypic Plasticity

Phenotypic plasticity, on the other hand, can often handle predictable and rapid changes in environmental conditions better than evolution can. Phenotypic plasticity refers to the tendency to produce different phenotypes (morphological, behavioral, or physiological) from a given genotype in response to different environments (figure 3.1). Phenotypic

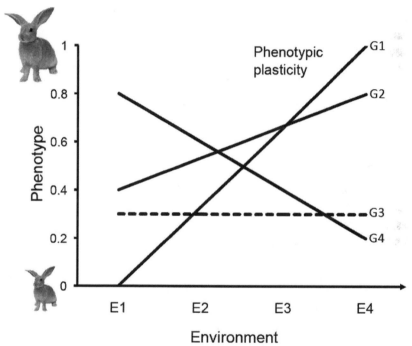

Figure 3.1. An example of phenotypic plasticity comes from the potential for genotype-by-environment interactions. Different degrees of phenotypic plasticity for a given trait are plotted as reaction norms in different environments. Plasticity occurs when the phenotype (rabbit size) produced by a given genotype (G1–G4), is determined by the environment. Note that in this example, genotypes G1, G2, and G4 converge on a similar phenotype at intermediate environments (E2 and E3) and are highly divergent at extreme environments (E1 and E4), thus representing genotype-by-environment interactions. Plasticity also varies, being greater in G1 (steeper slope) than in G2 or G4. Genotype G3 has no phenotypic plasticity and is said to be canalized. Modified from Garcia de Leaniz et al. 2007.

plasticity is best contrasted with the phenomenon of canalization, in which a particular trait will be expressed similarly under a wide range of natural environmental conditions (Stearns 1989). The role of plasticity under climate change will depend on whether the plasticity provides a fitness advantage that could allow for the persistence of populations until adaptive evolutionary change occurs (Price et al. 2003). Disentangling the relative effects of genetic background and phenotypic plasticity on behavioral or other adaptive responses can be difficult, though examples exist showing that plasticity can be used for adaptive responses under changing environmental conditions. For example, guppies adapted to low predation environments can still respond through phenotypic plasticity to predator-induced alarm pheromones (Huizinga et al. 2009).

The availability of predictive environmental cues at the correct time is crucial for the evolution and maintenance of phenotypic plasticity (Padilla and Adolph 1996, DeWitt et al. 1998). Behaviors are often immediately plastic. However, modifications in physiology and morphology require more time to occur relative to behavior. For example, geometrid moth larvae can match morphology and color to their diet and background over successive molts (Greene 1989).

Individuals are not infinitely plastic for all traits, and this is likely due to a lack of historically predictive cues or to the energy costs associated with maintaining the ability to be plastic (DeWitt et al. 1998). Nonetheless, some level of phenotypic plasticity is common for many traits in most organisms.

Evolutionary and Plastic Responses to Climate Change

Differentiating between evolutionary and plastic responses is an important component of our understanding of how and whether wildlife will adapt under climate change scenarios (Gienapp et al. 2008). Perhaps even more important, understanding the interplay of the two phenomena can lead to a richer appreciation of the ability of some species to cope with climate change. For example, phenotypic plasticity may allow species to respond immediately to change and facilitate the long-term evolutionary process (Rehfeldt et al. 2001, West-Eberhard 2003) by acting as a temporal buffer for adaptive evolution (Price et al. 2003). However, plasticity may not be able to sustain a directional response for long periods, as might be the case under long-term changes in climatic variables. Over longer periods, evolutionary change will likely be required for sustained directional responses, though at times the pace of evolution will also be insufficient for populations to sufficiently adapt (e.g., in long-lived organisms).

Plastic responses of animals under climate change will not always be adaptive, and may at times be detrimental. Such negative responses will be more likely when environmental conditions are novel and extreme. Plasticity will promote long-term population persistence under climate change if the plastic response is well matched to the optimal phenotype, and depending upon the rate at which nonheritable, environmentally induced variation can be converted to heritable variation (Ghalambor et al. 2007).

The extent of this theoretical role of plasticity in buffering populations from climate change is still unclear, though a few examples exist. Individual great tits (*Parus major*) that are highly plastic in their timing of reproduction in response to spring temperature and prey availability have had higher average fitness over a 30-year period. As a result, greater plasticity in an ecologically important trait (timing of reproduction) has led to natural selection acting for increased plasticity of laying date correlated with changing climate (Nussey et al. 2005). Examples of phenotypic plasticity in introduced species (e.g., anoles [Losos et al. 2000] and cane toads [Phillips and Shine 2006]) suggest an important role for plasticity in the invasion of new environments, including those habitats formed under climate change.

Simply documenting phenotypic change over time does not distinguish plasticity from evolution. Many studies focusing on phenotypic characters that purport to reflect rapid microevolution have been criticized for lacking evidence of genetic change (Gienapp et al. 2008). However, it is rarely feasible to measure the genetics of adaptively important traits (particularly in nonmodel organisms) and to differentiate genetic and environmental effects. Researchers can instead use organisms where controlled breeding or pedigree analysis is possible to examine the potential for adaptation to climatic changes through evolution and plasticity. Together with predictive modeling, these efforts will serve as a guide for predicting the responses of many other species to climate change. (van Asch et al. 2007, Ghalambor et al. 2007).

While phenotypic plasticity may be an essential mechanism for allowing populations to respond to climate change, alterations in the predictability of plasticity cues can have disastrous consequences. If formerly reliable cues become disassociated with a particular environment or resource, organisms may express a maladaptive phenotype in space and/or time. Environmental sex determination (ESD) is an example in which phenotypic plasticity (the determination of sex occuring often in response to temperature; Bull 1983) can lead to evolutionary traps (Schlaepfer et al. 2002). As environments change, ESD could increase the risk of extinction by resulting in large biases in sex ratio. Species with ESD (e.g., reptiles; Janzen 1994) are likely to be especially sensitive to global climate change (Walther et al. 2002). However, the detrimental effects of existing plasticity may be short-lived if selection acts quickly on nest site choice—for example, in response to skewed sex ratios (McGaugh et al. 2010). Another example is the use of photoperiod to predict environmental conditions, a widespread phenomenon in animals. The tendency of large, long-lived mammals to rely on photoperiod and other cues relative to smaller mammals could mean that the former will be less likely to adapt under climate change (Bronson 2010).

Limits to plasticity and evolution may exist partly because of internal and ecological limits to adaptation (figure 3.2) that are thought to be pervasive (DeWitt et al. 1998). Tadpoles that simply possessed the *ability* to be highly plastic in response to predators experienced some costs in mass, development, and survivorship (Relyea 2002). However, negative fitness costs associated with plasticity appear to be difficult to detect in natural systems (Van Buskirk and Steiner 2009).

From a conservation and management standpoint, species that express little plasticity and/or have low genetic variation will be the most vulnerable to long-term negative effects of directional climate change (Holt 1990). While predicting evolutionary responses may prove difficult, characteristics of species and populations can provide insight into the role that evolution and plasticity may play as climate continues to change. For example, species that already use a variety of habitats, foods, and other resources have demonstrated plasticity, which will likely be beneficial. Specialists and species that require narrow sets of resources will likely suffer. Small isolated populations, already of conservation concern, may be limited in their evolutionary adaptability, due to their low N_e and low standing genetic variation. The rate of climate change may also outpace the evolutionary response of long-lived organisms, particularly those with low fecundity.

The direct impact of plasticity and evolution within a wildlife conservation and management framework is still poorly documented. However, numerous related examples illustrate their importance. For example, naive prey can quickly adapt behaviorally to the reintroduction of predators in ecosystems (Berger et al. 2001). Habitat alteration and harvest regimes have been shown to alter selection regimes on phenotypes like size at first reproduction in fish (Fukuwaka and Morita 2008). And alterations in the seasonal behavior of red squirrels (*Tamiasciurus hudsonicus*) have been shown to have a heritable component, though most of the observed shift toward earlier breeding (18 days over a decade) was due primarily to plasticity (Reale et al. 2003).

a)

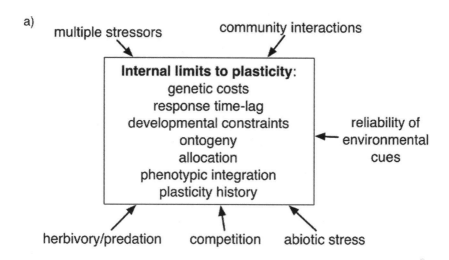

b)

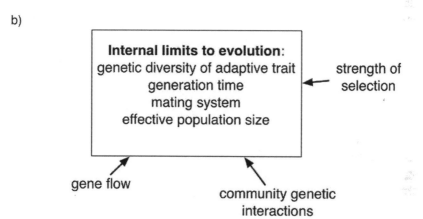

Figure 3.2. Multiple internal (*inside box*) and external (*outside box*) factors can influence the capacity of populations to respond to a given environmental factor via (a) plasticity and (b) natural selection. With plasticity, less is understood about the external or ecological limits than about the internal limits. These are the factors highly influenced by climate change. Both adaptation and plasticity are likely to be affected by dynamic changes to communities.

Applications of Evolution and Plasticity

The interpretation of climate change effects on wildlife, as well as the management strategies aimed at ameliorating those effects, can benefit by explicit consideration of phenotypic plasticity and evolutionary responses. These approaches are new to many managers and are increasingly important because adaptive change can happen rapidly, and in response to conservation schemes (Reznick and Ghalambor 2001, Smith and Bernatchez 2008).

Interpreting Recent Effects of Climate

While there have been numerous attempts to identify the fingerprint of contemporary climate change, only recently have investigators discovered whether the trends have occurred via evolution, plasticity, or both. A recent meta-analysis on the effect of human activities (e.g., introductions, translocations, harvesting) on rates of phenotypic change in 68 studies has demonstrated that most of the rapid phenotypic response has been largely due to plasticity rather than evolutionary change (Hendry et al. 2008), a result that may apply to interpreting climate-change effects as well (e.g., Ozgul et al. 2009).

Advances in modeling now provide a better understanding of the role of evolutionary responses and phenotypic plasticity in recent climate change. For example, Knight et al. (2008) extended stage-structured matrix models to incorporate evolutionary selective pressures on demographic traits of white trillium (*Trillium grandiflorum*). Coulson and Tuljakpurkar (2008) developed a novel framework that partitions the various types of selection and phenotypic plasticity within quantitative traits. Application of this model framework to a long-term dataset on red deer (*Cervus elaphus*) suggests that plasticity, rather than evolutionary responses to selection, has been a major driver in observed phenotypic change of birth weights. Such models could be applied to traits relevant to the effects of climate change, such as reproductive phenology, migration timing, or body size (Ozgul et al. 2009).

Forecasting Future Effects of Climate

Predicting how species will respond to future climate change and other rapidly occurring crises is increasingly becoming relevant not only for researchers but also for agencies, wildlife managers, and land planners. Most forecasting has focused on using bioclimatic envelopes of species-environment relationships and coupling them with projections for climate change (Lawler et al. 2009). These models have been adopted because they only require information on current distributions of species and climate variables at those locations, thereby allowing many species to be modeled across broad geographic ranges (Guisan and Thuiller 2005). Projections from these modeling attempts suggest devastating effects of climate on biodiversity in the coming decades. Although climate-envelope models of extinction probabilities have been criticized for ignoring adaptive response of organisms to perturbations (Dormann 2007,

Aitken et al. 2008), there is little for managers or policy makers to look to for guidance, particularly because adaptive responses to climate are not well understood (Davis et al. 2005) and few species have data that allow for more detailed mechanistic modeling of potential effects (but for an example of such modeling, see Rehfeldt et al. 1999).

How can these models incorporate evolution and phenotypic plasticity? Projections based on species distribution can be improved by incorporating dynamic processes and estimates of variation for data typical of species distribution models. For example, Oneill et al. (2008) incorporated population variation into modeling approaches, which substantially improved the model's performance by embracing the observed variation critical to evolutionary or plastic responses to climate change. Similarly, species distribution models that incorporate temporal variation in environmental relationships, such as variation in correlations of species occurrence in different years of variable climate, could further improve model performance and the ability to extrapolate to future climate conditions (e.g., Fletcher et al. 2011). Consequently, while it might be difficult or impossible to directly incorporate such complex processes as evolutionary response or phenotypic plasticity into data-sparse, phenomenological models of climate change, some of the dynamics necessary for such responses could be incorporated into model predictions, and could vastly improve them.

Biodiversity Conservation in a Community Context

The capacity for wildlife to respond to climate change through phenotypic plasticity does not only include the ability to respond to changing temperatures and precipitation. As climatic conditions shift, the assemblages of many plants and animals may change in a given area. A species may be able to cope with variation in temperature, but not to changes in the distribution or abundance of other species with which it interacts (e.g., Suttle et al. 2007, Post and Pedersen 2008). Thus, solely investigating the thermal tolerance of plants and animals will not give a complete picture of the potential for those species to adjust to climate change. For instance, the extinction of a major pollinator due to drought or higher temperatures may result in a domino effect, even for those plant species that themselves respond well to elevated temperatures (Hedhly et al. 2009). As communities shift in composition, selection pressures on individual species will likely change through changes in competition, predation, and mutualistic interactions. Changing selection pressures may be difficult to predict in all but the most extreme cases (e.g., obligate mutualists). Overall, understanding interactions among species will be an increasingly important part of wildlife conservation as habitats shift and novel interactions occur (e.g., introduced species). A considerable amount of research has shown that the ubiquity of plasticity in species' responses to one another will provide expanded adaptive potential and indirect effects on biodiversity (see Agrawal 2001, Miner et al. 2005).

Connectivity Conservation

A prominent approach to wildlife conservation is in facilitating connectivity (Crooks and Sanjayan 2006). Promoting connectivity of habitats in the context of future climate change is thought to be particularly important (Cross et al. this volume). This is because as climate changes, habitats will likely shift and organisms will putatively have to track those shifts to persist. Thus, promoting current and future connectivity of habitats as a "natural" way (as opposed to assisted migration; see below) to facilitate organisms to track changing conditions in space and time will potentially be crucial to minimizing the effects of climate change.

What does the potential for evolutionary responses and plasticity to climate change tell us about future connectivity conservation? While promoting connectivity will undoubtedly be a useful conservation measure, plasticity tells us that that some species may persist in the absence of tracking shifting habitat mosaics, at least for some period of time. As noted above, evolutionary perspectives also point to the role that gene flow can play in the changing ecological theater: in general, too much gene flow will swamp the potential for local adaptation, whereas too little may limit the variation available for selection to act upon.

These potential effects are complex, however, and may be highly dependent on numerous issues, including how the quality of habitats changes as climate continues to change. These changes could have profound effects on adaptive evolution. For example, theory suggests that high rates of immigration into sink habitats can facilitate adaptive niche evolution. Temporal variation in fitness, potentially driven by increased climate variation, can further facilitate adaptive evolution (Holt et al. 2004). In addition, the effects of expected asymmetric gene flow tracking favorable climate could be influenced by the quality of the habitats being connected. If these habitats function initially as source populations while previous source populations become sinks, natural selection should favor traits for the new source population (Kawecki and Holt 2002). Taken together, theoretical work on gene flow and spatiotemporal variation suggest that the effects of connectivity conservation in a changing climate may be profoundly influenced by how temporal changes in habitat quality progress.

Population genetics provides tools for estimating gene flow across changing landscapes and the role of connectivity, but the markers typically used are neutral ones, such that they are insufficient for interpreting adaptive evolutionary or phenotypic responses to connectivity conservation and climate change (Holderegger et al. 2006). Instead, interpreting these potential effects will require either a quantitative genetics approach or a focus on markers under selection (e.g., via genome scans; Holderegger and Wagner 2008).

Assisted Colonization

The past few years have seen increased focus on the merits of moving species that are potentially threatened by extinction from climate change (Mueller and Hellmann 2008; Hoegh-Guldberg et al. 2008; Popescu and Hunter, this volume). While we do not intend to contribute to the heated debate on the efficacy of assisted migrations here (Fazey and Fischer 2009), it is worth emphasizing the potential evolutionary implications associated with such strategies. As alluded to above, a high rate of migration can reduce the adaptation of recipient populations that are accustomed to different environmental conditions. This is particularly true when immigrants are from populations adapted differently from the host population. There are a number of genetic as well as demographic concerns about such management strategies, including the number and genetic makeup of colonizers, the frequency and amount of ongoing assisted migration that will be needed, and the landscape configuration of colonized populations. Quantitative approaches need to be adopted that focus on post-translocation monitoring in order to determine the factors that limit or drive successful translocations (e.g., Pelini et al. 2009).

Despite the potential positive effects of introducing novel genotypes from multiple source populations to a newly colonized small population, the disruption of coadapted gene complexes can result in overall reductions in fitness over time (Haig and Wagner 2001). Novel genotypes introduced through supplementation efforts have also been cited as a cause for native species becoming "more invasive" (Mueller and Hellmann 2008).

The viability of assisted migration efforts needs to be assessed prior to moving (or removing) individuals, to avoid detrimental impacts on source or target populations. This includes knowledge of the genetic and demographic resources available (i.e., across the native range) in order to help ascertain project viability (Haig and Wagner 2001). In other words, viability targets cannot be properly determined without an understanding of the basic population processes of intact populations. Landscape genetic studies of neutral and adaptive variation prior to assisted colonization will be critical to this end.

Conclusions

Understanding the effects of climate change on wildlife populations, and managing for such effects, will be insufficient for protecting and maintaining wildlife population viability without an appreciation of the potential for phenotypic plasticity and evolutionary responses of wildlife to climate change. These two phenomena provide the raw essentials for wildlife to respond to climate change, and they may allow populations to persist in a rapidly changing world. Nonetheless, there has been surprisingly little direct evidence that evolution and plasticity have reduced extinction rates in wild populations (Parmesan 2006). We attribute this lack of evidence, in part, to investigators not explicitly addressing these effects and not having the appropriate tools to do so. We caution, however, that not all populations may have the evolutionary or plastic potential necessary for adapting to rapid climate change. A major challenge in the coming years will be to better identify and interpret the roles of evolutionary responses and phenotypic plasticity for wildlife in a rapidly changing climate.

Addressing this challenge and adjusting management in light of evolutionary responses and phenotypic plasticity will require incorporating new tools and approaches. The development of monitoring designs that can rigorously and efficiently estimate effects of climate change, including the potential for estimating shifting species distributions and evolutionary and plastic responses, is needed. Targeted monitoring (sensu Nichols and Williams 2006) that both augments surveys with relevant data on individual variation (e.g., via mark-recapture, or with genetic markers) and targets locations for observing potential changes in spatial distribution may allow for better understanding of phenotypic change in space and time. Such monitoring strategies will need to be adaptive (Lindenmayer and Likens 2009) by updating monitoring designs as data become available regarding recent climate-change effects, especially observed effects during extreme climate events (e.g., extreme drought, heat waves). Such a focus on climatically relevant traits of species may further prove informative for interpreting the potential for evolutionary or plastic changes, but clearly more work is needed in this area as responses to climate change continue to be quantified. As the roles of evolutionary responses and phenotypic plasticity in observed and future effects of climate change are better understood, management and conservation strategies will become more effective at protecting biodiversity.

ACKNOWLEDGMENTS
This material is based on work supported by the National Science Foundation under grant no. NSF grant IOS-0926855 to CWM.

LITERATURE CITED

Agrawal, A. A. 2001. Phenotypic plasticity in the interactions and evolution of species. *Science* 294:321–26.

Aitken, S. N., S. Yeaman, J. A. Holliday, T. Wang, and S. Curtis-McLane. 2008. Adaptation, migration, or extirpation: Climate change outcomes for tree populations. *Evolutionary Applications* 1:95–111.

Antonovics, J. 1992. Toward community genetics. In *Plant Resistance to Herbivores and Pathogens: Ecology, Evolution, and Genetics*, edited by R. S. Fritz and E. L. Simms, 426–50. Chicago: University of Chicago Press.

Barrett, R. D. H., and D. Schluter. 2007. Adaptation from standing genetic variation. *Trends in Ecology and Evolution* 23:38–44.

Bell, G., and S. Collins. 2008. Adaptation, extinction and global change. *Evolutionary Applications* 1:3–16.

Berger, J., J. E. Swenson, and I.-L. Persson. 2001. Recolonizing carnivores and naïve prey: Conservation lessons from Pleistocene extinctions. *Science* 291: 1036–39.

Bronson, F. H. 2010. Climate change and seasonal reproduction in mammals. *Philosophical Transactions of the Royal Society B* 364:3331–40.

Bull, J. J. 1983. *Evolution of Sex Determining Mechanisms*. Benjamin-Cummings, Menlo Park, CA.

Coulson, T., and S. Tuljakpurkar. 2008. The dynamics of a quantitative trait in an age-structured population living in a variable environment. *American Naturalist* 172:599–612.

Crooks, K. R., and M. Sanjayan, eds. 2006. *Connectivity Conservation*. Cambridge University Press, Cambridge.

Davis, M. B., R. G. Shaw, and J. R. Etterson. 2005. Evolutionary responses to changing climate. *Ecology* 86:1704–14.

DeWitt, T. J., A. Sih, D. S. Wilson. 1998. Cost and limits of phenotypic plasticity. *Trends in Ecology and Evolution* 13:77–81.

Dormann, C. D. 2007. Promising the future? Global change projections of species distributions. *Basic and Applied Ecology* 8:387–97.

Etterson, J. R., and R. G. Shaw. 2001. Constraint to adaptive evolution in response to global warming. *Science* 294:151–54.

Fazey, I., and J. Fischer. 2009. Assisted colonization is a techno-fix. *Trends in Ecology and Evolution* 24:475.

Fletcher, R. J., Jr., J. S. Young, R. L. Hutto, A. Noson, and C. T. Rota. 2011. Insights from ecological theory on temporal dynamics and species distribution modeling. In *Predictive Modeling in Landscape Ecology*, edited by A. Drew, F. Huettmann, and Y. Wiersma, 91–107. New York: Springer.

Fukuwaka, M., and K. Morita. 2008. Increase in maturation size after the closure of a high seas gillnet fishery on hatchery-reared chum salmon *Oncorhynchus keta*. *Evolutionary Applications* 1:376–87.

Garcia de Leaniz, C., I. A. Fleming, S. Einum, E. Verspoor, W. C. Jordan, S. Consuegra, N. Aubin-Horth, D. Lajus, B. H. Letcher, A. F. Youngson, J. H. Webb,

L. A. Vøllestad, B. Villanueva, A. Ferguson, and T. P. Quinn. 2007. A critical review of adaptive genetic variation in Atlantic salmon: Implications for conservation. *Biological Reviews* 87:173–211.

Ghalambor, C. K., J. K. McKay, S. P. Carroll, and D. N. Reznick. 2007. Adaptive versus non-adaptive phenotypic plasticity and the potential for contemporary adaptation in new environments. *Functional Ecology* 21:394–407.

Gienapp, P., C. Teplitsky, J. S. Alho, J. A. Mills, and J. Merilä. 2008. Climate change and evolution: disentangling environmental and genetic responses. *Molecular Ecology* 17:167–78.

Gomulkiewicz, R., and D. Houle. 2009. Demographic and genetic constraints on evolution. *American Naturalist* 174:E218–29.

Greene, E. 1989. A diet-induced developmental polymorphism in a caterpillar. *Science* 243:643–46.

Groom, M. J., and N. Shumaker. 1993. Evaluating landscape change: patterns of worldwide deforestation and local fragmentation. In *Biotic Interactions and Global Change*, edited by P. M. Kareiva, J. G. Kingsolver, and R. B. Huey, 24–44. Sunderland, MA: Sinauer Associates.

Guisan, A. and W. Thuiller. 2005. Predicting species distribution: offering more than simple habitat models. *Ecology Letters* 8:993–1009.

Haig, S. M., and R. S. Wagner. 2001. Genetic considerations for introduced and augmented populations. Pages 444–51 in T. O'Neil and D. Johnson, eds., *Wildlife Habitats and Species Associations in Oregon and Washington: Building a Common Understanding for Management*. Oregon State University Press, Corvallis, OR.

Hedhly, A., J. I. Hormaza, and M. Herrero. 2009. Global warming and sexual plant reproduction. *Trends in Plant Science* 14:30–36.

Heino, J., Virkkala, R., and H. Toivonen. 2009. Climate change and freshwater biodiversity: detected patterns, future trends and adaptations in northern regions. *Biological Reviews* 84:39–54.

Hendry, A. P. 2004. Selection against migrants contributes to the rapid evolution of ecologically dependent reproductive isolation. *Evolutionary Ecology Research* 6:1219–36.

Hendry, A. P., T. J. Farrugia, and M. T. Kinnison. 2008. Human influences on rates of phenotypic change in wild animal populations. *Molecular Ecology* 17:20–29.

Hereford, J., T. F. Hansen, and D. Houle. 2004. Comparing strengths of directional selection: how strong is strong? *Evolution* 58:2133–43.

Hill, W. G. 1982. Rates of change in quantitative traits from fixation of new mutations. *Proceedings of the National Academy of Sciences USA* 79:142–45.

Holderegger, R., U. Kamm, and F. Gugerli. 2006. Adaptive vs. neutral genetic diversity: implications for landscape genetics. *Landscape Ecology* 21:797–807.

Holderegger, R. and H. H. Wagner 2008. Landscape genetics. *Bioscience* 58:199–208.

Hoegh-Guldberg, O., L. Hughes, S. McIntyre, D. B. Lindenmeyer, C. Parmesan, H. P. Possingham, and C. D. Thomas. 2008. Assisted colonization and rapid climate change. *Science* 321:345–46.

Holt, R. D. 1990. The microevolutionary consequences of climate change. *Trends in Ecology and Evolution* 5:311–15.

Holt, R. D., M. Barfield, and R. Gomulkiewicz. 2004. Temporal variation can facilitate niche evolution in harsh sink environments. *American Naturalist* 164:187–200.

Huizinga, M., C. K. Ghalambor, and D. N. Reznick. 2009. The genetic and environmental basis of adaptive differences in shoaling behaviour among populations of Trinidadian guppies, *Poecilia reticulata*. *Journal of Evolutionary Biology* 22:1860–66.

Janzen, F. J. 1994. Climate change and temperature-dependent sex determination in reptiles. *Proceedings of the National Academy of Sciences USA* 91:7487–90.

Kawecki, T. J., and R. D. Holt. 2002. Evolutionary consequences of asymmetric dispersal rates. *American Naturalist* 160:333–47.

Knight, T. M., M. Barfield, and R. D. Holt. 2008. Evolutionary dynamics as a component of stage-structured matrix models: An example using *Trillium grandiflorum*. *American Naturalist* 172:375–92.

Lawler, J. J., D. White, R. P. Neilson, and A. R. Blaustein. 2009. Predicting climate-induced range shifts: model differences and model reliability. *Global Change Biology* 12:1568–84.

Lindenmayer, D. B., and G. E. Likens. 2009. Adaptive monitoring: a new paradigm for long-term research and monitoring. *Trends in Ecology and Evolution* 24:482–86.

Losos, J. B., D. A. Creer, D. Glossip, R. Goellner, A. Hampton, G. Roberts, N. Haskell, P. Taylor, and J. Ettling. 2000. Evolutionary implications of phenotypic plasticity in the hindlimb of the lizard *Anolis sagrei*. *Evolution* 54:301–5.

McGaugh, S. E., L. E. Schwanz, R. M. Bowden, J. E. Gonzalez, and F. J. Janzen. 2010. Inheritance of nesting behaviour across natural environmental variation in a turtle with temperature-dependent sex determination. *Proceedings of the Royal Society B* 277:1219–26.

Mills, L. S., and F. W. Allendorf 1996. The one-migrant-per-generation rule in conservation and management. *Conservation Biology* 10:1509–18.

Miner, B. G., S. E. Sultan, S. G. Morgan, D. K. Padilla, and R. A. Relyea. 2005. Biological consequences of phenotypic plasticity. *Trends in Ecology and Evolution* 20:685–92.

Morin, X., and W. Thuiller. 2009. Comparing niche- and process-based models to reduce prediction uncertainty in species range shifts under climate change. *Ecology* 90:1301–13.

Mueller, J. M., and J. J. Hellmann. 2008. An assessment of invasion risk from assisted migration. *Conservation Biology* 22:562–67.

Nichols, J. D., and B. K. Williams. 2006. Monitoring for conservation. *Trends in Ecology and Evolution* 21:668–73.

Nussey, D. H., E. Postma, P. Gienapp, and M. E. Visser. 2005. Selection on heritable phenotypic plasticity in a wild bird population. *Science* 310:304–6.

Oneill, G. A., A. Hamann, and T. Wang. 2008. Accounting for population variation improves estimates of the impact of climate change on species' growth and distribution. *Journal of Applied Ecology* 45:1040–49.

Ozgul, A., S. Tuljapurkar, T. G. Benton, J. M. Pemberton, T. H. Clutton-Brock, and T. Coulson. 2009. The dynamics of phenotypic change and the shrinking sheep of St. Kilda. *Science* 352:464–67.

Padilla, D. K., and S. C. Adolph. 1996. Plastic inducible morphologies are not always adaptive: The importance of time delays in a stochastic environment. *Evolutionary Ecology* 10:105–17.

Parmesan, C. 2006. Ecological and evolutionary responses to recent climate change. *Annual Review of Ecology, Evolution, and Systematics* 37:637–69.

Parmesan, C., and G. Yohe. 2003. A globally coherent fingerprint of climate change impacts across natural systems. *Nature* 421:37–42.

Pelini, S. L., J. D. K. Dzurisin, K. M. Prior, C. M. Williams, T. D. Marsico, B. J. Sinclair, and J. J. Hellmann. 2009. Translocation experiments with butterflies reveal limits to enhancement of poleward populations under climate change. *Proceedings of the National Academy of Sciences USA* 106:11160–65.

Pigliucci, M., and K. Preston K, eds. 2004. Phenotypic Integration: *Studying the Ecology and Evolution of Complex Phenotypes*. Oxford: Oxford University Press.

Phillips, B. L., and R. Shine 2006. Spatial and temporal variation in the morphology (and thus, predicted impact) of an invasive species in Australia. *Ecography* 29:205–12.

Post, E. and C. Pedersen. 2008. Opposing plant community responses to warming with and without herbivores. *Proceedings of the National Academy of Sciences, USA* 105:12353–58.

Price, T. D., A. Qvarnstrom, and D. E. Irwin. 2003. The role of phenotypic plasticity in driving genetic evolution. *Proceedings of the Royal Society B* 270:1433–40.

Primack, R. B., I. Ibanez, H. Higuchi, S. D. Lee, A. J. Miller-Rushing, A. M. Wilson, and J. A. Silander. 2009. Spatial and interspecific variability in phonological responses to warming temperatures. *Biological Conservation* 142:2569–77.

Reale, D., A. G. McAdam, S. Boutin, D. Berteaux. 2003. Genetic and plastic responses of a northern mammal to climate change. *Proceedings of the Royal SocietyB* 270:591.

Rehfeldt, G. E., W. R. Wykoff, and C. C. Ying. 2001. Physiologic plasticity, evolution, and impacts of a changing climate on *Pinus contorta*. *Climatic Change* 50:355–76.

Rehfeldt, G. E., C. C. Ying, D. L. Spittlehouse, and D. A. Hamilton. 1999. Genetic responses to climate in *Pinus contorta*: Niche breadth, climate change, and reforestation. *Ecological Monographs* 69:375–407.

Relyea, R. A. 2002. Costs of Phenotypic Plasticity. *American Naturalist* 159:272–82.

Reznick, D. N. and C. K. Ghalambor. 2001 The population ecology of contemporary adaptations: what empirical studies reveal about the conditions that promote adaptive evolution. *Genetica* 112–13:183–98.

Roff, D. 2007. *Evolutionary Quantitative Genetics*. New York: Chapman & Hall.

Schlaepfer, M.A., M. C. Runge, and P. W. Sherman. 2002. Ecological and evolutionary traps. *Trends in Ecology and Evolution* 17:474–80.

Smith, T. B., and L. Bernatchez. 2008. Evolutionary change in human-altered environments. *Molecular Ecology* 17:1–8.

Solomon, S., D. Qin, M. Manning, R.B. Alley, T. Berntsen, N.L. Bindoff, Z. Chen, A. Chidthaisong, J.M. Gregory, G.C. Hegerl, M. Heimann, B. Hewitson, B.J. Hoskins, F. Joos, J. Jouzel, V. Kattsov, U. Lohmann, T. Matsuno, M. Molina, N. Nicholls, J. Overpeck, G. Raga, V. Ramaswamy, J. Ren, M. Rusticucci, R. Somerville, T.F. Stocker, P. Whetton, R.A. Wood and D. Wratt. 2007. Technical Summary. Pages 21–91 in S. Solomon, D. Qin, M. Manning, Z. Chen, M. Marquis, K. B. Averyt, M. Tignor and H. L. Miller, eds. *Climate Change 2007: The Physical Science Basis. Contribution of Working Group I to the Fourth Assessment Report of the Intergovernmental Panel on Climate Change*. Cambridge: Cambridge University Press.

Stearns, S. C. 1989. The evolutionary significance of phenotypic plasticity:

Phenotypic sources of variation among organisms can be described by developmental switches and reaction norms. *Bioscience* 39:436–45.

Suttle, K. B., M. A. Thomsen, and M. E. Power. 2007. Species interactions reverse grassland responses to changing climate. *Science* 315:640–42.

Van Asch, M., P. H. Tienderen, L. J. M. Holleman, and M. E. Visser. 2007. Predicting adaptation of phenology in response to climate change, an insect herbivore example. *Global Change Biology* 13:1596–1604.

Van Buskirk, J., and U. K. Steiner. 2009. The fitness costs of developmental canalization and plasticity. *Journal of Evolutionary Biology* 22:852–60.

Walther, G.-R., E. post, P. Convey, A. Menzel, C. Parmesan, T. J. C. Beebee, J.-M. Fromentin, O. Hoegh-Guldberg, and F. Bairlein. 2002. Ecological responses to recent climate change. *Nature* 416:389–95.

West-Eberhard, M. J. 2003. *Developmental Plasticity and Evolution*. Oxford: Oxford University Press.

Demographic Approaches to Assessing Climate Change Impact: An Application to Pond-Breeding Frogs and Shifting Hydropatterns

John H. Matthews, W. Chris Funk, and
Cameron K. Ghalambor

Applications of conservation science to aquatic ecosystems have traditionally focused on water quality and quantity issues, the establishment of protected areas, the need for restoring impaired or reduced hydrological function (including flow regime), and the improvement of connectivity that has been damaged by freshwater infrastructure (Brönmark and Hansson 2002; Brinson and Malvarez 2002). All of these issues will remain critically important in the future and in the face of climate change. However, anthropogenic climate change represents two new threats. First, traditional freshwater conservation practice assumes climate "stationarity" (Milly et al. 2008), which means that interventions risk producing poor results without reevaluation in light of realized and potential effects. Second, freshwater species have a diminished capacity to respond autonomously to climate change-induced shifts as a result of extensive human modifications of these ecosystems that are not climate-related (Matthews et al. 2009; Matthews and Wickel 2009). Indeed, management responses to climate change, such as building more "clean energy" hydropower facilities or increasing irrigation demands as a result of lower soil moistures, are likely to produce synergistic effects between historic threats and emerging climate conditions (Le Quesne et al. 2010).

How will freshwater species respond to these changes? Amphibians hold a special significance with regard to this question. Given their wide distribution globally and the ability of many species to connect both terrestrial and aquatic ecosystems through their life histories, amphibians may be especially vulnerable to climate change impacts in both sets of landscapes, and some researchers suggest that they may be the most threatened group of vertebrates globally (Stuart et al. 2004; IUCN et al. 2008), particularly as a result of climate change (Corn 2005; Pounds et al. 2006; McMenamin et al. 2008).

In this chapter, we focus on the gap between ecosystem-level shifts driven by climate change and the implications for population dynamics, connectivity, and management. A key assumption in this analysis is that amphibians may be sensitive to climate-change–related threats due to the coupling between many species' life histories and aquatic hydropattern, which is the "normal" cycling of high and low water in an annual period (typically represented by an annual hydrograph; Jackson 2006). Here, we will describe the importance of hydropattern as a key variable in freshwater ecosystems, discuss how variation in historic hydropatterns shape evolved and plastic reproductive strategies, describe a framework for how hydropattern may be shifting globally as a result of climate change, present a demographic model to examine the effects of hydropattern shifts on pond-breeding amphibians relative to management options to offset these impacts, posit the evolutionary implications of hydropattern shifts, and finally propose management approaches for amphibians and aquatic ecosystems that may encompass both traditional and climate-induced threats. In this chapter we attempt to demonstrate that demographic modeling combined with sufficient ecosystem and species knowledge can assist in the design of targeted conservation management interventions.

Shifting Hydropattern: An Emerging Focus for Amphibian Conservation

Phenology and demography are linked in many species. Recent research on fresh waters points to the critical role of hydroperiod (for standing water systems) and flow regime (for flowing water systems) in determining a wide range of species, communities, and ecosystem qualities. Hydroperiod and flow regime are collectively referred to as hydropattern. Poff (1997) has referred to hydropattern as the "master variable" for freshwater ecosystems, determining water quality and water quantity, life-history patterns for a wide variety of aquatic and semi-aquatic species, and the trophic structure of ephemeral/intermittent (fish-free) versus permanent/perennial (fish-present) systems. A growing body of literature has documented the importance of hydropattern in determining habitat suitability for aquatic taxa such as amphibians in tropical and temperate regions, particularly for phenologies such as breeding, egglaying, and life-stage transitions (Wiggins et al. 1980; Williams 1987; Williams 2006; see below).

Hydropatterns are determined in a particular region or ecosystem through a wide variety of factors, such as surface-to-volume ratio, flow hydraulics, and geomorphology. However, climate plays an especially critical role as hydropattern is determined by precipitation and groundwater inflows (including seasonality, intensity, quantity) and outflows (especially evapotranspiration or ET, reflecting air temperature and humidity: Winter and Woo 1990; Brinson 1993; Jackson 2006).

While air temperature is a critical variable when discussing terrestrial ecosystems, the relationship between air and water temperature is complex. Since liquid water has a greater thermal mass than the atmosphere, water temperatures may not be shifting in conjunction with air temperatures in some regions and/or some types of systems (Poole et al. 2004). Indeed, hydropattern-driven volume changes, such as changes in salinity, pH, and time until complete drying of a given water body (Williams 2006; Matthews 2011), are likely to play a more important role in water quality (Allen and Ingram 2002; Karl and Trenberth 2003; Lambert et al. 2004).

Historic Hydropatterns and the Evolved and Plastic Reproductive Strategies of Amphibians

Amphibians exhibit a diverse range of reproductive strategies that often reflect the historic hydropatterns and associated selective pressures under which they evolved. These evolved strategies are observed as the mean differences between populations and species in key life-history traits such as the number and size of offspring, the length of the larval period, and the age and size at maturity (reviewed in Morrison and Hero 2003). These life-history differences are correlated with environmental gradients, such as altitude, latitude, and hydropattern, that reflect variation in important ecological factors including predation risk, length of the breeding season, temperature, food availability, and the probability of pond desiccation (e.g., Wilber 1997; Altweg and Reyer 2003; Morrison and Hero 2003; Rose 2005; see below). For example, at higher latitudes and altitudes where the breeding season is short, temperatures are cool, and breeding ponds have longer hydroperiods, amphibians tend to produce smaller clutches of larger eggs, have longer larval periods, and take longer to mature despite being under selection for rapid growth to compensate for the relatively short and cool breeding season (Berven 1982a, b; Morrison and Hero 2003).

These same life-history traits also show extensive phenotypic plasticity in response to these ecological variables, and the degree of plasticity also varies between populations and species (e.g., Leips and Travis 1994; McCollum and VanBuskirk 1996; Denver et al. 1998; Laurila and Kujasalo 1999). For example, amphibians breeding in ponds with variable predation risk in time and space exhibit greater predator-induced plasticity in morphology and growth rate, presumably as an adaptive strategy to cope with the unpredictable nature of predation risk (Lardner 2000). Thus, as hydroperiods shift in response to climate change, amphibian species and populations may be more or less adapted to coping with environmental change depending on where they fall along these environmental gradients. Relatively little research has been done to identify specific life-history traits that would alter populations susceptibility to climate change, but it is reasonable to expect that populations that have historically experienced stable hydropatterns will be most vulnerable to disruption of the inflow and outflow of water.

For example, some amphibian species, such as spadefoot toads (genera *Scaphiopus* and *Spea*), have evolved striking patterns of phenotypic plasticity in their growth rates, ages, sizes at maturity, foraging modes as a means of dealing with highly variable hydropatterns (reviewed in Doughty and Reznick 2004). Yet there is also variation in these plastic responses that reflects the historic hydropatterns of a population or species. Further, the degree of plasticity in growth and stage of development in spadefoot toads increases with how ephemeral their hydropattern is (Morey and Reznick 2000). Again, the implications of this variation in plasticity for climate change remain unknown, but with regards to increasing severity and frequency of extreme events that alter hydropatterns, a reasonable hypothesis is that species and populations from ephemeral environments may be more resilient than their counterparts from more stable environments because they are more plastic.

A Framework for Describing Impacts on Amphibian Populations and Life-History Patterns

Many trends suggest that hydropatterns are shifting globally, perhaps even in the majority of freshwater ecosystems (Carney et al. this volume). However, the literature describing climate change interactions between hydropattern and amphibians is limited. Studies have tended to focus on what might be considered primary, direct climate change impacts, as in physiological tolerance studies (e.g., of phenological or range shifts: Beebee 2002; Parra-Olea et al. 2005), rather than on secondary or tertiary climate change impacts, as in studies of population demographics, interspecies relationships, community assemblages, and ecosystem tipping points—topics generally associated with multivariate and often highly uncertain processes.[1] Our purpose here is not to review the literature on the effects of climate change on amphibians, but rather to present new perspectives and modeling approaches on how phenology and studies of

1. Much of the discussion in recent years of climate change impact on amphibians has raged over the causes of observed global amphibian declines, such as over the relative importance of climate shifts on the distribution and abundance of invasive species or diseases (e.g., Pounds et al. 2006; Lips et al. 2008). In most cases, "the" global amphibian decline seems likely to be multifactorial rather than from a single cause, and we will avoid these debates in this analysis.

ecosystem-level impacts, landscape connectivity, and climate change can be integrated to understand the effects of climate change on amphibian demography. We refer the reader to recent reviews for a broad overview of the effects of climate change on amphibians (e.g., Corn 2005; Collins and Crump 2009).

Are shifts in hydropattern a current threat to amphibians? The climate science literature documents widespread impacts on hydropattern-relevant variables. Precipitation patterns have been shifting worldwide since at least 1945 (Dore 2005). Globally, annual precipitation is increasing (Wentz 2007), although there is much variation in regional and local precipitation patterns. In some areas, such as the Murray-Darling Basin of Australia, rates of evapotranspiration change are outpacing or reinforcing shifts in precipitation (Murphy and Timbal 2008). Thus, even when annual precipitation amounts may be increasing, given shifts in other climate variables, there may be a net decrease in surface water for aquatic species.

Given that key life-history events are keyed to hydropattern, how will climate change impact amphibian demographic patterns? Here we describe general geographic patterns of hydropattern alteration, and how the mode of climate change can influence hydropattern and amphibian demographic patterns.

At temperate and boreal latitudes, and at high altitudes in tropical and subtropical regions, hydropatterns are influenced by the timing and form of precipitation and ambient air temperature (Magnuson et al. 1997). Most freshwater ecosystems in these areas include a period of ice cover and water temperatures that do not sustain active movement and development for most amphibian species. In almost all temperate zones, ice-free periods are lengthening and more winter precipitation is falling as rain rather than snow, thus resulting in a reduced snowpack and a longer warm period (Magnuson et al. 1997; IPCC 2008). In many regions, summer temperatures and/or evapotranspiration rates are also increasing, thus raising the rate of outflows. For some water bodies, this may result in a transition from a long-hydropattern state (a permanent or perennial system) into a short-hydropattern state (an ephemeral, temporary, or intermittent system). Given the influence of water volume on water quality, not all water need leave a system to render it unviable for aquatic species: water temperatures are more likely to match high air temperatures at low volumes, dissolved oxygen rates can fall as temperatures increase, and pH and salinity can increase rapidly as solute concentration increases (Oviatt 1997; Poff et al. 2002; Issar 2003; Matthews 2011).

As suggested above, trophic structure can also shift dramatically in response to changes in state from short- to long-hydropattern regimes. At temperate and tropical latitudes, for instance, long-hydropattern systems tend to be dominated by vertebrate fish predators, while short-hydropattern systems are often dominated by invertebrate predators such as aquatic insects (e.g., Hecnar and M'Closkey 1997; Hero et al. 1998; Williams 2006). Regional shifts between long and short hydropatterns are likely to have profound impact on community assemblages and survivorship of amphibians.

In contrast, monsoonal precipitation patterns dominate influences on hydropattern at tropical and subtropical latitudes, as intra-annual temperatures do not vary as widely there as they do in temperate and boreal regions. Generally speaking, hydropatterns in these regions will be most sensitive to changes in monsoon timing, which can easily result in shifts between long (permanent) and short (ephemeral) hydropattern states. Although air temperature shifts are predicted to be more dramatic at middle or high latitudes, tropical species may be more sensitive than temperate species to small increases in temperature , because of a smaller difference between current and optimal temperatures (i.e., a smaller thermal safety margin; Deutsch et al. 2008).

Interestingly, these observations may also be relevant to species that are considered semiaquatic or dependent on very humid but nominally terrestrial ecosystems. Particularly in tropical rainforests and cloud forests, many amphibians have direct development, in which eggs are laid in leaf litter or on leaves; larval development occurs within the egg, and individuals hatch as miniature versions of adults (Duellman and Trueb 1994). These habitats are characterized by high humidity and abundant moisture even in the absence of a fluid environment. In effect, these species exist within a terrestrial "humidity envelope," so that they are subjected to pressures similar to those experienced by fully aquatic amphibian species. Indeed, high-elevation tropical amphibians in these habitats may be already in substantial decline due to the drying of the terrestrial humidity envelope (e.g., Pounds et al. 1999; Rovito et al. 2009)

While examination of broad latitudinal patterns is critical to understanding global amphibian demographic impacts from climate change, the mode through which regional or local climate shifts are expressed is also important to consider. Perhaps the three most important modes of climate change are shifts in "mean" climate conditions (e.g., annual precipitation increases), shifts in seasonality (e.g., more rain in winter, later monsoon arrival), and shifts in climate extremes (in effect, the climate variance around the mean, such as more intense tropical cyclones or more long droughts; see table 4.1 for an overview of all three modes;

Table 4.1. Modes of climate change and implications for ecosystems and management responses

Mode of climate shift	Relevant spatial and temporal scale	Implications for ecosystems	Implications for management	Evolutionary ecology impacts
Shifts in "mean" climate	Basin and seasonal/annual; slow change in "normal" weather "patterns	Persistent and usually slow movement of climate; probably the least harmful form of climate change for most species and ecosystems	Type of change most often predicted by circulation models, but not often shown in climate history	Unidirectional selection
Change in climate seasonality	Whole basin, monthly or below; seasonality changes already widespread	Flow regime and hydropattern shifts	Management for old or new flow regime extremely difficult to predict; best detected through retrospective analyses	Selection for changes in the timing of phenological events such as breeding and development time
More extreme weather	Basin to local, mostly subannual	Major changes in precipitation/ evapotranspiration intensity, and very hot/cold days that shock ecosystems for brief periods, destabilizing ecosystem processes and local species and populations	Water quality variability increasing due to intense precipitation and high evapotranspiration rates (ET) and temperature shocks	Bottlenecks, plasticity, microevolution

Le Quesne et al. 2010). Note that these three modes are not mutually exclusive and in many localities all three modes of change are occurring simultaneously (IPCC 2008).

Understanding the mode of change is important to understanding the climate pressures and potential for adaptive responses to those pressures. Gradual shifts in mean climate conditions, for instance, are probably the least acute of the three modes for amphibian populations over short time scales (roughly, decades to centuries). From an evolutionary perspective, a shift in mean climate is equivalent to persistent directional selection. However, not all shifts in mean climate are likely to be gradual. As discussed above, the paleoecological literature shows ample evidence that a cluster of climate variables describing a particular climate state or "plateaus" can cross thresholds or tipping points that result in abrupt transitions to new climate states. Many glacial-interglacial transitions, for instance, spanned only a few decades (Issar 2003; Anderson et al. 2007). To date, few state-level changes appear to have occurred as a result of anthropogenic climate change. Three possible large-scale candidates described in the current literature are alterations in ocean currents off the Pacific Northwest coast of North America (Grantham et al. 2004), the widespread disappearance of subtropical and tropical snowpacks and glaciers (Thompson et al. 2006), and rapid change in the precipitation-evapotranspiration regime in southeastern Australia (Murphy and Timbal 2008).

Shifts in seasonality present a different set of pressures, particularly for community-level interactions and life-history stages. Seasonal changes such as earlier springs, later winters, and increases in the number of frost-free days also are occurring (Magnuson et al. 1997; Schindler 1998). Monsoon timing is shifting in the tropics (Bueh et al. 2003; Mirza and Ahmad 2005), melting dates for snowpack are advancing (Nolin and Daly 2006; IPCC 2007), and ice-free periods for lakes and rivers are increasing in length as freeze dates recede and breakup dates advance (Prowse et al. 2006).

Phenological mismatches between species (e.g., butterfly-plant, parasite-host) and between triggering cues and significant life-history events are likely with seasonality shifts. In the latter case, dependence on cues that are proxies for climate variables (such as photoperiod) can present a significant "signaling" problem for an organism (Parmesan and Yohe 2003; Parmesan 2006). From an ecological perspective the ability to sense and appropriately respond to these cues is likely to occur initially through adaptive phenotypic plasticity, or the capacity of individuals to track environmental changes through altered developmental, physiological, and behavioral changes. How well amphibians can track phenological changes will in turn determine the strength of selection and the capacity for evolutionary changes to occur (Ghalambor et al. 2007). Given the importance of hydropattern signaling for amphibians and the responsiveness of hydropatterns to alterations in the precipitation-evapotranspiration regime, shifts in seasonality are likely to be a critical factor in shaping amphibian responses to climate change.

Likewise, changes in the frequency and severity of climate extremes can have a powerful impact on populations and communities if these shifts alter disturbance regimes. Weather extremes have also been increasing in frequency globally (Groismann et al. 1999; Easterling et al. 2000); they have been shown to have a powerful influence on populations and community assemblages (Parmesan et al. 2000; McLaughlin et al. 2003; Tews et al. this volume). Precipitation intensity is growing in many regions, impacting runoff, erosion, nutrient loading, and groundwater recharge processes (Zhang and Nearing 2005; Beighley et al. 2008). Greater climate variability is increasing the frequency of floods and droughts (IPCC 2007). Extreme events represent episodes of selection that in the short term may cause mass mortality and reproductive failure, and in the long term may impact the genetic structure of populations and the interactions between community members. Ultimately, these changes in extreme severity and frequency modify the ecological disturbance regime that over decades or centuries can alter key ecosystem qualities.

Demographic Modeling of Climate Change Impacts

A recent IPCC technical report (2008) on water and climate identified small, shallow lentic systems, such as ponds and other wetlands used by many amphibian species, as the most climate-change–threatened type of habitat globally, due to their rapid response to even minimal shifts in climate characteristics, with their low volume making them more likely to change from permanent to ephemeral (or from less ephemeral to more ephemeral). They are probably also one of the types of systems least likely to be monitored to observe these impacts.

Climate change can affect lentic-breeding amphibians directly, by reducing their survival and fecundity, or indirectly, by altering their habitats (figure 4.1). Much climate change research focuses on direct effects on fitness, although many species, including amphibians, are more likely to first be impacted by habitat changes (Collins and Crump 2009). Moreover, climate change can impact breeding, foraging, and/or overwintering habitats that are used by amphibians in different life-history stages (figure 4.1). These habitat changes, in turn, can reduce survival and fecundity rates, thereby reducing population growth rates and increasing extinction probabilities. Thus, predicting the effects of climate change on amphibian populations requires an understanding of which habitats will be impacted and how these changes will affect amphibian vital (birth and death) rates.

Here, we provide an example of how population projection matrix models can be used to assess the effects of climate change and management actions on the growth and extinction probabilities of amphibian populations. Similar modeling approaches have been used to assess the effects of shifting hydropattern, precipitation, and climate on the demog-

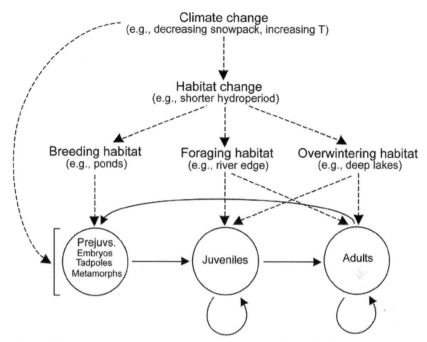

Figure 4.1. Conceptual diagram of the direct effects (on different life-history stages) and indirect effects (mediated through habitat changes) of climate change on pond-breeding amphibians. Solid arrows indicate life-history transitions; dashed arrows indicate climate change impacts.

raphy of plants (Smith et al. 2005), birds (Elderd and Nott 2008), and mammals (Saltz et al. 2006; Bakker et al. 2009). In a projection matrix model, the population is divided into discrete age or stage classes and the contributions of individuals in each class at one census to each stage at the next census are tracked (Morris and Doak 2002). In the context of climate change, projection matrix models can be used to predict how climate change impacts on specific vital rates will affect population growth rates (λ) and extinction probabilities (P_e). Moreover, projection matrix models can also be used to predict how management actions targeted at specific vital rates can offset the effects of climate change. Thus, one particular advantage of matrix models in climate change impact research and management is that they can be used to predict the effects of specific management actions while additional demographic data is collected to improve the model and thus improve future predictions (Bakker and Doak 2009).

We used projection matrix models to examine the effects of reduced tadpole survival (S_t) caused by reduced hydroperiods predicted under climate change on λ and P_e of amphibian populations. We had two specific questions. First, how will various scenarios of reduced hydroperiod affect λ and P_e? Second, how effective are different management strategies at offsetting the effects of reductions in S_t caused by shorter hydroperiods? Our goal, however, is not to make accurate predictions for a given population, but to demonstrate how projection matrix models can be used to address these questions for amphibians put at risk by climate change.

Our focal species for this analysis was the Columbia spotted frog (*Rana luteiventris*) in the northern Rocky Mountains of Montana. Columbia spotted frogs (hereafter referred to as CSFs) are a widely distributed pond-breeding species found in the Great Basin of Nevada, Utah, Idaho, and Oregon as well as northward into eastern Washington, northern Idaho, western Montana, northwestern Wyoming, western Alberta, British Columbia, and the southernmost reaches of the Yukon Territory. CSFs have a typical temperate-zone pond-breeding frog life history in that the adults lay eggs in the early spring, embryos hatch one to two weeks later, and tadpoles develop in ponds and metamorphose in the summer (figure 4.1; Stebbins 2003). These frogs become sexually mature in two or more years, depending on sex, latitude, altitude, and other local environmental conditions (Bull 2005).

Our analysis focuses on a low-elevation population of CSFs in Keeler Creek in the Cabinet Mountains of northwestern Montana. This population consists of eight breeding ponds (primarily beaver ponds) adjacent to Keeler Creek and connected to it by small inlet and outlet streams (Funk et al. 2005a). The breeding ponds are separated from each other by a maximum straight-line distance of 7 km. Genetic analysis using microsatellite loci showed that these eight ponds represent a single random mating population that is genetically distinct from nearby populations (Funk et al. 2005b).

We use CSFs from Keeler Creek as our focal population for this analysis for three main reasons. First, CSFs in the northern Rocky Mountains may be negatively impacted by climate change due to shortening in the length of the regional hydroperiod regime. Climate change is predicted not only to increase temperatures and therefore evapotranspiration, but also to reduce spring snowpack in the northern Rocky Mountains (Karl et al. 2009). This combination of increasing evapotranspiration and decreasing snowpack is predicted to result in shorter high-water periods (i.e., earlier drying) for ponds used by CSFs for breeding, since snowmelt is the main source of water in these ponds. Second, WCF collected detailed demographic data for CSFs in Keeler Creek between 2000 and 2003 as part of a dispersal study (Funk et al. 2005a), allowing us to parameterize a matrix model that quantitatively describes the demography of this population. Finally, during this demographic study, WCF observed CSF breeding ponds in Keeler Creek drying out completely before tadpoles were able to metamorphose, demonstrating that pond drying potentially already impacts this population, and providing a baseline estimate of the probability of pond drying prior to metamorphosis.

We constructed a female-based, post-birth pulse Lefkovitch matrix model (Burgman et al. 1993; Caswell 2000) with annual projection intervals representing a population with a life history consisting of three stages: prejuvenile (embryo, tadpole, and overwintering metamorph), juvenile, and reproductive adult. The matrix for this life history is

$$
\begin{bmatrix}
0 & [S_j \times \Psi_{ja} \times P_l \times C] & [S_a \times P_l \times C] \\
[S_e \times S_t \times S_j^{0.67}] & [S_j \times (1 - \Psi_{ja})] & 0 \\
0 & [S_j \times \Psi_{ja}] & [S_a]
\end{bmatrix}
$$

where S_e = embryo survival, S_t = tadpole survival, S_j = juvenile survival, S_a = adult survival, Ψ_{ja} = the probability that a juvenile becomes an adult (given that the juvenile has survived), P_l = the probability of laying eggs, and C = clutch size. S_j is taken to the power of 0.67 in the transition from prejuveniles to juveniles (matrix element a_{21}), to account for the fact that new juveniles metamorphose at the end of the summer and therefore only have to survive eight-twelfths (0.67) of a year until the next census. This matrix is the same as the one used in Biek et al. (2002) for temperate-zone pond-breeding frogs.

We parameterized our matrix model using vital rate means and temporal variances estimated for the low-elevation Keeler Creek population (table 4.2). S_j, S_a, and Ψ_{ja} were estimated in 2000, 2001, and 2002 using capture-mark-recapture (CMR) analysis (Funk et al. 2005a). Females were considered adults once they reached 50 mm snout-vent length, the minimum size of breeding females observed in this population. The probability of laying eggs (P) was assumed to be 1.00. This is likely a reasonable assumption for low-elevation CSFs. Most adult females were clearly gravid by the end of the summer and many were found breeding in consecutive years, suggesting that low-elevation females typically breed every year.

Clutch size (C) was estimated using a modified version of the sub-

Table 4.2. Vital rate estimates used in projection matrix model for Columbia spotted frogs from lower Keeler Creek, Montana

Vital rate symbol	Vital rate description	Mean	Temporal process variance
S_e	Embryo survival	0.9810	0.0006
S_t	Tadpole survival	0.0191	0.0000
S_j	Juvenile survival	0.4700	0.0570
S_a	Adult survival	0.7800	0.0470
P_l	Probability of laying eggs	1.0000	0.0000
C	Clutch size	434	1302
Ψ_{ja}	Probability of juvenile becoming adult	0.1000	0.0029

sampling-volume displacement method of Werner et al. (1999). In this method (1) the egg mass (clutch) volume is measured with a 1-liter beaker using volume displacement; (2) the volume of three different egg mass subsamples is measured with a 250 ml graduated cylinder; (3) the number of eggs in each subsample is counted; and (4) the average ratio of the number of eggs per milliliter for the three subsamples is used to estimate the number of eggs in the entire clutch. Counts of entire egg masses showed that this method provides accurate estimates of clutch size (Maxell and Funk, unpublished data). Clutch size estimates were divided by two to estimate the number of female embryos, assuming an equal sex ratio. Clutch sizes were estimated for 33, 45, and 61 egg masses in the low-elevation Keeler Creek population in 2001, 2002, and 2003, respectively.

Observations at the beginning of the study revealed that embryo survival approached 1.00 unless an entire clutch was stranded due to desiccation, in which case embryo survival was 0.00. Embryo survival (S_e) for a given breeding pond was thus estimated as the proportion of egg masses that did not dry out due to stranding. S_e was estimated at each breeding pond in 2001, 2002, and 2003.

Tadpole survival (S_t) was estimated for each pond as the estimated number of metamorphs (tadpoles that successfully metamorphosed into froglets) divided by the estimated number of hatchlings (embryos that hatched). The number of hatchlings was estimated as the number of egg masses $\times C \times S_e$. The number of metamorphs was estimated using Chapman's unbiased version of the Lincoln-Petersen closed-population CMR estimator of abundance (Seber 1982) and dividing by two to estimate the number of female metamorphs. S_t was also estimated at each breeding pond in 2001, 2002, and 2003.

In some years, ponds dried before any tadpoles metamorphosed due to a short hydroperiod. We defined P_{dry} as the probability of a pond drying before tadpoles were able to metamorphose. We estimated P_{dry} in 2001, 2002, and 2003 as the number of ponds that dried before metamorphosis divided by the total number of ponds in the population (eight). Ponds that dried before metamorphosis were not included in estimates of S_t so that baseline tadpole mortality ($1-S_t$) could be modeling separately from tadpole mortality caused by short hydroperiods.

For each vital rate, process variance among years (temporal variance) was estimated using White's (2000) method, which removes sampling variance from total variance to estimate process variance. Using process variance rather than total variance is important to avoid over-estimating P_e. We used Morris and Doak's (2002) program (white.m) in Matlab R2008b to estimate process variance. For our starting population size vector we assumed 100 adult females and a stable stage distribution.

We used stochastic projection matrix model simulations to estimate stochastic population growth rates (λ_s), the number of adult females at the end of simulations (only including simulated populations that did not decrease below the quasi-extinction threshold), and quasi-extinction probabilities (P_{qe}) under various hydroperiod and management scenarios for a single CSF breeding pond in the low-elevation Keeler Creek population. We defined P_{qe} as the probability of falling below a predefined threshold population size (10 adult females). Our matrix simulations are based on those of Morris and Doak (2002) as implemented in their Matlab program vitalsim.m. In our model, however, we did not include vital rate correlations, autocorrelations, or cross-correlations, as there are insufficient data to reliably estimate these parameters for CSFs.

We modeled projected reductions in hydroperiods in our simulations by incorporating a probability of complete reproductive failure for a given year ($S_t = 0$) due to pond drying before metamorphosis. Specifically, P_{dry} for each year of the simulation was determined by randomly drawing from a beta distribution of P_{dry}. The beta distribution of P_{dry} was determined from the mean and variance observed in the low-elevation Keeler Creek population from 2001 to 2003. If $P_{dry} \geq$ a uniform random number, then $S_t = 0$ for that year. If $P_{dry} <$ the uniform random number, then S_t was chosen from the baseline S_t beta distribution. To model an increasing probability of P_{dry} over the next 40 years, the mean of the beta distribution was increased. For example, in a scenario of a 50% increase in P_{dry} over the next 40 years and a baseline P_{dry} of 0.20, mean P_{dry} would be 0.25

in year 20 [= 0.20 × (1+ (20 × (0.50/40)))]. For each scenario, we ran five independent runs of 5,000 iterations each for 40 years. Our Matlab code is available from WCF upon request.

Climate change projections for the northwest United States (which includes Keeler Creek) predict a 10% to 30% decrease in spring snowpack by mid-century and a 1.7 to 5.6° C increase in temperature by the end of the century (Karl et al. 2009). The impact of these climatic changes on hydroperiods of the ephemeral, semipermanent, and permanent ponds used by CSFs for breeding is unknown, although a conservative estimate is that these changes in climate will result in a reduction in hydroperiod equivalent to the reduction in snowpack (10%–30%) since snowmelt is the main source of water in these ponds.

Also uncertain is the relationship between hydroperiod length and the probability of a pond drying before CSF tadpoles metamorphose (P_{dry} as defined above). Field observations by WCF show that in many cases, CSF tadpoles metamorphose "just in time," within a few days before ponds dry out completely. This leads to the prediction that a small reduction in hydroperiod length may result in a large increase in P_{dry}. Thus we hypothesize that the relationship between P_{dry} and hydroperiod length is sigmoidal, with small reductions in hydroperiod resulting in dramatic increases in P_{dry} (figure 4.2).

Accordingly, we considered climate change scenarios with large increases in P_{dry} over the next 40 years of 50%, 100%, and 200%. We feel that these are reasonable possibilities for lower Keeler Creek. We also ran simulations with no probability of pond drying before metamorphosis, and with the current, baseline P_{dry} for comparison. In addition, we modeled the effects of three different management scenarios to counteract the negative effects of a 100% increase in P_{dry}. First, we simulated the effects of introducing 10, 50, 100, or 200 metamorphs from a different source population to the focal population once every five years. In Keeler Creek and other nearby CSF populations, WCF observed considerable variation in recruitment among ponds from 2001 to 2003. Thus, in any given year, there were usually some ponds with high recruitment that could serve as source populations for translocations. Translocations would therefore be a relatively straightforward management option. Second, we considered the effects of increasing connectivity so that five metamorphs immigrate to the focal population each year. One possible method of increasing connectivity for amphibians is the construction of road underpasses to facilitate movement between populations on opposite sides of roads (Lesbarreres et al. 2004). Finally, we examined the effects of increasing

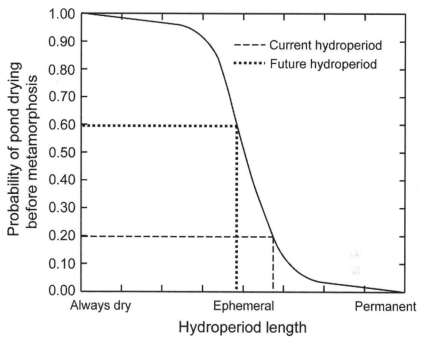

Figure 4.2. Hypothesized relationship between the probability of pond drying before tadpoles can metamorphose (P_{dry}) and hydroperiod length. In the example shown, a small future reduction in hydroperiod results in a 200% increase in P_{dry}.

juvenile survival by 10%, for example, by controlling native or introduced predators or by using road underpasses to reduce juvenile mortality.

Output from Demographic Models

The projection matrix model simulations indicate that reproductive failure due to pond drying is predicted to have a significant negative impact on population growth rates and population persistence (table 4.3; figure 4.3). With the current rate of pond drying (P_{dry} = 0.1984), the stochastic population growth (λ_s) is 4.3% lower than it would be without any pond drying. The final number of adults is 74% lower with current pond drying compared to what it would be with no pond drying. The probability of quasi-extinction (P_{qe}) is also 214% higher with the current rate of pond drying than with no drying. With the current rate of pond drying, λ_s is slightly greater than one, thus indicating that the population is growing at a slow rate. With increases in P_{dry} over the next 40 years, however, population growth dips below one, indicating a population de-

Table 4.3. Results of projection matrix model simulations for Columbia spotted frogs from lower Keeler Creek, Montana, under different climate change and management scenarios. P_{dry} = baseline probability of a pond drying before tadpoles can metamorphose; $IncP_{dry}$ = proportional increase in P_{dry} over the next 40 years due to climate change; $IncP_{dry}$ = proportional increase metamorphs are introduced or immigrating; $IncS_j$ = increase in juvenile survival through management actions; λ_s = stochastic growth rate for a given scenario; number of adults = number of adult females after 40 years for simulated populations that have remained above the quasi-extinction threshold of 10 or fewer adults. Means and 95% confidence intervals (CIs) are based on five independent runs of 5,000 iterations per run for each scenario.

Scenario	Impacts		Management			λ_s		Number of adults	
	P_{dry}	$IncP_{dry}$	N_{met}	MetFreq	$IncS_j$	Mean	95% CI	Mean	95% CI
No pond drying	0.00	0.00	0	N.A.	0.00	1.0534	1.0527–1.0541	998	970–1,026
Baseline (current) pond drying	0.1984	0.00	0	N.A.	0.00	1.0082	1.0076–1.0088	258	249–67
50% increase in pond drying	0.1984	0.50	0	N.A.	0.00	0.9960	0.9953–0.9967	178	177–79
100% increase in pond drying	0.1984	1.00	0	N.A.	0.00	0.9834	0.9830–0.9838	129	127–31
200% increase in pond drying	0.1984	2.00	0	N.A.	0.00	0.9567	0.9562–0.9572	69	68–70
Introduce 10 metamorphs every 5 years	0.1984	1.00	10	Every 5 years	0.00	0.9871*	0.9868–0.9874*	128	125–31
Introduce 50 metamorphs every 5 years	0.1984	1.00	50	Every 5 years	0.00	0.9942*	0.9936–0.9948*	132	126–38
Introduce 100 metamorphs every 5 years	0.1984	1.00	100	Every 5 years	0.00	1.0000*	0.9997–1.0003*	137	133–41
Introduce 200 metamorphs every 5 years	0.1984	1.00	200	Every 5 years	0.00	1.0078*	1.0074–1.0082*	156	153–59
Increase connectivity	0.1984	1.00	5	Every year	0.00	0.9909*	0.9903–0.9915*	128	123–33
Increase juvenile survival by 10%	0.1984	1.00	0	N.A.	0.10	1.0265	1.0260–1.0270	403	393–413

*For the metamorph introduction and increased connectivity management scenarios, estimates of λ_s are expected to approximate 1, because when there is a fixed introduction of a given number of individuals to a population that otherwise has $\lambda_s < 1$, the population will be stable (D. F. Doak, personal communication).

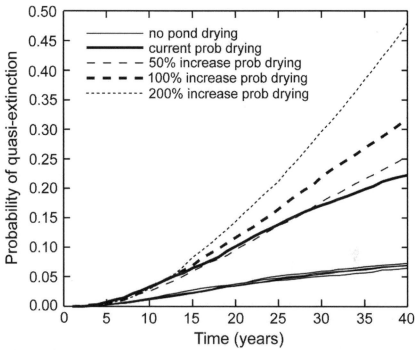

Figure 4.3. Cumulative distribution of quasi-extinction probability (P_{qe}) for scenarios with different probabilities of pond drying before tadpoles can metamorphose (P_{dry}). The five different runs for the "no pond drying" scenario are shown as an example of the amount of variability typically observed among runs.

cline. The number of adults is predicted to decrease 31%, 50%, and 73%, and P_{qe} is predicted to increase 14%, 45%, and 118% over the current level in 40 years with a 50%, 100%, and 200% increase in P_{dry}, respectively.

The three management scenarios counteract the negative effects of a 100% increase in P_{dry} to varying degrees (table 4.3; figure 4.4). A relatively ineffective management scenario is increased connectivity resulting in five juvenile immigrants per year. This strategy only increases λ_s by 0.8% and reduces P_{qe} in 40 years by 13%. It has essentially no effect on the final number of adults. The second most effective management strategy is introduction of metamorphs once every five years. For example, introduction of 100 metamorphs once every five years increases λ_s by 1.7%, increases the number of adults by 6.2%, and reduces P_{qe} by 34%. The most effective option is increasing juvenile survival by 10%, which results in an increase in λ_s of 4.4%, an increase in the number of adults of 212%, and a reduction in P_{qe} by 63%.

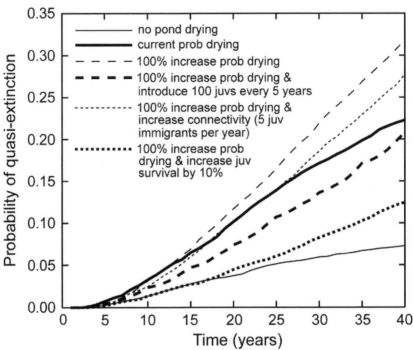

Figure 4.4. Cumulative distribution of quasi-extinction probability (P_{qe}) for different management scenarios assuming a 100% increase over the next 40 years in the probability of pond drying before tadpoles can metamorphose (P_{dry}).

These projection matrix model simulations show that, despite generally low sensitivity of λ to variation in S_t for pond-breeding frogs (Biek et al. 2002), periodic mass mortality of tadpoles due to pond drying can have a significant negative impact on λ_s, the final number of adults, and P_{qe} for CSFs in the low-elevation Keeler Creek population. Moreover, although historical P_{dry} is unknown, λ_s and the final number of adults are substantially lower and P_{qe} is much higher with the current rate of P_{dry} than in the case of $P_{dry} = 0$, thus suggesting that this population may already be impacted by pond drying. Indeed, McMenamin et al. (2008) have recently found evidence that CSFs and two other amphibian species have declined in Yellowstone National Park in the last 16 years due at least in part to pond desiccation. Increases in P_{dry} will only exacerbate the predicted impact on the lower Keeler Creek population of CSFs. A large increase in P_{dry} is required, however, for a significant impact on this population. A 50% increase in P_{dry} is predicted to have a minimal effect on λ_s, the number of adults, and P_{qe}. But impacts on λ_s, the number of adults, and P_{qe} increase substantially if P_{dry} increases by 100% or 200% in 40 years.

Fortunately, the simulations also demonstrate that management actions can buffer this population from the negative impacts of brief high-water periods and earlier drying, although different strategies vary substantially in their effectiveness. If all of these management actions were equally feasible and cost the same, then increasing S_j would clearly be the best option because it would increase λ_s and the number of adults and decrease P_{qe} the most. For a given species and population, however, some strategies may not be possible at all, or some may be more expensive than others. In a real management situation, feasibility and cost-effectiveness clearly should be included in the analysis.

There are several data and model limitations in the current analysis. The first data limitation is that our vital rate estimates are based on data from a limited number of years (four years for this population). This is often the case with projection matrix models. Second, although climate change models clearly show that hydroperiods in the northern Rocky Mountains will become shorter, the magnitude of this reduction is uncertain. Third, the relationship between P_{dry} and hydroperiod length is unknown although, as explained earlier, the observation that many tadpoles metamorphose a few days before complete desiccation of ponds suggests that a short reduction in hydroperiod may result in a large increase in P_{dry}. Finally, the effect of climate change on other vital rates is unknown for CSFs. For example, Scherer et al. (2008) found that adult survival of boreal toads (*Bufo boreas*) in Colorado was positively related to winter temperatures. Similarly, McCaffery and Maxell (2010) found a positive relationship between survival and winter temperatures and between breeding probabilities and winter temperatures in high-elevation populations of CSFs in the Bitterroot Mountains of Montana. Thus, increased temperatures from climate change may also have some positive effects on amphibian populations.

Our simulations are limited in that they do not include vital rate correlations, autocorrelations, cross-correlations, density dependence, or demographic stochasticity, all of which could change our predictions. Collecting additional years of data will allow proper estimation of these parameters. A second limitation is that the models do not incorporate plasticity or evolution, which we recognize can buffer populations over shorter and longer time periods (see above). For example, tadpoles may be able to develop and complete metamorphosis faster, either through phenotypic plasticity within generations or by adaptation to selection imposed by shorter hydroperiods across generations. Furthermore, populations may vary in their capacity to exhibit plasticity or in their response to selection, thus making it difficult for us to make universal generalizations. Any demographic model that does not include plasticity or adaptation is likely to predict a greater impact than a model that does include these factors, thus resulting in somewhat "pessimistic" results.

Several gaps in the current understanding of climate change and amphibian demography need to be filled before population models can accurately predict climate change impacts and the effectiveness of management actions for amphibians. First, much more research is needed to understand future hydroperiod changes of amphibian breeding ponds, the relationship between P_{dry} and hydroperiod, and the effect of climate change on multiple vital rates. Second, long-term demographic studies are needed to estimate vital rates, vital rate correlations, autocorrelations, and cross-correlations, and also to understand the importance of other demographic processes, such as density dependence, in amphibian population dynamics. Third, demographic studies should be conducted on amphibian species with different reproductive modes and life histories to determine the variation in their responses to climate change. In addition, amphibian demographic research and amphibian management should be conducted in an adaptive management framework in which model predictions are tested, data is collected to improve those predictions, and management practices are adjusted on the basis of the improved models (Bakker and Doak 2009).

Recommendations for Management

Perhaps the most important contribution of the modeling efforts presented here is the implication that physiological ecology approaches to climate change, such as bioclimate envelope modeling, are likely to be insufficient for capturing the complex suite of climate variables that describe hydropattern. The phenological mismatches induced by advancing drying times present a severe, unforgiving threat for many amphibians in an aquatic stage. Thus, observational data for regional drying trends is critical to determine whether permanent ecosystems are becoming intermittent, or whether intermittent systems are becoming more frequently intermittent. Such an assessment can identify habitats, regions, and individual water bodies that may be at special risk from climate change, even in the absence of more detailed natural history information or high-confidence global circulation model downscaling to discern trends in precipitation impacts (Matthews and Wickel 2009).

Closely related to this is the need for a formal consideration of connectivity within a network of aquatic ecosystems (cf. Cross et al. this volume). Ideally, demographic modeling can provide guidance about the critical or sensitive stages of amphibian life history to guide management interventions. For instance, are there "traditional" barriers to dispersal between habitat patches, such as water infrastructure? Is climate change altering these barriers—for example, is it increasing the frequency of terrestrial fires? Likewise, can water infrastructure be managed to buffer shifts in hydroperiod for critically endangered habitat or species, and to buffer ecosystems from some climate change impacts (Le Quesne et al. 2010)? By extension, the importance of including an area of management focus large enough to encompass significant microclimate variation could be useful in preventing more frequent extreme weather events from eliminating multiple populations from a network and disrupting their connectivity over large areas.

Over longer time scales, amphibian management should also consider the evolutionary implications of shifts in mean climate and seasonality, both of which are likely to be experienced as directional selection. What are the limits to adaptation and phenotypic plasticity in responses to climatic change (see Austin et al. this volume)? Is there evidence of potential critical mismatches in phenology between other members of the trophic network or the climate-regulated portions of the disturbance regime? Does change appear to advance much faster than species can respond? Is there genetic evidence for multigenerational connectivity between disjunct populations? A decision framework has recently been proposed for assisted migration (Hoegh-Goldberg et al. 2008; Popescu and Hunter, this volume); it may be necessary to consider this for taxa, such as amphibians, that inhabit isolated habitat patches embedded in semiarid landscapes, with little potential for autonomous connectivity.

Here we have focused on the importance of hydropattern as an aspect of freshwater ecosystems that are rapidly shifting in response to anthropogenic climate change, especially as a function of shifts in the timing and amount of precipitation. Hydropattern has a critical relationship with the life-history patterns of many amphibians, particularly pond-dwelling species. Shifts in ponds from long to short hydropattern regimes are likely to be especially dire for amphibian populations that have been acclimated to relatively stable hydrological conditions. Matrix modeling is one method for untangling the relative impacts of climate change from other demographic influences, such as loss of connectivity, and it can be helpful in pursuing sound, adaptive management interventions. Ideally, matrix modeling can assist resource managers in trying to explore future impacts. However, some amphibian species demonstrate complex reactions to shifting hydropatterns; and species- or population-specific research that focuses on estimates of major demographic variables, evolutionary implications of demographic changes across population networks, and the possibility of phenotypically plastic interactions with hydropattern should also be considered.

LITERATURE CITED

Allen, M. R., and W. J. Ingram. 2002. Constraints on future changes in climate and the hydrologic cycle. *Nature* 419:224–32.

Altwegg, R., and H.U. Reyer. 2003. Patterns of natural selection on size at metamorphosis in water frogs. *Evolution* 57:872–82.

Anderson, D., A. Goudie, and A. Parker. 2007. *Global Environments through the Quaternary*. Oxford: Oxford University Press.

Bakker, V. J., and D. F. Doak. 2009. Population viability management: Ecological standards to guide adaptive management for rare species. *Frontiers in Ecology and the Environment* 7:158–65.

Bakker, V. J., D. F. Doak, G. W. Roemer, D. K. Garcelon, T. J. Coonan, S. A. Morrison, C. Lynch, K. Ralls, and R. Shaw. 2009. Incorporating ecological drivers and uncertainty into a demographic population viability analysis for the island fox. *Ecological Monographs* 79:77–108.

Beighley, R. E., T. Dunne, and J. M. Melack. 2008. Impacts of climate variability and land use alterations on frequency distributions of terrestrial runoff loading to coastal waters in southern California. *Journal of the American Water Resources Association* 44: 62–74.

Beebee, T. J. C. 2002. Amphibian phenology and climate change. *Conservation Biology* 16:1454.

Berven, K. A.. 1982a. The genetic basis of altitudinal variation in the wood frog *Rana sylvatica*. 1. An experimental analysis of life history traits. *Evolution* 36:962–83.

———. 1982b. The genetic basis of altitudinal variation in the wood frog *Rana sylvatica*. 2. An experimental analysis of larval development. *Oecologia* 52: 360–69.

Biek, R., W. C. Funk, B. A. Maxell, and L. S. Mills. 2002. What is missing in amphibian decline research: Insights from ecological sensitivity analysis. *Conservation Biology* 16:728–34.

Brinson, M. M. 1993. Changes in the functioning of wetlands along environmental gradients. *Wetlands* 13:65–74.

Brinson, M. M., and A. I. Malvarez. 2002. Temperate freshwater wetlands: Types, status, and threats. *Environmental Conservation* 29:115–33.

Brönmark, C., and L. Hansson. 2002. Environmental issues in lakes and ponds: current state and perspectives. *Environmental Conservation* 29:290–306.

Bueh, C., U. Cubasch, and S. Hageman. 2003. Impacts of global warming on changes in the east Asian monsoon and the related time-slice experiment. *Climate Research* 24:47–57.

Bull, E. L.. 2005. *Ecology of the Columbia Spotted Frog in Northeastern Oregon*. USDA Forest Service, Pacific Northwest Research Station, Portland, OR.

Burgman, M. A., S. Ferson, and H. R. Akçakaya. 1993. *Risk Assessment in Conservation Biology*. New York: Chapmann and Hall.

Caswell, H.. 2000. Matrix population models: *Construction, Analysis, and Interpretation*. Sunderland, MA: Sinauer Associates.

Colins, J. P., and M. L. Crump. 2009. Extinction in Our Times: *Global Amphibian Declines*. Oxford: Oxford University Press.

Corn, P. S.. 2005. Climate change and amphibians. *Animal Biodiversity and Conservation* 28:59–67.

Denver, R. J., N. Mirhadi, and M. Philips. 1998. Adaptive plasticity in amphibian metamorphosis: Response of *Scaphiopus hammondi* tadpoles to habitat dessication. *Ecology* 79:1859–72.

Deutsch, C. A., J. J. Tewksbury, R. B. Huey, K. S. Sheldon, C. K. Ghalambor, D. C. Haak, and P. R. Martin. 2008. Impacts of climate warming on terrestrial ectotherms across latitude. *Proceedings of the National Academy of Sciences of the United States of America* 105:6668–72.

Dore, M. H. I. 2005. Climate change and changes in global precipitation patterns: What do we know? *Environment International* 31:1167–81.

Doughty, P., and D. N. Reznick. 2004. Patterns and analysis of adaptive phenotypic plasticity in animals. In *Phenotypic Plasticity: Functional and Conceptual Approaches*, edited by T. J. DeWitt and S. M. Scheiner, 126–50. Oxford: Oxford University Press.

Duellman, W. E., and L. Trueb. 1994. Biology of Amphibians. Baltimore: Johns Hopkins University Press.

Easterling, D. R., J. L. Evans, P. Y. Groisman, T. R. Karl, K. E. Kunkel, and P. Ambenje. 2000. Observed variability and trends in extreme climate events: A brief review. *Bulletin of the American Meteorological Society* 81:417–25.

Elderd, B.D., and M.P. Nott. 2008. Hydrology, habitat change and population demography: an individual-based model for the endangered Cape Sable seaside sparrow *Ammodramus maritimus mirabilis*. *Journal of Applied Ecology* 45:258–68.

Funk, W. C., M. S. Blouin, P. S. Corn, B. A. Maxell, D. S. Pilliod, S. Amish, and F. W. Allendorf. 2005b. Population structure of Columbia spotted frogs (*Rana luteiventris*) is strongly affected by the landscape. *Molecular Ecology* 14:483–96.

Funk, W. C., A. E. Greene, P. S. Corn, and F. W. Allendorf. 2005a. High dispersal in a frog species suggests that it is vulnerable to habitat fragmentation. *Biology Letters* 1:13–16.

Ghalambor, C. K., J. K. McKay, S. Carroll, and D. N. Reznick. 2007. Adaptive versus non-adaptive phenotypic plasticity and the potential for contemporary adaptation to new environments. *Functional Ecology* 21:394–407.

Grantham, B. A., F. Chan, K. J. Nielsen, D. S. Fox, J. A. Barth, A. Huyer, J. Lubchenco, and B. A. Menge. 2004. Upwelling-driven nearshore hypoxia signals ecosystem and oceanographic changes in the northeast Pacific. *Nature* 429:749–54.

Groisman, P. Y., T. R. Karl, D. R. Easterling, R. W. Knight, P. F. Jamason, K. J. Hennessy, R. Suppiah, C. M. Page, J. Wibig, K. Fortuniak, V. N. Razuvaev, A. Douglas, E. Føland, and P. M. Zhai. 1999. Changes in the probability of heavy precipitation: important indicators of climatic change. *Climatic Change* 42:243–83.

Hecnar, S. J., and R. T. M'Closkey. 1997. The effects of predatory fish on amphibian species richness and distribution. *Biological Conservation* 79:123–31.

Hero, J. M., C. Gascon, and W. E. Magnusson. 1998. Direct and indirect effects of predation on tadpole community structure in the Amazon rainforest. *Australian Journal of Ecology* 23:474–82.

Hoegh-Goldberg, O., L. Hughes, S. McIntyre, D. B. Lindemayer, C. Parmesan, H. P. Possingham, and C. D. Thomas. 2008. Assisted migration and rapid climate change. *Science* 321:345–46.

Intergovernmental Panel on Climate Change. 2007. *Climate Change 2007: Impacts, Adaptation, and Vulnerability*, edited by M. Parry, O. Canziani, J. Palutikoff, P. van der Linden, and C. Hanson. Cambridge: Cambridge University Press.

Intergovernmental Panel on Climate Change. 2008. Climate change and water, edited by B. Bates, Z. Kundzewicz, S. Wu, and J. Palutikoff. Technical paper of the Intergovernmental Panel on Climate Change, IPCC Secretariat, Geneva.

Issar, A.S.. 2003. *Climate Changes during the Holocene and their Impact on Hydrological Systems*. Cambridge: Cambridge University Press.

IUCN, Conservation International, and NatureServe. 2008. An analysis of amphibians on the 2008 IUCN Red List <www.iucnredlist.org/amphibians>. Accessed August 4, 2009.

Jackson, C. R.. 2006. Wetland hydrology. In *Ecology of Freshwater and Estuarine Wetlands*, edited by D. P. Batzer and R. R. Sharitz , 43–81. Berkeley: University of California Press.

Karl, T. R., J. M. Melillo, and T. C. Peterson. 2009. *Global Climate Change Impacts in the United States*. Cambridge: Cambridge University Press.

Karl, T. R., and K. E. Trenberth. 2003. Modern global climate change. *Science* 302:1719–23.

Lambert, F. H., P. A.Stott, M. R. Allen, and M. A. Palmer. 2004. Detection and attribution of changes in 20th century land precipitation. *Geophysical Research Letters* 31: L10203.

Lardner, B.. 2000. Morphological and life history responses to predators in larvae of seven anurans. *Oikos* 88:169–80.

Laurila, A., and J. Kujasalo. 1999. Habitat duration, predation risk, and phenotypic plasticity in common frog (*Rana temporaria*) tadpoles. *Journal of Animal Ecology* 68:1123–32.

Leips, J., and J. Travis. 1994. Metamorphic responses to changing food levels in 2 species of hylid frogs. *Ecology* 75:1345–56.

Le Quesne, T., J. H. Matthews, C. von der Heyden, A. J. Wickel, R. Wilby, J. Hartmann, G. Pegram, E. Kistin, G. Blate, G. Kimura de Freitas, E. Levine, C. Guthrie, C. McSweeney, and N. Sindorf. 2010. *Flowing Forward: Freshwater Ecosystem Adaptation to Climate Change in Water Resources Management and Biodiversity Conservation*. Water working note no. 28. Washington: World Bank Group.

Lesbarreres, D., T. Lode, and J. Merila. 2004. What type of amphibian tunnel could reduce road kills? *Oryx* 38:220–23.

Lips, K. R., J. Diffendorfer, J. R.Mendelson III, and M. W. Sears. 2008. Riding the wave: reconciling the roles of disease and climate change in amphibian declines. *Public Library of Science Biology* 6:441–54.

Magnuson, J., K. E. Webster, R. A. Assel, C. J. Bowser, P. J. Dillon, J. G. Eaton, H. E. Evans, E. J. Fee, R .I. Hall, L. R. Mortsch, D. W. Schindler, and F. H. Quinn. 1997. Potential impacts of climate change on aquatic systems: Laurentian great lakes and the Precambrian shield region. *Hydrological Processes* 11:825–71.

Matthews, J. H.. 2011. Anthropogenic climate change impacts on aquatic macroinvertebrates: A thermal mass perspective. In *Odonate Conservation: Climate Change Impacts and Monitoring Strategies*, edited by J. Ott. Sofia, Bulgaria: Pensoft Publishers.

Matthews, J. H., A. Aldous, and A. J. Wickel. 2009. Managing water in a shifting climate. *Journal of the American Water Works Association* 101:28–29, 99.

Matthews, J. H., and A. J. Wickel. 2009. Embracing uncertainty in climate change adaptation: a natural history approach. *Climate and Development* 1:269–79.

McCaffery, R. M, and B. A. Maxell. 2010. Decreased winter severity increases viability of a montane frog population. *Proceedings of the National Academy of Sciences* 107:8644–49.

McCollum, S. A., and J. VanBuskirk. 1996. Costs and benefits of a predator-induced polyphenism in the gray treefrog *Hyla chrysoscelis*. *Evolution* 50:583–93.

McLaughlin, J. F., J. J. Hellmann, C. L. Boggs, and P. R. Ehrlich. 2003. Climate change hastens population extinctions. *Proceedings of the National Academy of Sciences* 99:6070–74.

McMenamin, S. K., E. A. Hadly, and C. K. Wright. 2008. Climate change and wetland desiccation cause amphibian decline in Yellowstone National Park. *Proceedings of the National Academy of Sciences USA* 105:16988–93.

Milly, P., J. Betancourt, M. Falkenmark, R. Hirsch, Z. Kundzewicz, D. Lettenmaier, and R. Stouffer. 2008. Stationarity is dead: Whither water management? *Science* 319:573–74.

Mirza, M. M. Q., and Q. K. Ahmad. 2005. *Climate Change and Water Resources in South Asia*. London: Taylor and Francis.

Morey, S., and D. Reznick. 2000. A comparative analysis of plasticity in larval development in three species of spadefoot toads. *Ecology* 81:1736–49.

Morris, W. F., and D. F. Doak. 2002. Quantitative conservation biology. Sunderland, MA: Sinauer Associates.

Morrison, C., and Hero J.-M.. 2003. Geographic variation in life-history characteristics of amphibians: A review. *Journal of Animal Ecology* 72:270–79.

Murphy, B. F., and B. Timbal. 2008. A review of recent climate variability and climate change in southeastern Australia. *International Journal of Climatology* 28:859–79.

Nolin, A. W., and C. Daly. 2006. Mapping "at risk" snow in the Pacific Northwest. *Journal of Hydrometeorology* 7:1164–71.

Oviatt, C. G.. 1997. Lake Bonneville fluctuations and global climate change. *Geology* 25:155–58.

Parmesan, C. 2006. Ecological and evolutionary responses to recent climate change. *Annual Reviews of Ecology and Evolution* 37:637–69.

Parmesan, C., T. L. Root, and M. R. Wilig. 2000. Impacts of extreme weather and climate on terrestrial biota. *Bulletin of the American Meteorological Society* 81:443–50.

Parmesan, C., and G. Yohe. 2003. A globally coherent fingerprint of climate change impacts across natural systems. *Nature* 421:37–42.

Parra-Olea, G., E. Martínez-Meyer, and G.P.-P. de Leon. 2005. Forecasting climate change effects on salamander distribution in the highlands of central Mexico. *Biotropica* 37:202–8.

Poff, N. L. 1997. The natural flow regime: A paradigm for river conservation and restoration. *BioScience* 47:769–84.

Poff, N. L., M. M. Brinson, and J. W. Day. 2002. *Aquatic Ecosystems and Global Climate Change: Potential Impacts on Inland Freshwater and Coastal Wetland Ecosystems in the United States.* Arlington, VA: Pew Center on Global Climate Change.

Poole, G. C., J. B. Dunham, D. M. Keenan, S. T. Sauter, D. A. McCullough, C. Mebane, J. C. Lockwood, D. A. Essig, M. P. Hicks, D. J. Sturdevant, E. J. Materna, S. A. Spalding, J. Risley, and M. Deppman 2004. The case for regime-based water quality standards. *BioScience* 54:155–61.

Pounds, J. A, M. R. Bustamante, L. A. Coloma, J. A. Consuegra, M. P. L. Fogden, P. N. Foster, E. La Marca, K. L. Master, A. Merino-Viteri, R. Puschendorf, S. R. Ron, G. A. Sánchez-Azofeifa, C. J. Still, and B. E. Young. 2006. Widespread amphibian extinctions from epidemic disease driven by global warming. *Nature* 439:161–67.

Pounds, J. A., M. P. Fogden, and J. H. Campbell. 1999. Biological response to climate change on a tropical mountain. *Nature* 398:611–15.

Prowse, T. D., F. J. Wrona, J. D. Reist, J. J. Gibson, J. E. Hobbie, L. M. J. Levesque, and W. F. Vincent. 2006. Climate change effects on hydroecology of arctic freshwater ecosystems. *Ambio* 35:347–58.

Rose, C. S.. 2005. Integrating ecology and developmental biology to explain the timing of frog metamorphosis. *Trends in Ecology and Evolution* 20:129–35.

Rovito, S. M., G. Parra-Olea, C. R. Vasquesz-Almazan, T. J. Papenfuss, and D. B. Wake. 2009. Dramatic declines in neotropical salamander populations are an important part of the global amphibian crisis. *Proceedings of the National Academy of Sciences* 106:3231–36.

Saltz, D., D. I. Rubenstein, and G. C. White. 2006. The impact of increased environmental stochasticity due to climate change on the dynamics of asiatic wild ass. *Conservation Biology* 20:1402–9.

Scherer, R. D., E. Muths, and B. A. Lambert. 2008. Effects of weather on survival in populations of boreal toads in Colorado. *Journal of Herpetology* 42:508–17.

Schindler, D. 1998. Sustaining aquatic ecosystems in boreal regions. *Conservation Ecology* 2:18.

Seber, G. A. F. 1982. *The Estimation of Animal Abundance and Related Parameters.* New York: Macmillan.

Smith, M., H. Caswell, and P. Mettler-Cherry. 2005. Stochastic flood and precipitation regimes and the population dynamics of a threatened floodplain plant. *Ecological Applications* 15:1036–52.

Stebbins, R. C.. 2003. *A Field Guide to Western Reptiles and Amphibians.* Boston: Houghton-Mifflin.

Stuart, S.N., J. S. Chanson, N. A. Cox, B. E. Young, A. S. L. Rodrigues, D. L. Fischman, and R. W. Waller. 2004. Status and trends of amphibian declines and extinctions worldwide. *Science* 306:1783–86.

Thompson, L. G., E. Mosley-Thompson, H. Brcher, M. Davis, B. Leon, D. Les, P. N. Lin, T. Mashiotta, and K. Mountain. 2006. Abrupt tropical climate change: Past and present. *Proceedings of the National Academy of Sciences* 103:10536–43.

Wentz, F. 2007. How much more rain will global warming bring? *Nature* 317:233–35.

Werner, J. K., J. Weaselhead, and T. Plummer. 1999. The accuracy of estimating eggs in anuran egg masses using weight or volume measurements. *Herpetological Review* 30:30–31.

White, G. C. 2000. Population viability analysis: Data requirements and essential analyses. In *Research Techniques in Animal Ecology: Controversies and Consequences*, edited by Boitani, L., and T. K. Fuller. New York: Columbia University Press.

Wiggins, G. B., R. J. Mackay, and I. M. Smith. 1980. Evolutionary and ecological strategies of animals in annual temporary ponds. *Archiv für Hydrobiologie* 58 (Suppl): 97–206.

Wilbur, H. M.. 1997. Experimental ecology of food webs: Complex systems in temporary ponds: The Robert MacArthur Award Lecture. *Ecology* 78:2279–2302.

Williams, D. D.. 2006. *The Biology of Temporary Waters*. Oxford: Oxford University Press.

———. 1987. *The Ecology of Temporary Waters*. Caldwell, NJ: Blackburn Press.

Winter, T. C., and M. K. Woo. 1990. Hydrology of lakes and wetlands. In *Surface Water Hydrology*, edited by M. G. Wolman and H. C. Riggs, 159–87. Boulder, CO: Geological Society of America.

Zhang, X. C., and M. A. Nearing. 2005. Impact of climate change on soil erosion, runoff, and wheat productivity in central Oklahoma. *Catena* 61:185–95.

5

Modeling Range Shifts for Invasive Vertebrates in Response to Climate Change

Damien A. Fordham, H. Resit Akçakaya,
Miguel Araújo, and Barry W. Brook

Invasive species and climate change are often viewed as important but separate threats to biodiversity that affect species interactions and alter ecosystem processes. Consequently, it is unsurprising (Brook 2008) that most studies of vertebrate responses to climate change have tended to focus on native organisms, for which conservation is the primary goal (e.g., Lovejoy and Hannah 2005). Yet climate change will also both facilitate new invasions and alter the ranges and ecological effects of already established aliens in their invasive range, and this may have strong ecological and economic implications (Hellmann et al. 2008).

Invasive organisms, wildlife exploitation, and habitat destruction and other forms of land use change are the major historical drivers of range reductions (Myers et al. 2000), directly causing severe biodiversity loss at local scales (e.g., Brook 2008) and indirectly limiting the scope for sufficient ecological and evolutionary adaptation to future environmental change (Fordham and Brook 2010). Mitigation of these human-mediated impacts is a central focus of global change ecology and conservation biology, but with an increasing acknowledgement that these deterministic and stochastic extinction drivers are usually interacting and self- or mutually reinforcing (Brook et al. 2008). In particular, the mechanisms by which these historical threats interact with the new overarching threat of climate change are becoming increasingly apparent (e.g., Benning et al. 2002), leading to a greater awareness of the problem of synergistic feedbacks. For example, recent research on insect and plant invasions under climate change (e.g., Ward and Masters 2007, Peterson et al. 2008) provide prime examples of the type of nonlinear, mutually reinforcing ecological processes that are expected to drive ecosystem change and biological homogenization in this century (MEA 2005).

Past efforts to forecast the effect of climate change on the distribution of invasive organisms have focused on bioclimatic envelope approaches (e.g., Peterson et al. 2008, Ficetola et al. 2009) or nonstatistical procedures that model the relationship between species' physiology and climatic requirements (e.g., Kearney et al. 2008). These methods do not, however, account for potentially important demographic and multispecies responses (Brook et al. 2009, Brodie et al. this volume). These responses can influence distribution, population structure, and extinction risk at local scales. In this chapter we describe a framework for investigating the dynamics of range edges for metapopulations under climate change, highlighting that (1) range limits are an inherently population level problem for which a demographically explicit approach is needed but is rarely applied, and (2) invasive species provide good candidate models to clarify why some species spread rapidly across novel environments while others expand their ranges only incrementally.

Although synergistic feedbacks between the impacts of introduced species and climate change on native biota have been established for plants and invertebrates (Smith et al. 2000, Stachowicz et al. 2002), there are scant real-world projections of future vertebrate invasions (or contractions) under climate change (Brook 2008). A notable exception is research on the South American endemic cane toad, *Chaunus* [*Bufo*] *marinus*, which has invaded northern Australia (e.g., Sutherst et al. 1996, Urban et al. 2007), and on slider turtles, *Trachemyz scripta*, in Italy (Ficetola et al. 2009). Here we use predictions of species' responses to climate change that incorporate metapopulation dynamics and elements of dispersal to explore the geographical range margin dynamics of an infamous Australian invasive species—the European rabbit, *Oryctolagus cuniculus*—in response to future climate change.

Invasiveness in Rapidly Changing Environments

The detrimental impacts of invasive species on native communities and ecosystems has been documented for decades (e.g., Wilcove et al. 1998). Invasive organisms often act as novel direct stressors to resident species as competitors, parasites, or predators (Lockwood et al. 2007). They can also cause indirect effects by modifying physical elements of their environment, which alter habitat suitability and can provoke subsequent invasions (Byers 2002). Recent evidence suggests that future mean global surface temperature will affect the range and abundance of many organisms (Thomas et al. 2004), including already established invasive species (Dukes and Mooney 1999). However, it has been speculated (Bradley 2009) that rather than just increasing invasion risk, climate change may in some instances reduce invasive competitiveness, leading to range contraction and potential extirpation of invasive species.

Anthropogenic warming during the last 50 years has been rapid in comparison to long-term historical changes, and it is forecast to accelerate, with carbon dioxide concentrations. Without substantive intervention, global mean temperatures are projected to increase by 2.4° C to 6.4° C by the year 2100 (IPCC 2007). This will bring with it drastic environmental alterations, producing conditions with no current climate analog and the complete disappearance of some extant climates (Williams et al. 2007). Rapid change is likely to result in the formation of new microhabitats, increased introduction rates of alien species, and decreased resistance to invasion for populations of native species (Dukes and Mooney 1999).

These abrupt global changes threaten to push species out of the environmental and geographic space that have defined their evolutionary history (Parmesan 2006), creating new competitive situations that are just as novel for noninvasive species as they are for invasives. This will have direct implications for how we define an invasive species under global warming, forcing new challenges upon conservation practitioners and policy makers (Pyke et al. 2008). For the purpose of this chapter, we adopt a pragmatic biological definition of invasions, considering a species to be invasive if it has been introduced recently and exerts a strong negative effect on native biota.

Climate Change and the Invasion Pathway

To become invasive, species must pass through a set of environmental filters. Success at any of these well-established phases—initial introduction, colonization, establishment, and spread—depends upon a distinct set of mechanisms, many of which are likely to be influenced by future climate (Sutherst 2000). Global warming may accelerate the *introduction* of invasive species by linking geographic areas, thus enabling species to overcome dispersal barriers. For example, in regions where future water temperatures are expected to shift, or more frequent or intense flooding is forecast in response to global warming, invasive aquatic vertebrates may be able to advance their range due to the loss of dispersal barriers (Rahel et al. 2008). In the Arctic, warming has changed the migration patterns of some fish species, increasing the likelihood that fishless lakes will be colonized (Post et al. 2009). Once new invasive species are transported into a novel environment, climate change can enhance *colonization* by removing the climatic constraints on population processes such as survival and fecundity. New arrivals may have a greater competitive advantage over native species under altered climates, thus enhancing *establishment*. The movement, in response to global warming, of native organisms out of conditions to which they are adapted is likely to weaken competitive resistance, enabling invaders to more easily overcome biotic constraints on population growth and persistence (Stachowicz et al. 2002). Indeed, paleoecological evidence associated with late Quaternary global warming suggests that altered climates are more likely to cause the range dynamics of species to shift individualistically, forming novel community assemblages with different species associations, rather than causing existing communities to migrate together as tightly coevolved units (Graham and Grimm 1990). The consequences of such ecological reorganization are likely to be severe (O'Dowd et al. 2003). For example, the elimination of formerly dominant or keystone species because their climate tolerances are exceeded will force a reorganization of the structure of many ecological communities; this can be expected to provoke cascades of future invasions.

Once established, invasive species may *spread* across the landscape via both continuing long-distance dispersal from foreign sources and short-distance dispersal, thus leading to lateral expansion of the existing population (Sakai et al. 2001). The rate of spread is influenced by system-specific and species-specific factors such as connectivity of viable habitat patches, demographic and dispersal dynamics, and evolutionary history (Hastings et al. 2005). Invasive species often display excellent adaptations for dispersal (Phillips et al. 2008a), which has been shown to be an important trait in the successes of vertebrate invasions (e.g., O'Connor et al. 1986, Alford et al. 2009) and to influence range margins (expansion and contraction) under climate change (Anderson et al. 2009). For example, range expansion of cane toads in Australia over the last 30 years has been linked to increased climatic suitability (rise in maximum temperature), dispersal capacity, and evolutionary adaptation (Urban et al. 2008).

Considering that temperature and hydrological constraints are predicted to strongly influence the invasion process of both vertebrate and invertebrate species (Stachowicz et al. 2002, Urban et al. 2007), identifying those characteristics that predispose species to becoming better invaders under climate change will assist in targeting future management and conservation efforts (Ward and Masters 2007). The ability to use trait-based generalities to indicate future invasiveness has been a long-held goal of invasive species research. However, determining range shifts for invasive as well as noninvasive species under climate change is challenging, because of uncertainties arising from alternative climatic and species-distributional modeling techniques, idiosyncratic species characteristics, and uncertainties about future climate (Beaumont et al. 2008). We now discuss the pros and cons of different approaches to investigating the impact of global warming on future invasiveness.

Predicting the Future Spread of Invasive Species

Range margins occur along the boundary where individual populations are no longer sustained by local recruitment or dispersal from adjacent areas, because local deaths exceed local births and immigrations. Causes of species' range limits include unfavorable abiotic conditions (e.g., temperature extremes, precipitation, dispersal barriers), biotic interactions (e.g., competitors, predators, lack of essential mutualists), demographic-genetic processes (e.g., Allee effects, genetic diversity), and evolution history, all of which often occur in unison (Gaston 2003). Given that climate is one of the primary constraints on species distributions and ecosystem function, ecologists are faced with an enormously complex challenge in trying to realistically forecast ecological responses to climate change. In this context, models are the conceptual tools for linking demographic processes (population growth, dispersal, and trending vital rates) and ecosystem processes (ecological interactions and feedbacks) to an organism's perception and use of landscape spatial features, thus allowing us to predict ranges and persistence under different human impact scenarios (Akçakaya et al. 2004).

Habitat and Bioclimatic Niche Models

Species' ranges are traditionally modeled by linking occupancy or abundance data directly to environmental variables. The resulting bioclimatic model (also known as a habitat, niche, or species distribution model) can then be projected onto current or future landscapes to identify geographic regions with potentially suitable ecological conditions. However, the implicit simplifying assumption, that current climatic and topographic constraints act to define a species distribution and reflect its biophysical preferences, ignores (1) interactions between the distribution and the spatial structure of suitable habitat, (2 ecosystem processes derived from species interactions, and (3) how life history and morphological and physiological traits (e.g., longevity, dispersal ability, body size, and heat exchange) interact with shifting ranges (Botkin et al. 2007). Moreover, when we model ecological responses to environmental conditions outside the range of current data or historical experience, errors can arise owing to the behavior of fitted models in extrapolated space (Menke et al. 2009). Model transferability is highly unpredictable due to inherent nonlinearities, and it depends on both the type of model and the number and combination of predictor variables (Elith et al. 2006).

Thus, although bioclimatic-niche models have a role to play in supporting efforts to protect current patterns of biodiversity (Araújo and Williams 2000, Peterson 2003), and although they have been used to project future range shifts for invasive species (e.g., Urban et al. 2007, Peterson et al. 2008), these predictions should be viewed at best as only a first approximation of the potential impact of climate-induced landscape change on biodiversity (Pearson and Dawson 2003). This is because niche models are potentially limited in their ability to predict species occurrences in novel environmental space, and because forecasting range shifts under global warming often requires extrapolating the model into new climate space (see also Paull and Johnson, this volume).

Mechanistic Models

Approaches that explicitly model the mechanistic links between functional traits of organisms and their environments have been used to explain current species distributions and project future ecological trends (Kearney and Porter 2009). Unlike bioclimatic-niche models, which attempt to predict the realized niche of the species, these "first principle" approaches do not use occurrence data. Instead, they try to capture the ecophysiological limits of a species' tolerance of environmental factors (i.e., its estimated fundamental niche on the basis of controlled experiments establishing its physiological responses to environmental parameters. This approach provides reasonably good projections of distribution when constrained to regions within which the species are known a priori to occur, and when sufficient detail is known about their ecophysiological traits (Gaston and Fuller 2009). It can be expected that as physiological knowledge becomes better integrated into species distribution type modeling, our capacity to make more robust predictions of range shifts in novel or nonequilibrium contexts such as climate induced invasions will increase (Kearney and Porter 2009).

Model development essentially follows a three-step process: (1) spatial data for macroclimatic conditions and topography are converted into the microclimatic settings actually experienced by the species; (2) using principles of biophysical physiology, details are established on how species traits interact with microclimatic settings to change the state of the organism; and (3) the relationship between important dependent variables (e.g., a species trait such as body temperature) and fitness is used to identify areas where populations or individuals of the target species can and cannot survive (Phillips et al. 2008b). These biophysical models are then used to make predictions of range shifts under future climate scenarios. Research on cane toads in Australia provides a good example of where a process-based approach has been used to explore range shifts in an invasive vertebrate. Thermal constraints on the locomotor potential of adult cane toads, together with limitations on the availability of water for the larval stage, were shown to constrain the cane toad's continued colonization of Australia (Kearney et al. 2008).

CLIMEX is a generically designed software package that functions by attempting to mimic the mechanisms that limit species' geographical distributions (Sutherst et al. 2007). It is frequently used to model the impacts of climate change on invasive plant and insect species (e.g., Sutherst and Maywald 2005, van Klinken et al. 2009); and to investigate future range expansion in vertebrates such as cane toads (Sutherst et al. 1996). However, the use of CLIMEX or more fundamental biophysical models to predict range shifts among invasive species has its limitations. These models fail to account for the ability of invasive vertebrates and other organisms to adapt their physiology (either through evolution or plasticity) in response to environmental change (Phillips et al. 2008a). They also fail to incorporate metapopulation and dispersal dynamics. Inconsistencies found in the climate spaces occupied by invasive species on different continents and the conditions that prevail in those same species' original habitats point to the importance of nonclimatic factors in regulating range and abundance (Duncan et al. 2009).

Spatial Population Models

Spatial population models integrate landscape features with demographic and ecosystem processes, thus enabling more accurate forecasting of species' persistence and distribution (Brook et al. 2009). The utility of these models has been somewhat constrained because it is often intractable to gather detailed demographic and dispersal data for a whole community of species. However, recent computational and statistical developments have meant that it is now tractable to use spatial population models to investigate the dynamics of range margins for metapopulations under forecasted landscape change (Anderson et al. 2009). The remainder of this chapter will focus on an advanced framework we are developing for invasive risk assessment under climate change. This framework accounts for spatially explicit interactions between demography and climate and their simultaneous effect on habitat distribution and population viability (figure 5.1). We make the point that invasive vertebrates provide particularly useful candidate model organisms for advancing our understanding of the interplay between range dynamics and climate change, reducing model uncertainty in future projections of extinction, because (1) invasive species are, by definition, species shifting their ranges; (2) invasive species tend not to be rare or cryptic in their non-native environment; (3) the initial invasion, establishment, and spread is often well documented; (4) the population ecology and biology of invasive species, especially those that cause economic loss, is often well studied; and (5) they are not a focus of conservation, and so are amenable to manipulation. Because the spread of alien species is often well documented, temporal steps in the invasion process provide good scope for assessing how well models can predict species range shifts (Ficetola et al. 2010). This provides a framework to test different modeling approaches and, in turn, to test whether data-intensive approaches provide more reliable projections than models with less complexity. Model validation can be tested through retrospective projections.

Range Limits: A Population-Level Problem

Because factors that regulate species distributions, operating singly or in combination, ultimately influence vital rates (survival and reproduction) and associated population traits (Holt and Keitt 2005), integrating bioclimatic approaches with spatial population models will provide more realistic forecasts of species invasions under climate change. The model architecture that we suggest (figure 5.1) can be cohort- or individual-based, incorporating explicit and dynamic spatial structure. Habitat and vegetation modeling approaches are used to quantify the interaction between landscape and species ranges. Demographic and ecosystem responses to temporal and spatial landscape variation are simulated by developing predictive models that link rates and patterns of human impacts (e.g., decreased habitat connectivity and altered fire regimes) to ecological data (i.e., survival and dispersal rates; see Fordham and Brook 2010 for further details). When applied to many case studies, this integrative model architecture offers the prospect of developing generalizations about the traits that make species vulnerable to climate change (Brook et al. 2009). This in turn will provide the opportunity to better understand patterns of invasion, their underlying mechanism, and the role played by the environment. However, success will depend on minimizing uncertainty that arises from the choice of species distribution model (SDM), global climate model (GCM), and greenhouse gas emission scenario (GES; Buisson et al. 2009). Further, it is important to develop methods to better reflect differences in vital rates at the core of a species range because the marginal areas are often the most relevant in climate change risk assessment (Gaston 2003).

Global Climate Model Ensembles

Most attempts to model climate change impacts on invasive species and on biodiversity in general have used only a single GCM and GES to generate forward projections. Yet this simplification masks considerable

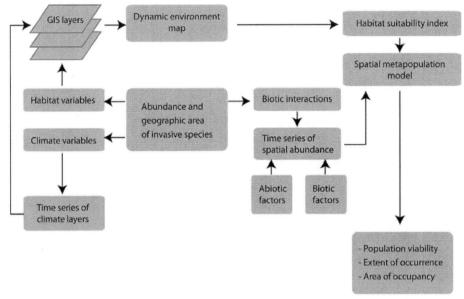

Figure 5.1. Next-generation modeling for extinction risk under climate change, incorporating dynamic environmental change, demographic variation, general metapopulation processes and structure, and range margin dynamics.

uncertainty; the choice of GCM and GES influences the expected spatial spread of invasive organisms (Mika et al. 2008), owing to alternative GCM structures and a range of plausible possibilities for carbon mitigation efforts. To account for this variation, we suggest a novel (at least for the ecological modeling discipline) approach that uses multiple runs of a suite of GCMs to make ensemble forecasts of climate change under multiple GESs. However, current GCM output grids tend to be coarse in resolution (typically at least 250 × 250 km) compared to biological processes, and this limits their usefulness for making predictions using species distribution models (Seo et al. 2009). As such, GCM projections need to be downscaled to spatial resolutions similar to those available for present-day climate data (i.e., approximately 1–5 km2). This can be done using bilinear interpolation to initially downscale the GCM output (approximately 50 × 50 km), and then overlaying the climate change projection on top of fine-resolution baseline climate data, as derived from smoothed historical meteorological station data (figure 5.2). Although dynamical downscaling, which involves a regional climate model (RCM) nested within a GCM, provides an alternative method for downscaling climate scenarios, forecasts are limited by the skill of the driving GCM and the strength of regional forcing (Beaumont et al. 2008).

Ensemble Distribution Modeling

Predictions of shifts in species distribution are sensitive to the initial environmental conditions, model type, parameters, and boundary conditions (Araújo and New 2007). Although empirical examination of input parameters and outputs can enable informed judgment of model performance under current conditions (Elith et al. 2006), it is difficult to determine what a good model is when forecasting species distributions under climate change, since the future is unknowable (Araújo and Rahbek 2006). A potential solution is to take intermodal variability into account by using large numbers of simulations across a range of model types and combining them into a consensus forecast (Araújo and New 2007). The output is a probability density function (figure 5.2), rather than mean-point projections. This approach is gaining much support, with computer freeware now available for generating bioclimatic ensembles of moderate size (Thuiller et al. 2009).

Spatially Structured Population Models

Variations in population abundance and changing demographic rates along the edge of the range, for subpopulations within the total area of occupancy and ecotones along the range limits, should be some of the first and most sensitive signs of broader species response to environmental change. Unfortunately, most population models treat vital life-history

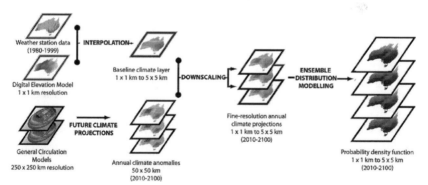

Figure 5.2. Removing uncertainty from climate envelope projections: downscaling climate projections by splicing surface response models of meteorological station data to multiple global circulation models (GCMs) with good regional skill and grid-cell linear interpolation to achieve spatial smoothing.

traits such as survival and dispersal as stochastic, but spatially invariant and nontrending. Demographic models of population and metapopulation dynamics used in this integrated approach provide enormous potential to account for fine-scale spatial variance in vital rates. For example dispersal can be modeled as being dependent on the size, shape, location, and number of populations, dynamically changing as a result of shifting suitability in response to predicted climate change.

Future Rabbit Rrange and Abundance in Australia

The European rabbit has shown considerable ability to establish and flourish in Australia and many other regions across the globe (Tablado et al. 2009), causing large-scale economic loss and environmental degradation. Rabbits were introduced to mainland Australia in 1859 and spread quickly through the southern part of the continent. Today, the limits to their distribution in Australia appear to be determined chiefly by climatic variables. The northern range boundary is probably constrained by the influence that the seasonal timing of pasture growth and warren (burrow network) temperature has on reproductive success (Cooke 1977). The fecundity of rabbits is largely governed by the interaction between rainfall and pasture growth (Poole 1960), and high temperatures directly inhibit their reproduction by causing lactation problems (Cooke 1977). The altitudinal limit to the distribution is determined by the number of days snow lies on the ground (Myers et al. 1975)—a limited consideration in Australia. Rabbits also do not tolerate very wet environments, owing to frequent flooding of their burrows, which causes increased mortality (Bowen and Read 1998). Landscape variables such as soil type and vegetation cover also influence their spatial abundance. Rabbits prefer sandy soils, which easily allow warren construction (Parer and Libke 1985), and they avoid heavy forest (Hamilton et al. 2006) and highly urbanized regions.

The persistence and recolonization of lagomorph populations following climate-related local extinction has been shown to be influenced by metapopulation processes and dispersal dynamics in Europe (Anderson et al. 2009). In this context investigating the impact of climate change on the invasive range and abundance of rabbits in Australia would seem to require a dynamic, spatially explicit metapopulation approach. Following figure 5.1:

1. Four climate variables were identified a priori as having the
   greatest influence on rabbit distribution: maximum temperature
   in the hottest month, minimum temperature in the coolest month,

annual rainfall, and summer rainfall. An annual time series of climate change layers for each variable was generated according to two GES scenarios: a fossil-fuel-intensive reference scenario (WRE750) and a more conservative scenario that assumes substantive intervention (LEV1; Wigley et al. 1996, Wigley et al. 2009). A coupled gas cycle/aerosol/climate model called MAGICC/SCENGEN 5.3 (http://www.cgd.ucar.edu/cas/wigley/magicc), used in the IPCC Fourth Assessment Report (IPCC 2007), was used to generate future climate anomalies using an ensemble of nine GCMs, chosen according to their superior skill in reproducing the Australian baseline climate (1980–99).

2. Meteorological weather station and elevational data were spatially smoothed using thin-plate spline models (Hutchinson 1995), and overlain on the GCM projections (50 × 50 km resolution) to produce 5 km downscaled yearly projections (2010–2100) of regional climate change (figure 5.2).

3. The Bioensembles software (Diniz-Filho et al. 2009) was used to generate two annual time series of climate envelopes (2010–2100) from presence- type survey data using a large number of forecasts obtained from generating models with 10 alternative cross-validated samples, 6 alternative combinations of variables (i.e., the full factorial combination of the 3 selected climatic variables), and 7 bioclimatic-niche models (BIOCLIM, Mahalanobis distances, Euclidian distances, generalized linear models, random forest, maximum entropy, and GARP). Pseudo-absences were randomly chosen from locations with low climatic suitability (Wisz and Guisan 2009), having been selected from outside the observed 90th percentile for each climate variable. We assumed a climate suitability threshold of 0.3, based on CLIMEX projections of present-day, physiologically suitable rabbit habitat (Brian Cooke, Invasive Animals CRC, unpublished data).

4. A habitat suitability index was derived by combining climate envelopes with static habitat variables. Heavily forested (with average tree basal area in a 5 × 5 km grid cell in exceess of 20 m) and urban regions were treated as unviable habitat andexcluded via a mask (*mask*). On the basis of expert knowledge, sandy soils (*soil*) were modeled as having a strong positive influence on rabbits' carrying capacity (Greg Mutze, Department of Water, Land and Biodiversity Conservation, personal commmunication). The habitat relationship was modeled using the equation

Habitat suitability (HS) =   [eq. 1]
$mask$*max ($soil$,0.5)*thr ($clim$, 0.3)

where max is the maximum of two arguments, thr is a threshold function (0 below the threshold value, 1 above it), and *clim* is the climate envelope probability of occurrence (see 3). The function calculates the suitability for each location (cell) on the basis of environmental input maps (Akçakaya 2005).

The habitat suitability values were linked to the metapopulation model (see below) by the equation

Carrying capacity (K) = thr (THS*10000, 3250)   [eq. 2]

where *THS* is total habitat suitability in patch clusters or groups of nearby cells that have HS values higher than or equal to a threshold value of 0.25. Values were chosen through an iterative process based on published population estimates and expert advice.

5. Spatially explicit demographic models of population and metapopulation dynamics were constructed in RAMAS GIS (Akçakaya 2005). Population models were scalar (not age- or stage-structured), and are also commonly referred to as count-type models (Dunham et al. 2006). We used a Ricker-type equation to model a density-dependent process of worsening returns, whereby as population size increases, the amount of available resources per individual decreases (Akçakaya 2005). Time-series data for eight populations of rabbits, accessed from the Global Population Dynamics Database (www3.imperial.ac.uk/cpb/research/patternsandprocesses/gpdd), were used to calculate the maximum annual rate of increase (R_{max} = 1.5) and its standard deviation (SD = 0.95) according to the Ricker model. Published estimates of vital rates such as dispersal (Hamilton et al. 2006) were directly incorporated into the model. Owing to present computational limitations in trying to simulate the entire continental distribution at such a fine scale, our models focused on the northern range boundary north of −25.5° latitude. The model results we describe below are the average of 1,000 stochastic simulations using the RAMAS Metapop module.

Ensemble forecasts from Bioensembles predict a southern contraction of areas of high climate suitability for rabbits in Australia, under both the mitigation policy (LEV1) and reference scenario (WRE750); the former occurs at a less substantial rate (figure 5.3). Accordingly, total habitat

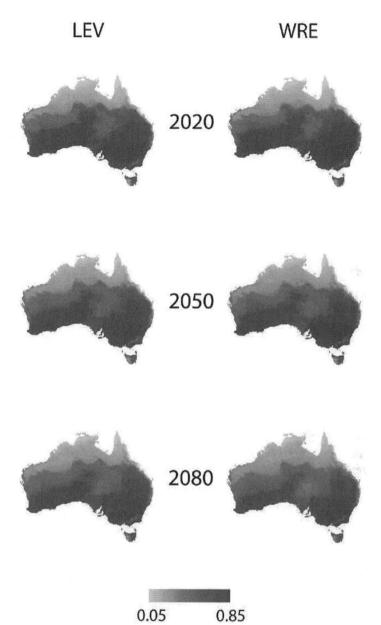

Figure 5.3. Future climate suitability for invasive European rabbits (*Oryctolagus cuniculus*) in Australia (5 km × 5 km resolution), according to a low-emission "policy" scenario (LEV1) and a high-emission "reference" scenario (WRE750). The scale is a probability density function, calculated using ensemble forecasts from seven species-distribution modeling techniques.

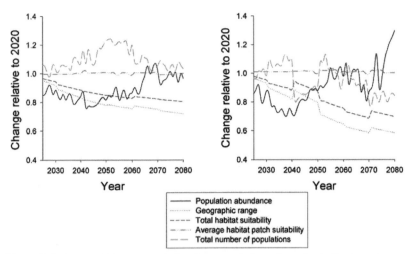

Figure 5.4. Future rabbit range and abundance on their northern Australian boundary (for the years 2025–80) according to two climate change projections: (a) the LEV1 policy scenario, and (b) the WRE750 reference scenario. Forecasts are relative to a 2020 mean baseline estimate of population size, calculated from model projection (2015–25). Spatially explicit metapopulation models were used to estimate population abundance (total number of individuals) under a medium rate of dispersal (1,000 simulated iterations). Geographic range is the total number of cells that can be occupied at a given point in time; their suitability is described by the average habitat patch suitability index (range = 0–1 for each landscape cell). Total habitat suitability is the sum of suitability across all cells in the focal region. Total number of populations is the number of patches (clusters or groups of nearby cells higher than or equal to a habitat suitability threshold) defined by the species-habitat relationship.

suitability and future geographic range (measured as the number of environmentally suitable cells) are predicted to decline along the northern range boundary (figure 5.4).

Although rabbits in Australia may experience a future range contraction in a southerly direction, the demographic model suggests that it is unlikely that the reduction in geographic range will be matched by a steady reduction in future rabbit abundance (figure 5.4). Our spatially explicit metapopulation projections indicate that mean rabbit abundance in the northern range margin will undergo systematic temporal and spatial fluctuations not connected with environmental stochasticity, including periods of substantial increase and rapid decline. In fact, population abundance in 2080 may exceed the 2020 mean baseline projection (2015–25) under the reference scenario. These fluctuations in abundance are not directly explained by concordant shifts in the average suitability of occupied patches (figure 5.4); instead, they are probably a complex response to the spatial configuration and number of viable habitat patches. For example, the general increase in population abundance following 2040 (under both scenarios) corresponds to a highly significant fragmentation event—the largest population (accounting for approximately 90% of total occupied habitat) splits into two, on or around the year 2040. Moreover, for models developed under the reference WRE750 scenario, the breakup of the "super population" in 2040 also coincides with a decrease in the number of suitable habitat patches, and this leads in turn to a substantial rise in relative population abundance. Increased fragmentation can cause increased local extinctions, owing to demographic and environmental stochasticity. A second steep rise in relative population abundance, after 2075 under the reference scenario, requires further investigation to pin down its cause; however, it is interesting to note that this is preceded by a period of low population numbers and an abrupt jump in the area of suitable habitat (figure 5.4).

Dispersal rate also influences projections of relative population abundance, especially after 2040, when the "super population" fragments (figure 5.5). A high rate of long-distance dispersal (5 in every 100 rabbits can move 10 km over a one-year period) had a noticeably positive influence on abundance, following the moderate climate change scenario (figure 5.5A). This rate of dispersal had a pronounced influence on relative rabbit abundance up to 2065 under the business-as-usual emission scenario, after which the influence was diminished. It is again interesting to note that this period is preceded by a period of relatively few populations (figure 5.4) and a trend towards increased relative abundance. A reduction in the influence of dispersal rate on population abundance and a concurrent increase in population size would be expected if the decline in population numbers was matched by less connectivity between habitat patches (i.e., less chance of long-distance migration under the high- and low-dispersal scenarios), which would reduce the risk of rabbits moving into marginal habitat patches and thereby increase the likelihood that they become isolated and locally extinct.

Conversely, it is also plausible that the formation of fewer, less geographically isolated populations could explain the weakened influence of dispersal, due to increased connectivity. This would seem a less likely explanation, however, given that under the low-dispersal scenario fewer than one rabbit in a thousand can move more than five km, thus meaning that the populations would need to be in very close proximity. Both these hypotheses require further investigation.

Our rabbit modeling illustrates the influence of spatial structure and demographic parameters (such as dispersal) on forecasts of species range

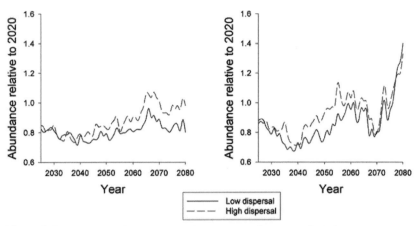

Figure 5.5. The impact of varied rates of dispersal on forecasts of relative rabbit abundance in Australia (2025–80) for two climate change projections: (a) the LEV1 policy scenario and (b) the WRE750 reference scenario. Abundance is relative to a 2020 mean baseline estimate of population size, calculated from model projection (2015–25). Dispersal is modeled as low or high, accordingly influencing rates of long-distance dispersal.

and abundance in a changing climate. Dispersal often depends on the spatial configuration of viable habitats (size, shape, location, and number of patches) which can change in time as a result of shifting climates. Using a spatially structured population model, global climate change is forecast to reduce rabbit habitat suitability on the northern Australian range boundary, influencing its abundance in ways that would not be detected by less sophisticated approaches which do not model the relationship between dispersal and the spatial context of the landscape. Future research should focus on determining the consequences of different parameter values and/or model structure on forecasts of rabbit range and abundance.

Summary

Predicting the rate at which invasive species shift their distribution in response to climate change is important because such organisms are a key component of the synergistic threats to biodiversity caused by global change, and result in a substantial financial burden to agriculture and biodiversity conservation. There have been few efforts to model the effect of climate change on the distribution of invasive vertebrates, and those that have been made have failed to account for important dynamics in demography and multispecies response, which in turn influence distribution, population structure, and viability. We have presented an integrative framework that will strengthen future forecasts of the range and abundance of invasive species, and which will thus enable land managers and conservation practitioners to better manage alien vertebrates under climate and landscape change. A caveat is that these models are data-intensive, requiring a strong understanding of the focal species' population dynamics, as well as of distributional data. Although invasion processes and the ecology of alien species are often well documented (especially in developed countries), robust ecological data will not be available for all invasive species. Further research is needed to compare the validity of less data-intensive approaches to forecasting future invasions (e.g., niche and bioclimatic models) with more demanding techniques (e.g., spatial population models; Brook et al. 2009). Some alien species may provide the key for critically evaluating different model types—that is, if the spatial and temporal dynamics of the invasions are well recorded.

To our knowledge, the targeted modeling on rabbits presented in this chapter provides a precedent: never has the future range and abundance of an invasive vertebrate been explored using an integration of climatic, geophysical, and demographic processes. The uncertainty of our modeling approach was reduced by using downscaled global climate model ensembles and multiple types of species distribution models, and by incorporating demography and metapopulation and dispersal dynamics.

ACKNOWLEDGMENTS
We are very grateful to David Berman and Brian Cooke for providing access to survey data on rabbits and knowledgeable insights into the potential drivers of their range and abundance. We are also thankful to Camille Mellin for helping generate the figures. The research was supported by the Australian Research Council (LP0989420).

LITERATURE CITED
Akçakaya, H. R. 2005. *RAMAS GIS: Linking Spatial Data with Population Viability Analysis, Version 5*. Setauken, NY: Applied Biomathematics.
Akçakaya, H. R., V. C. Radeloff, D. J. Mlandenoff, and H. S. He. 2004. Integrating landscape and metapopulation modeling approaches: Viability of the sharp-tailed grouse in a dynamic landscape. *Conservation Biology* 18:526–37.
Alford, R. A., G. P. Brown, L. Schwarzkopf, B. L. Phillips, and R. Shine. 2009. Comparisons through time and space suggest rapid evolution of dispersal behaviour in an invasive species. *Wildlife Research* 36:23–28.
Anderson, B. J., H. R. Akçakaya, M. B. F. Araújo, D.A., E. Martinez-Meyer, W. Thuiller, and B. W. Brook. 2009. Dynamics of range margins for metapopulations under climate change. *Proceedings of the Royal Society of London Series B: Biological Sciences* 276:1415–20.
Araújo, M. B., and M. New. 2007. Ensemble forecasting of species distributions. *Trends in Ecology & Evolution* 22:42–47.
Araújo, M. B., and C. Rahbek. 2006. How does climate change affect biodiversity? *Science* 313:1396–97.

Araújo, M. B., and P. H. Williams. 2000. Selecting areas for species persistence using occurrence data. *Biological Conservation* 96:331–45.

Beaumont, L. J., L. Hughes, and A. J. Pitman. 2008. Why is the choice of future climate scenarios for species distribution modelling important? *Ecology Letters* 11:1135–46.

Benning, T. L., D. LaPointe, C. T. Atkinson, and P. M. Vitousek. 2002. Interactions of climate change with biological invasions and land use in the Hawaiian Islands: Modeling the fate of endemic birds using a geographic information system. *Proceedings of the National Academy of Sciences of the United States of America* 99:14246–49.

Botkin, D. B., H. Saxe, M. B. Araújo, R. Betts, R. H. W. Bradshaw, T. Cedhagen, P. Chesson, T. P. Dawson, J. R. Etterson, D. P. Faith, S. Ferrier, A. Guisan, A. S. Hansen, D. W. Hilbert, C. Loehle, C. Margules, M. New, M. J. Sobel, and D. R. B. Stockwell. 2007. Forecasting the effects of global warming on biodiversity. *Bioscience* 57:227–36.

Bowen, Z., and J. Read. 1998. Population and demographic patterns of rabbits (*Oryctolagus cuniculus*) at Roxby Downs in arid South Australia and the influence of rabbit haemorrhagic disease. *Wildlife Research* 25:655–62.

Bradley, B. A. 2009. Regional analysis of the impacts of climate change on cheatgrass invasion shows potential risk and opportunity. *Global Change Biology* 15:196–208.

Brook, B. W. 2008. Synergies between climate change, extinctions and invasive vertebrates. *Wildlife Research* 35:249–52.

Brook, B. W., H. R. Akçakaya, D. A. Keith, G. M. Mace, R. G. Pearson, and M. B. Araújo. 2009. Integrating bioclimate with population models to improve forecasts of species extinctions under climate change. *Biology Letters* 5:723–25.

Buisson, L., W. Thuiller, N. Casajus, S. Lek, and G. L. Grenouillet. 2009. Uncertainty in ensemble forecasting of species distribution. *Global Change Biology*. 2010. doi: 10.1111/j.1365–2486.2009.02000.x.

Byers, J. E. 2002. Impact of non-indigenous species on natives enhanced by anthropogenic alteration of selection regimes. *Oikos* 97:449–58.

Cooke, B. 1977. Factors limiting the distribution of the wild rabbit in Australia. *Proceedings of the Ecological Society of Australia* 10:113–20.

Diniz-Filho, J. A., L. M. Bini, T. F. L. B. Rangel, R. D. Loyola, C. Hof, D. Nogués-Bravo, and M. B. Araújo. 2009. Partitioning and mapping uncertainties in ensembles of forecasts of species turnover under climate changes. *Ecography* 32:1–10.

Dukes, J. S., and H. A. Mooney. 1999. Does global change increase the success of biological invaders? *Trends in Ecology & Evolution* 14:135–39.

Duncan, R. P., P. Cassey, and T. M. Blackburn. 2009. Do climate envelope models transfer? A manipulative test using dung beetle introductions. *Proceedings of the Royal Society B: Biological Sciences* 276:1449–57.

Dunham, A. E., H. R. Akçakaya, and T. S. Bridges. 2006. Using scalar models for precautionary assessments of threatened species. *Conservation Biology* 20:1499–1506.

Elith, J., C. H. Graham, R. P. Anderson, M. Dudik, S. Ferrier, A. Guisan, R. J. Hijmans, F. Huettmann, J. R. Leathwick, A. Lehmann, J. Li, L. G. Lohmann, B. A. Loiselle, G. Manion, C. Moritz, M. Nakamura, Y. Nakazawa, J. M. Overton, A. T. Peterson, S. J. Phillips, K. Richardson, R. Scachetti-Pereira, R. E. Schapire, J. Soberon,

S. Williams, M. S. Wisz, and N. E. Zimmermann. 2006. Novel methods improve prediction of species' distributions from occurrence data. *Ecography* 29:129–51.

Ficetola, G. F., L. Maiorano, A. Falcucci, N. Dendoncker, L. Boitani, E. Padoa-Schioppa, C. Miaud, and W. Thuiller. 2010. Knowing the past to predict the future: Land-use change and the distribution of invasive bullfrogs. *Global Change Biology* 16:528–37.

Ficetola, G. F., W. Thuiller, and E. Padoa-Schioppa. 2009. From introduction to the establishment of alien species: bioclimatic differences between presence and reproduction localities in the slider turtle. *Diversity and Distributions* 15:108–16.

Fordham, D. A., and B. W. Brook. 2010. Why tropical island endemics are acutely susceptible to global change. *Biodiversity and Conservation* 19:329–42.

Gaston, K. J. 2003. *The Structure and Dynamics of Geographic Ranges*. New York: Oxford University Press.

Gaston, K. J., and R. A. Fuller. 2009. The sizes of species' geographic ranges. *Journal of Applied Ecology* 46:1–9.

Graham, R., and E. Grimm. 1990. Effects of global climate change on the patterns of terrestrial biological communities. *Trends in Ecology & Evolution* 5:289–92.

Hamilton, G. S., P. B. Mather, and C. Wilson. 2006. Habitat heterogeneity influences connectivity in a spatially structured population. *Journal of Applied Ecology* 43:219–26.

Hastings, A., K. Cuddington, K. F. Davies, C. J. Dugaw, S. Elmendorf, A. Freestone, S. Harrison, M. Holland, J. Lambrinos, U. Malvadkar, B. A. Melbourne, K. Moore, C. Taylor, and D. Thomson. 2005. The spatial spread of invasions: new developments in theory and evidence. *Ecology Letters* 8:91–101.

Hellmann, J. J., J. E. Byers, B. G. Bierwagen, and J. S. Dukes. 2008. Five potential consequences of climate change for invasive species. *Conservation Biology* 22:534–43.

Holt, R. D., and T. H. Keitt. 2005. Species' borders: A unifying theme in ecology. *Oikos* 108:3–6.

Hutchinson, M. F. 1995. Interpolating mean rainfall using thin plate smoothing splines. *International Journal of GIS* 9:305–403.

Intergovernmental Panel on Climate Change. 2007 The AR4 synthesis report. Accessed at www.ipcc.ch/.

Kearney, M., B. L. Phillips, C. R. Tracy, K. A. Christian, G. Betts, and W. P. Porter. 2008. Modelling species distributions without using species distributions: the cane toad in Australia under current and future climates. *Ecography* 31:423–34.

Kearney, M., and W. Porter. 2009. Mechanistic niche modelling: Combining physiological and spatial data to predict species ranges. *Ecology Letters* 12:334–50.

Keith, D. A., H. R. Akçakaya, W. Thuiller, G. F. Midgley, R. G. Pearson, S. J. Phillips, H. M. Regan, M. B. Araújo, and T. G. Rebelo. 2008. Predicting extinction risks under climate change: Coupling stochastic population models with dynamic bioclimatic habitat models. *Biology Letters* 4:560–63.

Lockwood, J. L., M. F. Hoopes, and M. P. Marchetti. 2007. *Invasion Ecology*. Oxford: Blackwell.

Lovejoy, T. E., and L. Hannah. 2005. *Climate Change and Biodiversity*. New Haven: Yale University Press.

Menke, S. B., D. A. Holway, R. N. Fisher, and W. Jetz. 2009. Characterizing and predicting species distributions across environments and scales: Argentine ant occurrences in the eye of the beholder. *Global Ecology and Biogeography* 18:50–63.

Mika, A. M., R. M. Weiss, O. Olfert, R. H. Hallett, and J. A. Newman. 2008. Will climate change be beneficial or detrimental to the invasive swede midge in North America? Contrasting predictions using climate projections from different general circulation models. *Global Change Biology* 14:1721–33.

Millennium Ecosystem Assessment. 2005. Ecosystems and human wellbeing: Synthesis of the Mellenium Ecosystem Assessment. Accessed at www.millennium assessment.org/en/synthesis.aspx.

Myers, K., B. S. Parker, and J. D. Dunsmore. 1975. Changes in number of rabbits and their burrows in a subalpine environment in South-eastern New South Wales. *Wildlife Research* 2:121–33.

Myers, N., R. A. Mittermeier, C. G. Mittermeier, G. A. B. da Fonseca, and J. Kent. 2000. Biodiversity hotspots for conservation priorities. *Nature* 403:853–58.

O'Connor, R. J., M. B. Usher, A. Gibbs, and K. C. Brown. 1986. Biological characteristics of invaders among bird species in Britain. *Philosophical Transactions of the Royal Society of London: Series B, Biological Sciences* 314:583–98.

O'Dowd, D. J., P. T. Green, and P. S. Lake. 2003. Invasional "meltdown" on an oceanic island. *Ecology Letters* 6:812–17.

Parer, I., and J. A. Libke. 1985. Distribution of rabbit, *Oryctolagus-cuniculus*, warrens in relation to soil type. *Australian Wildlife Research* 12:387–405.

Parmesan, C. 2006. Ecological and evolutionary responses to recent climate change. *Annual Review of Ecology Evolution and Systematics* 37:637–69.

Pearson, R. G., and T. P. Dawson. 2003. Predicting the impacts of climate change on the distribution of species: Are bioclimate envelope models useful? *Global Ecology and Biogeography* 12:361–71.

Peterson, A. T. 2003. Predicting the geography of species' invasions via ecological niche modeling. *Quarterly Review of Biology* 78:419–33.

Peterson, A. T., A. Stewart, K. I. Mohamed, and M. Araújo. 2008. Shifting global invasive potential of Europe plants with climate change. *PLoS One* 3:e2441.

Phillips, B. L., G. P. Brown, J. M. J. Travis, and R. Shine. 2008a. Reid's paradox revisited: The evolution of dispersal kernels during range expansion. S34–48.

Phillips, B. L., J. D. Chipperfield, and M. R. Kearney. 2008b. The toad ahead: Challenges of modelling the range and spread of an invasive species. *Wildlife Research* 35:222–34.

Poole, W. E. 1960. Breeding of the wild rabbit, *Orctolagus cuniculus*, in relation to the environment. *Australian Wildlife Research* 5:21–43.

Post, E., M. C. Forchhammer, M. S. Bret-Harte, T. V. Callaghan, T. R. Christensen, B. Elberling, A. D. Fox, O. Gilg, D. S. Hik, T. T. Hoye, R. A. Ims, E. Jeppesen, D. R. Klein, J. Madsen, A. D. McGuire, S. Rysgaard, D. E. Schindler, I. Stirling, M. P. Tamstorf, N. J. C. Tyler, R. van der Wal, J. Welker, P. A. Wookey, N. M. Schmidt, and P. Aastrup. 2009. Ecological dynamics across the Arctic associated with recent climate change. *Science* 325:1355–58.

Pyke, C. R., R. Thomas, R. D. Porter, J. J. Hellmann, J. S. Dukes, D. M. Lodge, and G. Chavarria. 2008. Current practices and future opportunities for policy on climate change and invasive species. *Conservation Biology* 22:585–92.

Rahel, F. J., B. Bierwagen, and Y. Taniguchi. 2008. Managing aquatic species of conservation concern in the face of climate change and invasive species. *Conservation Biology* 22:551–61.

Sakai, A. K., F. W. Allendorf, J. S. Holt, D. M. Lodge, J. Molofsky, K. A. With, S. Baughman, R. J. Cabin, J. E. Cohen, N. C. Ellstrand, D. E. McCauley, P. O'Neil, I. M. Parker, J. N. Thompson, and S. G. Weller. 2001. The population biology of invasive species. *Annual Review of Ecology Evolution and Systematics* 32:305–32.

Seo, C., J. H. Thorne, L. Hannah, and W. Thuiller. 2009. Scale effects in species distribution models: Implications for conservation planning under climate change. *Biology Letters* 5:39–43.

Smith, S. D., T. E. Huxman, S. F. Zitzer, T. N. Charlet, D. C. Housman, J. S. Coleman, L. K. Fenstermaker, J. R. Seemann, and R. S. Nowak. 2000. Elevated CO_2 increases productivity and invasive species success in an arid ecosystem. *Nature* 408:79–82.

Stachowicz, J. J., J. R. Terwin, R. B. Whitlatch, and R. W. Osman. 2002. Linking climate change and biological invasions: Ocean warming facilitates nonindigenous species invasions. *Proceedings of the National Academy of Science* 99:15497–500.

Sutherst, R. W. 2000. Climate change and invasive species: a conceptual framework. In *Invasive Species in a Changing World*, edited by H. A. Mooney and R. J. Hobbs, 211–240. Washington: Island Press.

Sutherst, R. W., R. B. Floyd, and G. F. Maywald. 1996. The potential geographical distribution of the cane toad, *Bufo marinus L* in Australia. *Conservation Biology* 10:294–299.

Sutherst, R. W., and G. Maywald. 2005. A climate model of the red imported fire ant, *Solenopsis invicta Buren* (Hymenoptera: Formicidae): Implications for invasion of new regions, particularly Oceania. *Environmental Entomology* 34:317–35.

Sutherst, R. W., G. F. Maywald, and D. Kriticos. 2007. *CLIMEX Version 3 User's Guide*. Melbourne: CSIRO.

Tablado, Z., E. Revilla, and F. Palomares. 2009. Breeding like rabbits: Global patterns of variability and determinants of European wild rabbit reproduction. *Ecography* 32:310–20.

Thomas, C. D., A. Cameron, R. E. Green, M. Bakkenes, L. J. Beaumont, Y. C. Collingham, B. F. N. Erasmus, M. F. de Siqueira, A. Grainger, L. Hannah, L. Hughes, B. Huntley, A. S. van Jaarsveld, G. F. Midgley, L. Miles, M. A. Ortega-Huerta, A. T. Peterson, O. L. Phillips, and S. E. Williams. 2004. Extinction risk from climate change. *Nature* 427:145–48.

Thuiller, W., B. Lafourcade, R. Engler, and M. B. Araújo. 2009. BIOMOD: A platform for ensemble forecasting of species distributions. *Ecography* 32:369–73.

Urban, M. C., B. L. Phillips, D. K. Skelly, and R. Shine. 2007. The cane toad's (*Chaunus [Bufo] marinus*) increasing ability to invade Australia is revealed by a dynamically updated range model. *Proceedings of the Royal Society B: Biological Sciences* 274:1413–19.

Urban, M. C., B. L. Phillips, D. K. Skelly, and R. Shine. 2008. A toad more travelled: The heterogeneous invasion dynamics of cane toads in Australia. *American Naturalist* 171:E134–48.

Van Klinken, R. D., B. E. Lawson, and M. P. Zalucki. 2009. Predicting invasions in Australia by a neotropical shrub under climate change: The challenge of novel climates and parameter estimation. *Global Ecology and Biogeography* 18:688–700.

Ward, N. L., and G. J. Masters. 2007. Linking climate change and species invasion: An illustration using insect herbivores. *Global Change Biology* 13:1605–15.

Wigley, T. M. L., L. E. Clarke, J. A. Edmonds, H. D. Jacoby, S. Paltsev, H. Pitcher, J. M. Reilly, R. Richels, M. C. Sarofim, and S. J. Smith. 2009. Uncertainties in climate stabilization. *Climatic Change* 97:85–121.

Wigley, T. M. L., R. Richels, and J. A. Edmonds. 1996. Economic and environmental choices in the stabilization of atmospheric CO2 concentrations. *Nature* 379:240–43.

Wilcove, D. S., D. Rothstein, J. Dubow, A. Phillips, and E. Losos. 1998. Quantifying threats to imperilled species in the United States. *Bioscience* 48:607–15.

Williams, J. W., S. T. Jackson, and J. E. Kutzbacht. 2007. Projected distributions of novel and disappearing climates by 2100 AD. *Proceedings of the National Academy of Sciences of the United States of America* 104:5738–42.

Wisz, M. S., and A. Guisan. 2009. Do pseudo-absence selection strategies influence species distribution models and their predictions? An information-theoretic approach based on simulated data. *BMC Ecology* 9:8.

6

Can We Predict Climate-Driven Changes to Disease Dynamics? Applications for Theory and Management in the Face of Uncertainty

Sara H. Paull and Pieter T. J. Johnson

How climate change will affect diseases is rapidly becoming one of the most pressing and challenging questions for epidemiologists and conservationists. Advances in modeling techniques and climate science since publication of the first assessment of global climate change in 1990 have led to increasingly reliable predictions about temperature and precipitation changes (IPCC 2007). Corresponding theoretical developments regarding ecological effects of climate on disease have also occurred, but consensus remains elusive (Wilson 2009). While there is a strong body of theoretical work exploring potential climate-disease interactions, there has been little consideration of the potential synergistic effects of climate change and disease on the resilience of wildlife populations and communities. This will be a key concept for identifying effective management strategies in the face of multiple interacting threats and uncertainty (Hoegh-Guldberg and Bruno 2010). We begin this chapter by highlighting the debate over the role of climate in the spread of disease, using human malaria and amphibian chytridiomycosis as case studies. Next, we discuss the mechanisms through which climate change can alter host-pathogen physiology, distribution, interactions, and evolution, emphasizing empirical examples that illustrate the predominant trends. Finally, we discuss current statistical and empirical methods used to evaluate climate-disease linkages before proposing novel methods for studying, predicting, and managing the problems associated with climate-driven variations in disease.

Complexities of Predicting Climate-Driven Changes in Disease Dynamics

The effects of climate change on the distribution and severity of diseases is currently a source of vigorous debate (Lafferty 2009, Randolph 2009, Wilson 2009, Rohr et al. 2011). Multiple, interacting global change drivers, the complex interaction networks of many infectious diseases, and uncertainties in the specifics of temperature and precipitation predictions preclude simplistic generalizations about disease changes resulting from climate change (Tylianakis et al. 2008, Rohr et al. 2011). Given these complexities, it is not surprising that efforts to identify climate signatures in disease patterns have yielded equivocal results. We illustrate this point with two case studies, one involving a human disease (malaria) and one involving a wildlife disease (amphibian chytridiomycosis).

Early attempts at projecting the consequences of climate change for malaria risk based on biological models of vectors predicted large range expansions for human disease risk, igniting more than a decade of debate on the topic (e.g., Martens et al. 1995). Rogers and Randolph (2000) criticized initial models as overly simplistic and argued that they overestimated future malaria distributions. Using current malaria distributions to statistically model future ranges under climate change scenarios, they predicted little change in malaria risk. Ostfeld (2009), however, pointed out that such forecasts underrepresented the climatic tolerances for malaria because they excluded regions of targeted malaria eradication.

Similar controversies have arisen over the role of climate change in the emergence and spread of *Batrachochytrium dendrobatidis* (*Bd*), a chytridiomycete pathogen that has caused dramatic amphibian population declines and extinctions (Skerratt et al. 2007). To explain the decline of tropical montane frogs, Pounds et al. (2006) proposed the"chytrid-thermal-optimum hypothesis," which suggests that climate-mediated increases in cloud cover shifted temperatures towards the growth-optimum for *Bd*. Lips et al. (2008), on the other hand, found no support for climate-driven *Bd* outbreaks in Central and South America, showing instead that disease patterns were consistent with the epidemic spread of a recently introduced pathogen. An analysis by Rohr et al. (2008) further argued that correlations between temperature and frog declines do not necessarily imply that climate change has caused these species declines.

The case studies of malaria and *Bd* emphasize some of the fundamental problems that plague the debate over climate-driven effects on diseases, particularly those of urgent human health or conservation concern. Given the variety and complexity of disease systems and the uncertainties inherent in making climate predictions, it is prudent to question at what scale we can accurately forecast climate-mediated changes in disease. The answer is vitally important because the success of mitigation strategies ultimately depends on the efficient allocation of limited resources to regions most at risk of disease increases. To assess the feasibility of predicting climate change effects on disease dynamics, we first discuss the mechanisms through which climate change can influence host-pathogen physiology, distributions, interactions, and evolution.

How Does Climate Change Affect Patterns of Disease?
Physiological Changes
Climate change can influence the physiology of hosts, vectors, and pathogens in different ways, introducing intriguing shifts in disease patterns. Nonlinear responses of pathogens to rising temperature could have a major impact on their abundance. In one example, recent warming shifted the development time of arctic nematode larvae from two years to a single season, thereby increasing the infection risk experienced by musk oxen (Kutz et al. 2005). If warming temperatures consistently accelerate the development of parasites more than that of their hosts, climate change could dramatically increase parasite abundance and host infection. Alongside changes in mean temperature, shifts in climate variability will also affect disease dynamics by changing pathogen development rates or host immune responses relative to constant temperatures (Paaijmans et al. 2009, Rohr and Raffel 2010).

Climate change will alter mortality rates as well as developmental rates, however, and the balance between these changes for hosts, vectors, and pathogens will influence disease severity in a system (figure 6.1). In temperate zones, warmer winters could enhance the overwintering survival of some pathogens and vectors (Harvell et al. 2002, Canto et al. 2009), but higher metabolic rates and temperatures that exceed thermal tolerance limits can also reduce vector and parasite survival (Lafferty 2009, Snall et al. 2009). Nor will the effects of climate change on disease necessarily be consistent across the distribution of a pathogen. For instance, elevated precipitation can reduce water salinity, and therefore the survival of *Vibrio* bacteria, in mesic areas while increasing cholera risk in drier areas, possibly due to lower water availability and increased concentration of the pathogen in available water sources (Pascual et al. 2002).

A final complication of physiological influences on disease patterns is the influence of climate on host immunity, particularly for ectothermic species. Warming temperatures may either increase or decrease host immunity (e.g., Harvell et al. 2002, Canto et al. 2009). Other climatic changes, such as prolonged drought or increased atmospheric carbon dioxide concentrations, can also alter host resistance to pathogens (Garrett et al. 2006). These observations collectively suggest that the net effect of climate change on disease will depend on how the physiology of different hosts, vectors, and pathogens, responds to temperature and precipitation changes.

Range Shifts
Climate models predict a greater rise in minimum temperatures than in maxima, such that temperatures may be more likely to approach the ther-

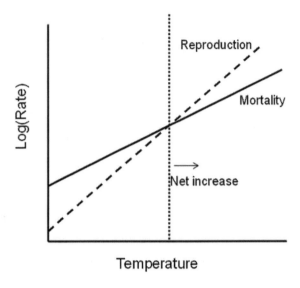

Figure 6.1. Hypothetical trade-offs between temperature-dependent growth and mortality rates of a pathogen. As temperatures approach a pathogen's thermal optimum, the reproductive rate (*dashed line*) may increase more quickly with temperature than the mortality rate (*solid line*). In this scenario, increasing the temperature past the point where the two lines intersect (*dotted line*) will result in a net increase in pathogen populations. If the slopes of these two lines are reversed, as might happen when temperatures approach the thermal maximum, pathogen spread will decline.

mal optima of many organisms, thus leading to predictions that diseases will expand their ranges as temperate areas warm (Ostfeld 2009). Many plant and animal pathogens and vectors may shift poleward in latitude or upward in elevation with climate change (e.g., Pascual et al. 2006; figure 6.2). For example, bluetongue virus has spread northward into Europe since 1998, likely as a result of northward expansion of its vector and increased overwinter virus persistence (Purse et al. 2005). However, many organisms face barriers to dispersal and physiological limitations that prevent range expansions (Root et al. 2003, Lafferty 2009). For instance, Randolph and Rogers (2000) projected a net range reduction of tick-borne encephalitis (TBE) in Europe owing to its dependency on infected nymphal ticks feeding in close proximity to larval ticks—an event that only occurs in particular climatic conditions. Poleward movements of pathogens and vectors could have a strong effect on the immunologically naïve host populations they encounter, regardless of whether pathogens quantitatively expand their ranges. For instance, if malaria shifts its distribution upwards in elevation, it will move into the most populous regions of Africa and South America such that a net decrease in range could still translate into an increase in human impact (Pascual and Bouma 2009). An elevational increase in avian malaria in Hawaii could have devastating effects on highly susceptible native bird populations (Atkinson et al. 2009). Proactive disease regulation measures (e.g., mosquito control) should be considered for areas where the latitudinal and elevational boundaries of pathogens seem to be determined by climatic factors rather than dispersal barriers.

Biotic Interactions

Local differences in species' physiological responses to climate change will scale up to influence biotic interactions that can have consequences for disease (Gilman et al. 2010). For instance, climate-driven changes to

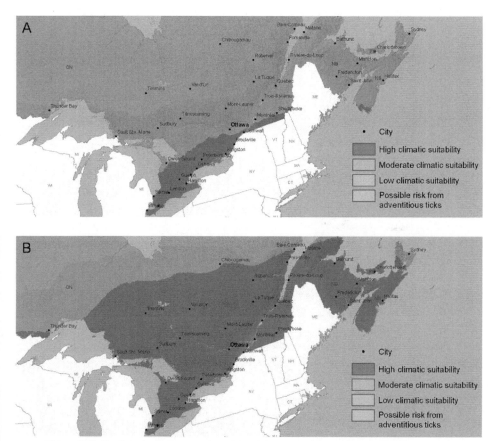

Figure 6.2. Predicted range expansion of the tick vector of Lyme disease, *Ixodes scapularis*, under future projected climate conditions. These risk maps predict the distribution of I. *scapularis* (a) between 2000 and 2019 and (b) between 2080 and 2109. Predictions are based on climate predictions from the CGCM2 model under emissions scenario A2. Figure reproduced with permission from Ogden et al. 2008.

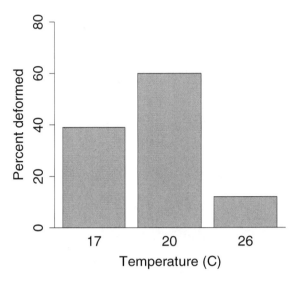

Figure 6.3. A greater percentage of amphibians emerged deformed as a result of infection with the parasite *Ribeiroia ondatrae* at 20° C than at 17 or 26 ° C. This mid-temperature peak in infection likely occurred because although high temperature exacerbated parasites' infectivity, it also reduced tadpoles' vulnerability by causing them to develop more quickly into stages at which they were substantially less likely to become deformed as a result of infection (Paull unpublished data).

host-parasite interactions can affect host pathology. One such example involves the trematode parasite, *Ribeiroia ondatrae*, which sequentially infects snails, amphibians, and birds to complete its life cycle. A laboratory experiment demonstrated that faster parasite development at elevated temperatures increased the pathology experienced by snail hosts infected with *R. ondatrae* by reducing their fecundity (Paull and Johnson 2011). A separate experiment on amphibian hosts showed that temperature-driven increases in parasite infectivity combined with faster development of tadpoles out of vulnerable developmental stages leads to a mid-temperature peak in pathology, measured as the percentage of tadpoles that metamorphose with deformities (figure 6.3). The combination of these temperature-driven changes in host susceptibility, parasite infectivity, and host-parasite reproduction, development and mortality rates across multiple hosts will likely lead to further shifts in the timing and consequence of this host-parasite interaction (Paull and Johnson 2011). Climate change can also shift host-parasite infection dynamics. For example, an analysis of a 30-year dataset of chytrid-diatom dynamics revealed that milder winters reduced chytrid fungal infections of the diatom host *Asterionella formosa* because the diatoms became infected before they bloomed, thus reducing population sizes of both species (Ibelings et al. 2011). Temperature-driven changes in predator-prey relationships can also affect disease transmission. Hall et al. (2006) showed that regulation of fungal epidemics in *Daphnia* by predatory fish may be stronger at warmer temperatures, because the fish respond more strongly to temperature. Climate-driven changes in the compositions of host and parasite communities will also lead to novel host-parasite interactions. For example, hosts may gain or lose parasites as a result of interspecific differences in movement rates (Garrett et al. 2006, Brooks and Hoberg 2007, Harvell et al. 2009). These novel parasite communities can influence host pathology via competition, facilitation, and predation among different parasite species (Pederson and Fenton 2007). Novel communities could thus result in interactions between parasites that complicate efforts to predict disease patterns.

Evolutionary Responses

Changing climates will shift the selective pressures operating on pathogens and their hosts, thus providing a catalyst for evolutionary change. Pathogens tend to have short generation times and high mutation rates, which facilitate adaption to changing environmental conditions (Koelle et al. 2005). Range shifts may also enhance the evolution and spread of drug-resistant pathogens by aiding host movement and gene flow (Criscione and Blouin 2004, Bonizzoni et al. 2009). Strong selection for use of new hosts by pathogens may lead to disease emergence in previously unaffected populations (Marcogliese 2001, Brooks and Hoberg 2007). A greater understanding of pathogen evolutionary processes can be achieved using a phylodynamic approach to characterize the drivers of pathogen evolutionary dynamics across multiple scales (Holmes and Grenfell 2009). Pathogen evolution and changing climate will stimulate host and vector adaptations as well, although evolutionary constraints may restrict the rate at which this occurs (Austin et al., this volume). Climate is changing at a rate unprecedented in the last 50 million years (IPCC 2007). Adaptation will be constrained in most cases by the rate of microevolution, as well as by antagonistic genetic correlations, which may not proceed at rates fast enough for the predicted pace of climate change (Etterson and Shaw 2001, Visser 2008). While mobile organisms may be able to escape regions of increasing parasitism, those with slow migration rates, such as plants, will be forced to rely more on adaptation to changing threats resulting from pathogens (Garrett et al. 2006). Reductions in plant genetic diversity resulting from local adaptation to rapid climate change could also reduce disease resistance in some populations (Jump and Peñuelas 2005).

Current Methods for Studying
Climate-Driven Changes in Disease

Current techniques for forecasting the influence of climate change on disease risk include correlative studies along temporal and spatial climate gradients, synthetic meta-analyses, predictive models, and experimental investigations. We discuss the relative merits and weaknesses of each technique before suggesting novel strategies that can enhance their implementation. Because of the urgency of the problem, the complexity of host-parasite-climate systems, and the paucity of baseline data for most diseases, a combination of approaches is likely to generate the most reliable results for forecasting.

Statistical approaches attempt to link temporally or spatially variable climatic patterns with disease incidence. These approaches take two main forms: tests for connections between regional warming trends and disease, and correlations between the El Niño Southern Oscillation (ENSO) or North Atlantic Oscillation (NAO) indices and disease. Studies linking disease incidence to ENSO or NAO indices can be useful for forecasting disease severity up to a year in advance (Chaves and Pascual 2006). Interpreting the results requires some caution, however, as such models often lack mechanistic components, and correlations involving large-scale, monotonic increases in both variables can be misleading (Rohr et al. 2008). Other intrinsic factors, such as changes in the number of immune hosts, could cause cyclical changes in disease, underscoring the need to consider the relative importance of extrinsic climatic forcing as opposed to internal drivers (Dobson 2009, Harvell et al. 2009).

Meta-analyses and other synthetic approaches can distill large-scale patterns from the synthesis of numerous small-scale experiments or surveys. By quantitatively summarizing current knowledge, meta-analyses have provided compelling evidence to support theoretical predictions about biotic responses to climate change (e.g., Parmesan and Yohe 2003, Root et al. 2003). Meta-analyses can be problematic, however, because working with published data can introduce bias (e.g., null results often go unpublished) or nonindependent studies into the analysis (Lei et al. 2007). While these problems can be minimized with a careful literature selection process, the results typically do not mechanistically explain patterns and mask the variability among studies that is essential for fine-scale prediction (Lei et al. 2007).

There are two main types of bioclimatic models: mechanistic models, which use physiological parameters to infer changes to disease distributions, and statistical models, which use the climatic parameters associated with the current range of the disease to forecast future range shifts (Jeschke and Strayer 2008). The utility of bioclimatic models lies in their ability to generate quantitative hypotheses about disease shifts resulting from climate change that can be used in directing management efforts (Jeschke and Strayer 2008). Many bioclimatic models, however, depend on assumptions that are frequently violated, such as the assumption that species ranges are not limited by dispersal or biotic interactions, and that a species' genetic climate tolerance does not vary through space or time (Peterson and Shaw 2003, Jeschke and Strayer 2008, Gilman et al. 2010, Brodie et al., this volume). Incorporating other aspects known to influence disease risk, such as anticipated changes in host population size or limitations to dispersal will strengthen the predictive power of such models (Peterson and Shaw 2003).

Carefully designed experiments provide the most mechanistic evidence of how climate affects disease dynamics. While short-term laboratory studies are typically limited to one component of a complex disease system, they help clarify the mechanisms underlying climate-disease interactions and can be used to parameterize mechanistic predictive models. For example, Terblanche et al. (2008) experimentally measured the thermal tolerance of tsetse flies, which are vectors for human and animal trypanosomiases, and inferred that future warming would exceed their upper thermal limits and lead to a reduction in their geographic range. Field experiments are needed to explore the effects of climate change on disease in a more realistic larger-scale context. For instance, field studies that examine elevated temperatures of plant-pathogen systems generally find that pathogens respond uniquely to the warming treatment, with some increasing in number and others decreasing (e.g., Wiedermann et al. 2007), thus suggesting that large-scale patterns may not be fully elucidated by smaller single-system experiments.

Where Do We Go from Here? Predicting and Mitigating the Effects of Climate Change on Disease

It is critical that we focus attention on developing more quantitative predictions, greater mechanistic understanding, and explicit management advice regarding the effects of climate on disease. Here, we summarize novel strategies in the areas of *modeling*, *empirical research*, and *disease management*, which can be used in conjunction with existing tools to provide an informative framework for mitigating the effects of climate change on disease. Although forecasts about climate-driven changes in disease dynamics will always be plagued by the complexity of the issue, unpredictable stochastic forces, and variation in disease response across scales, a greater mechanistic understanding of the processes involved, including more cross-disciplinary research, and a focus on climate-sensitive aspects of disease transmission (figure 6.4) will enhance our ability to respond to changing disease risks.

Climate effects on host-pathogen interactions

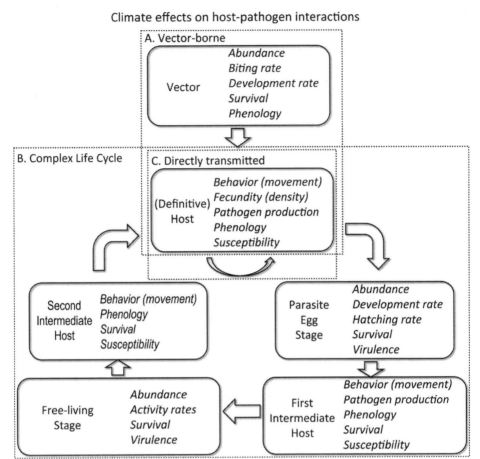

Figure 6.4. Potential direct influence of climate change at different stages of pathogen transmission for hypothetical vector-borne, complex-life-cycle, and directly transmitted diseases. Changes that occur in the dynamics of any disease will be affected by whether the organisms involved are ectothermic or endothermic, as well as by indirect interactions with the community and environment. Attempts to understand climate-driven changes in disease dynamics will be further complicated by evolutionary changes and distributional shifts of hosts, vectors, and pathogens.

Physiology Meets Ecology: Using Novel Modeling Strategies to Enhance Disease Forecasting

Mechanistic models based on physiological parameters of specific host-pathogen systems and regional climate forecasts can facilitate management plans at the local level. Dynamic energy budget (DEB) models describe organismal physiologies in detail, and can be adapted to model interactions between organisms at the population and community levels (e.g., Kooijman 2001, Vasseur and McCann 2005). Biophysical models have strong predictive power relative to correlative methods because they can detect effects across multiple scales and can distinguish between hypothesized drivers of disease change (Helmuth 2009, Kearney et al. 2009). For example, a biophysical model that incorporated the microclimatic effects of different water storage containers demonstrated that water storage had a larger impact than climate change on the distribution of dengue-carrying mosquitoes (Kearney et al. 2009). Such mechanistic models are rare, however, owing to the increase in parameters that occurs when bioenergetics models are extended to the community level (Vasseur and McCann 2005). In the absence of such detailed physiological data, qualitative predictions about host-parasite interactions can still be made by considering easily measured physiological parameters such as thermal windows, which describe the *range* of temperatures across which an organism can maintain stable performance, and the Q_{10} coefficient, which describes a change in a given physiological *rate* for every 10° C change in temperature. These measures may be useful in predicting the direction of changes in host-pathogen interactions (figure 6.5). An increased emphasis on physiological models for predicting climate-driven changes in biotic interactions within specific disease systems could complement the current use of bioclimatic models to provide a more complete overall picture.

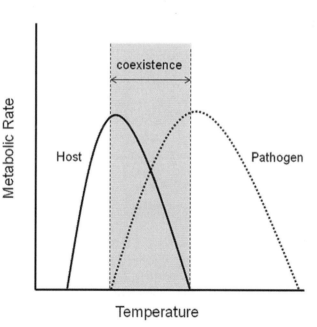

Figure 6.5. Depiction of the thermal tolerance window of a hypothetical host (*solid line*) and pathogen (*dotted line*). The shaded region is the range of temperatures at which the host and parasite can coexist. As temperatures rise they become more favorable for the parasite, thus shifting the character of the relationship.

Experimental Climate Change: Using Novel Empirical Approaches to Address Disease Effects

Empirical research on the linkage between climate and disease has been hindered by the logistical complications of experimentally modifying environmental temperatures. We suggest two forms of empirical study to help advance research on climate change: (1) spatial gradients in temperature and (2) experimental mesocosm and field studies. Because elevation gradients offer a range of climatic conditions over relatively short distances, correlations between climate and disease patterns across differing altitudes will have fewer covariates to influence the results (Fukami and Wardle 2005). Relatively few studies have correlated biotic responses to climate change using available spatial gradients, despite the value of such geographical gradients in temperature for understanding climate-driven changes in disease risk (Hudson et al. 2006, Altizer et al. 2006). Seminatural mesocosm experiments are also an effective way to test theories about climate-driven changes in disease transmission. Designs for large outdoor mesocosms can range from expensive computerized aquatic systems to simple Plexiglas heat-trapping structures in terrestrial or aquatic environments (Liboriussen et al. 2005, Netten et al. 2008). Such use of simplified host-parasite systems can further facilitate experimental testing of hypotheses about disease reactions to climate change. The Arctic is a promising region for such studies, due to its low anthropogenic influence, its low level of biodiversity, and its predicted large climatic shifts (Kutz et al. 2009).

Further study of the direct and indirect climatic drivers, aside from temperature, of disease dynamics is also necessary. Climate change involves alteration of a suite of variables beyond mean temperature, including precipitation, diurnal and seasonal temperature ranges, and the frequency and severity of extreme weather events. For example, in aquatic systems, changes in ice cover, acidification, eutrophication, lake mixing regimes, and ultraviolet exposure will also affect hosts and parasites (Marcogliese 2001). Changes in terrestrial ecosystems are expected in response to climate-driven alteration of snow cover, wildfire disturbance regimes, the frequency of extreme weather events, and carbon dioxide concentrations (Garrett et al. 2006, Schumacher and Bugman 2006, Jentsch et al. 2009). Studies that test factors other than changes in mean temperature and precipitation on disease systems are needed to assess whether climatic variability or indirect drivers may also play a role in disease dynamics.

Technological Advances for Research and Collaboration

Interdisciplinary collaboration will be important in developing novel research methods to explore climate-driven changes in disease. For instance, molecular PCR-based techniques can improve disease surveillance methods (Polley and Thompson 2009). Recent breakthroughs in DNA sequencing techniques will also pave the way for phylodynamic approaches to epidemiology that could provide key insights into evolutionary responses of pathogens to climate change (Holmes and Grenfell 2009). Technological advances in sequencing and identification techniques have also made the direct analysis of paleoparasitological changes associated with changes in paleoclimate data an increasingly useful avenue of research (Dittmar 2009). Geographic information systems (GIS) offer another powerful epidemiological tool that can be used for mapping potential climate-driven changes to disease risk (Ostfeld et al. 2005). For example, remote sensing technology can be used to characterize regions with high disease risk (Glass et al. 2007). Collectively, these tools should be applied toward developing a large-scale, interactive, and publicly accessible database of disease distribution and pathology that would provide an invaluable resource for global-scale analyses of climate-driven changes in disease (Semenza and Menne 2009). Because the study of climate change and disease spans disciplines ranging from atmospheric science to epidemiology, multidisciplinary efforts are key to developing the creative research and data acquisition strategies necessary for effective disease management.

Planning for the Unpredictable: Management and Surveillance Tools

Despite our best efforts to predict disease responses to climate change, the complexities and uncertainties within these systems ensure that ecological "surprises" will occur. Acting in the face of such uncertainties requires effective use of management strategies in combination with enhanced disease treatment and surveillance methods. Managing populations to increase their resilience will be critical given the uncertainties in predicting climate-driven changes to ecosystems (Hoegh-Guldberg and Bruno 2010). To increase the resilience of populations to climate-driven disease susceptibility, managers should focus their efforts on maintaining high genetic and species diversity while reducing other environmental stressors (Evans and Perschel 2009, Hoegh-Guldberg and Bruno 2010). Simple changes in animal husbandry practices, including grazing, housing, and shearing, can also reduce disease risk in domestic animals and the spillover of infections into wildlife populations (Morgan and Wall

2009). In the case of diseases in humans or in threatened or economically valuable species, vaccination against infection may be necessary (Hampson et al. 2009). Flexibility will be a key component of plans for managing climate-driven changes in disease. Adaptive management is one such strategy that involves development of experimental management programs to test alternative hypotheses for effective resource management (Walters and Holling 1990). Another emerging strategy for dealing with stochastic events that cause unpredictable regime shifts is to develop alternative management plans for multiple future scenarios to cope with unexpected changes quickly and effectively (Bennett et al. 2003).

Linking Climate Change, Disease, and Conservation

Ultimately, the success of predictions about climate-disease interactions will be measured by their utility in mitigating the negative consequences of diseases for human and wildlife populations. Although low host population sizes can often reduce disease persistence, diseases can still cause species extinctions by driving populations to unstably small numbers or by residing in reservoir hosts (de Castro and Bolker 2005). Even if a disease does not cause extinction of a particular species, it can lead to genetic homogenization that reduces its ability to cope with other environmental stressors (Smith et al. 2006). An overall assessment of climate-driven changes to disease risk should use these characteristics to determine which diseases are most likely to pose a threat to wildlife populations.

Finally, the conservation of parasites and pathogens themselves may be an appropriate management consideration. Parasites play a vital role in many ecosystems, and their disappearance could have a cascading effect on other processes. Parasites serve as key links in food webs and as regulators of host populations (Dobson et al. 2008). Specialist parasites may have a higher extinction risk, which could allow generalist parasites (typically associated with higher pathological effects) to increase as a result of competitive release (Dunn et al 2009). Parasites and pathogens have a dramatic influence on the health of human and wildlife populations. Changes in their dynamics and abundance will be among the most important consequences of global climate change.

Conclusions

Given the complexities involved in disease dynamics and the uncertainties surrounding climate change predictions, the debate over whether climate change will increase or reduce global disease risk will continue. Diseases and species tend to respond idiosyncratically to climate change, with range shifts, physiological changes, phenological changes, and evolutionary rates differing among species. Nevertheless, the risk of changing diseases to human health and wildlife conservation is great enough that action is required. Our predictive abilities can be enhanced through biophysical modeling, increased use of experimental manipulations incorporating the direct and indirect effects of climate change, and collaborative efforts that capitalize on recent technological innovations. Our forecasting capabilities may be limited to predicting the general behavior of specific, well-parameterized host-pathogen systems, or to identifying the factors that are important in driving the dynamics of specific classes of pathogens. The best approach for maximizing our predictive capabilities is to combine mechanistic empirical and modeling approaches. The uncertainty arising from the complexities of host-parasite-climate interactions, particularly in conjunction with other global change drivers, underscores the need for managing wildlife populations to increase resilience within a framework of flexible adaptive management practices.

ACKNOWLEDGMENTS
SHP was supported by the United States Environmental Protection Agency (EPA) under the Science to Achieve Results (STAR) Graduate Fellowship Program (SHP). The EPA has not officially endorsed this publication and the views expressed herein may not reflect the views of the EPA. PTJJ was supported by a fellowship from the David and Lucile Packard Foundation and a grant from NSF (DEB-0841758).

LITERATURE CITED
Altizer, S., A. Dobson, P. Hosseini, P. Hudson, M. Pascual, and P. Rohani. 2006. Seasonality and the dynamics of infectious diseases. *Ecology Letters* 9:467–84.
Atkinson, C. T., and D. A. LaPointe. 2009. Introduced avian diseases, climate change, and the future of Hawaiian honeycreepers. *Journal of Avian Medicine and Surgery* 23:53–63.
Bennett, E. M., S. R. Carpenter, G. D. Peterson, G. S. Cumming, M. Zurek, and P. Pingali. 2003. Why global scenarios need ecology. *Frontiers in Ecology and the Environment* 1:322–29.
Bonizzoni, M., Y. Afrane, F. N. Baliraine, D. A. Amenya, A. K. Githeko, and G. Yan. 2009. Genetic structure of *Plasmodium falciparum* populations between lowland and highland sites and antimalarial drug resistance in Western Kenya. *Infection, Genetics and Evolution* 9:806–12.
Brooks, D. R., and E. P. Hoberg. 2007. How will global climate change affect parasite-host assemblages? *Trends in Parasitology* 23:571–74.
Canto, T., M. A. Aranda, and A. Fereres. 2009. Climate change effects on physiology and population processes of hosts and vectors that influence the spread of hemipteran-borne plant viruses. *Global Change Biology* 15:1884–94.
Chaves, L. F., and M. Pascual. 2006. Climate cycles and forecasts of cutaneous leishmaniasis, a nonstationary vector-borne disease. *Plos Medicine* 3:1320–28.

Criscione, C. D., and M. A. Blouin. 2004. Life cycles shape parasite evolution: Comparative population genetics of salmon trematodes. *Evolution* 58:198–202.

De Castro, F., and B. Bolker. 2005. Mechanisms of disease-induced extinction. *Ecology Letters* 8:117–26.

Dittmar, K. 2009. Old parasites for a new world: The future of paleoparasitological research. A review. *Journal of Parasitology* 95:365–71.

Dobson A. 2009. Climate variability, global change, immunity, and the dynamics of infectious diseases. *Ecology* 90:920–27.

Dobson, A., K. D. Lafferty, A. M. Kuris, R. F. Hechinger, and W. Jetz. 2008. Homage to Linnaeus: How many parasites? How many hosts? *Proceedings of the National Academy of Sciences of the United States of America* 105:11482–89.

Dunn, R. R., N. C. Harris, R. K. Colwell, L. P. Koh, and N. S. Sodhi. 2009. The sixth mass coextinction: Are most endangered species parasites and mutualists? *Proceedings of the Royal Society B: Biological Sciences* 276:3037–45.

Etterson, J. R., and R. G. Shaw. 2001. Constraint to adaptive evolution in response to global warming. *Science* 294:151–54.

Evans, A. M., and R. Perschel. 2009. A review of forestry mitigation and adaptation strategies in the Northeast U.S. *Climatic Change* 96:167–83.

Fukami, T., and D. A. Wardle. 2005. Long-term ecological dynamics: reciprocal insights from natural and anthropogenic gradients. *Proceedings of the Royal Society B: Biological Sciences* 272:2105–15.

Garrett, K. A., S. P. Dendy, E. E. Frank, M. N. Rouse, and S. E. Travers. 2006. Climate change effects on plant disease: Genomes to ecosystems. *Annual Review of Phytopathology* 44:489–509.

Gilman, S. E., M. C. Urban, J. Tewksbury, G. W. Gilchrist, and R. D. Holt. 2010. A framework for community interactions under climate change. *Trends in Ecology and Evolution* 25:325–31.

Glass, G. E., T. Shields, B. Cai, T. L. Yates, and R. Parmenter. 2007. Persistently highest risk areas for hantavirus pulmonary syndrome: Potential sites for refugia. *Ecological Applications* 17:129–39.

Hall, S. R., A. J. Tessier, M. A. Duffy, M. Huebner, and C. E. Caceres. 2006. Warmer does not have to mean sicker: Temperature and predators can jointly drive timing of epidemics. *Ecology* 87:1684–95.

Hampson, H., J. Dushoff, S. Cleaveland, D. T. Haydon, M. Kaare, C. Packer, and A. Dobson. 2009. Transmission dynamics and prospects for the elimination of canine rabies. *PloS Biology* 7:462–71.

Harvell, C. D., C. E. Mitchell, J. R. Ward, S. Altizer, A. P. Dobson, R. S. Ostfeld, and M. D. Samuel. 2002. Ecology: Climate warming and disease risks for terrestrial and marine biota. *Science* 296:2158–62.

Harvell, D., S. Altizer, I. M. Cattadori, L. Harrington, and E. Weil. 2009. Climate change and wildlife diseases: When does the host matter the most? *Ecology* 90:912–20.

Helmuth, B. 2009. From cells to coastlines: How can we use physiology to forecast the impacts of climate change? *Journal of Experimental Biology* 212:753–60.

Hoegh-Guldberg, O., and J. F. Bruno. 2010. The impact of climate change on the world's marine ecosystems. *Science* 328:1523–28.

Holmes, E. C., and B. T. Grenfell. 2009. Discovering the phylodynamics of RNA Viruses. *PloS Computational Biology* 5:1–6.

Hudson, P. J., M. Cattadori, B. Boag, and A. P. Dobson. 2006. Climate disruption and parasite-host dynamics: Patterns and processes associated with warming and the frequency of extreme climatic events. *Journal of Helminthology* 80:175–82.

Ibelings, B. W., A. S. Gsell, W. M. Mooij, E. van Donk, S. van den Wyngaert, and S. N. de Senerpont Domis. 2011. Chytrid infections and diatom spring blooms: Paradoxical effects of climate warming on fungal epidemics in lakes. *Freshwater Biology* 56:754–66.

IPCC. 2007. Climate Change 2007: The Physical Science Basis. Contribution of Working Group I to the Fourth Assessment Report of the Intergovernmental Panel on Climate Change, edited by S. Solomon, D. Qin, M. Manning, Z. Chen, M. Marquis, K. B. Averyt, M. Tignor and H. L. Millers. New York: Cambridge University Press.

Jentsch, A., J. Kreyling, J. Boettcher-Treschkow, and C. Beierkuhnlein. 2009. Beyond gradual warming: Extreme weather events alter flower phenology of European grassland and heath species. *Global Change Biology* 15:837–49.

Jeschke, J. M., and D. L. Strayer. 2008. Usefulness of bioclimatic models for studying climate change and invasive species. *Year in Ecology and Conservation Biology* 2008 1134:1–24.

Johnson, P. T. J., K. B. Lunde, E. G. Ritchie, and A. E. Launer. 1999. The effect of trematode infection on amphibian limb development and survivorship. *Science* 284:802–4.

Jump, A. S., and Peñuelas J. 2005. Running to stand still: Adaptation and the response of plants to rapid climate change. *Ecology Letters* 8:1010–20.

Kearney, M., W. P. Porter, C. Williams, S. Ritchie, and A. A. Hoffmann. 2009. Integrating biophysical models and evolutionary theory to predict climatic impacts on species' ranges: the dengue mosquito *Aedes aegypti* in Australia. *Functional Ecology* 23:528–38.

Koelle K., M. Pascual, and M. Yunus. 2005. Pathogen adaptation to seasonal forcing and climate change. *Proceedings of the Royal Society B: Biological Sciences* 272:971–77.

Kooijman, S. A. L. M. 2001. Quantitative aspects of metabolic organization: a discussion of concepts. *Philosophical Transactions of the Royal Society of London Series B: Biological Sciences* 356:331–49.

Kutz, S. J., E. P. Hoberg, L. Polley, and E. J. Jenkins. 2005. Global warming is changing the dynamics of Arctic host-parasite systems. *Proceedings of the Royal Society B: Biological Sciences* 272:2571–76.

Kutz, S. J., E. J. Jenkins, A. M. Veitch, J. Ducrocq, L. Polley, B. Elkin, and S. Lair. 2009. The Arctic as a model for anticipating, preventing, and mitigating climate change impacts on host-parasite interactions. *Veterinary Parasitology* 163:217–28.

Lafferty, K. D. 2009. The ecology of climate change and infectious diseases. *Ecology* 90:888–900.

Lei X., Peng C., Tian D., and Sun J.. 2007. Meta-analysis and its application in global change research. *Chinese Science Bulletin* 52:289–302.

Liboriussen, L., F. Landkildehus, M. Meerhoff, M. E. Bramm, M. Sondergaard, K. Christoffersen, K. Richardson, M. Sondergaard, T. L. Lauridsen, and E. Jeppesen. 2005. Global warming: Design of a flow-through shallow lake mesocosm climate experiment. *Limnology and Oceanography–Methods* 3:1–9.

Lips, K. R., J. Diffendorfer, J. R. Mendelson III., and M. W. Sears. 2008. Riding the wave: Reconciling the roles of disease and climate change in amphibian declines. *PloS Biology* 6:441–54.

Marcogliese, D. J. 2001. Implications of climate change for parasitism of animals in the aquatic environment. Canadian Journal of Zoology / *Revue canadienne de zoologie* 79:1331–52.

Martens, W. J. M., L. W. Niessen, J. Rotmans, T. H. Jetten, and A. J. Mcmichael. 1995. Potential impact of global climate change on malaria risk. *Environmental Health Perspectives* 103:458–64.

Morgan, E. R., and R. Wall. 2009. Climate change and parasitic disease: Farmer mitigation? *Trends in Parasitology* 25:308–13.

Netten, J. J. C., E. H. van Nes, M. Scheffer, and R. M. M. Roijackers. 2008. Use of open-top chambers to study the effect of climate change in aquatic ecosystems. *Limnology and Oceanography–Methods* 6:223–29.

Ogden, N. H., L. St-Onge, I. K. Barker, S. Brazeau, M. Bigras-Poulin, D. F. Charron, C. M. Francis, A. Heagy, L. R. Lindsay, A. Maarouf, P. Michel, F. Milord, C. J. O'Callaghan, L. Trudel, and R. A. Thompson. 2008. Risk maps for range expansion of the Lyme disease vector, *Ixodes scapularis*, in Canada now and with climate change. *International Journal of Health Geographics* 7:1–15.

Ostfeld, R. S. 2009. Climate change and the distribution and intensity of infectious diseases. *Ecology* 90:903–5.

Ostfeld, R. S., G. E. Glass, and F. Keesing. 2005. Spatial epidemiology: An emerging (or re-emerging) discipline. *Trends in Ecology and Evolution* 20:328–36.

Paaijmans, K. P., A. F. Read, and M. B. Thomas. 2009. Understanding the link between malaria risk and climate. *Proceedings of the National Academy of Sciences of the United States of America* 106:13844–49.

Parmesan, C., and G. Yohe. 2003. A globally coherent fingerprint of climate change impacts across natural systems. *Nature* 421:37–42.

Pascual, M., J. A. Ahumada, L. F. Chaves, X. Rodo, and M. Bouma. 2006. Malaria resurgence in the East African highlands: Temperature trends revisited. *Proceedings of the National Academy of Sciences of the United States of America* 103:5829–34.

Pascual, M., and M. J. Bouma. 2009. Do rising temperatures matter? *Ecology* 90:906–12.

Pascual, M., M. J. Bouma, and A. P. Dobson. 2002. Cholera and climate: Revisiting the quantitative evidence. *Microbes and Infection* 4:237–45.

Paull, S. H., and P. T. J. Johnson. 2011. High temperature enhances host pathology in a snail-trematode system: Possible consequences of climate change for the emergence of disease. *Freshwater Biology* 56:767–78.

Pedersen, A. B., and A. Fenton. 2007. Emphasizing the ecology in parasite community ecology. *Trends in Ecology & Evolution* 22:133–39.

Peterson, A. T., and J. Shaw. 2003. *Lutzomyia* vectors for cutaneous leishmaniasis in Southern Brazil: Ecological niche models, predicted geographic distributions, and climate change effects. *International Journal for Parasitology* 33:919–31.

Polley, L., and R. C. A. Thompson. 2009. Parasite zoonoses and climate change: Molecular tools for tracking shifting boundaries. *Trends in Parasitology* 25: 285–91.

Pounds, J. A., M. R. Bustamante, L. A. Coloma, J. A. Consuegra, M. P. L. Fogden, P. N. Foster, E. La Marca, K. L. Masters, A. Merino-Viteri, R. Puschendorf, S. R. Ron,

G. A. Sanchez-Azofeifa, C. J. Still, and B. E. Young. 2006. Widespread amphibian extinctions from epidemic disease driven by global warming. *Nature* 439:161–67.

Purse, B. V., P. S. Mellor, D. J. Rogers, A. R. Samuel, P. P. C. Mertens, and M. Baylis. 2005. Climate change and recent emergence of bluetongue in Europe. *Nature Reviews Microbiology* 3:171–81.

Randolph, S. E. 2009. Perspectives on climate change impacts on infectious diseases. *Ecology* 90:927–31.

Randolph, S. E., and D. J. Rogers. 2000. Fragile transmission cycles of tick-borne encephalitis virus may be disrupted by predicted climate change. *Proceedings of the Royal Society of London Series B: Biological Sciences* 267:1741–44.

Rogers, D. J., and S. E. Randolph. 2000. The global spread of malaria in a future, warmer world. *Science* 289:1763–66.

Rohr, J. R., A. P. Dobson, P. T. J. Johnson, A. M. Kilpatrick, S. H. Paull, T. R. Raffel, D. Ruiz-Moreno, and M. B. Thomas. 2011. Frontiers in climate change-disease research. *Trends in Ecology and Evolution* 26:270–77.

Rohr, J. R., and T. R. Raffel. 2010. Linking global climate and temperature variability to widespread amphibian declines putatively caused by disease. *Proceedings of the National Academy of Sciences of the United States of America* 107:8269–74.

Rohr J. R., T. R. Raffel, J. M. Romansic, H. McCallum, and P. J. Hudson. 2008. Evaluating the links between climate, disease spread, and amphibian declines. *Proceedings of the National Academy of Sciences of the United States of America* 105:17436–41.

Root, T. L., J. T. Price, K. R. Hall, S. H. Schneider, C. Rosenzweig, and J. A. Pounds. 2003. Fingerprints of global warming on wild animals and plants. *Nature* 421: 57–60.

Schumacher, S., and H. Bugmann. 2006. The relative importance of climatic effects, wildfires and management for future forest landscape dynamics in the Swiss Alps. *Global Change Biology* 12:1435–50.

Semenza, J. C., and B. Menne. 2009. Climate change and infectious diseases in Europe. *Lancet Infectious Diseases* 9:365–75.

Skerratt, L. F., L. Berger, R. Speare, S. Cashins, K. R. McDonald, A. D. Phillott, H. B. Hines, and N. Kenyon. 2007. Spread of chytridiomycosis has caused the rapid global decline and extinction of frogs. *Ecohealth* 4:125–34.

Smith, K. F., D. F. Sax, and K. D. Lafferty. 2006. Evidence for the role of infectious disease in species extinction and endangerment. *Conservation Biology* 20: 1349–57.

Snall, T., R. E. Benestad, and N. C. Stenseth. 2009. Expected future plague levels in a wildlife host under different scenarios of climate change. *Global Change Biology* 15:500–507.

Terblanche, J. S., S. Clusella-Trullas, J. A. Deere, and S. L. Chown. 2008. Thermal tolerance in a south-east African population of the tsetse fly *Glossina pallidipes* (Diptera, Glossinidae): Implications for forecasting climate change impacts. *Journal of Insect Physiology* 54:114–27.

Tylianakis, J. M., R. K. Didham, J. Bascompte, and D. A. Wardle. 2008. Global change and species interactions in terrestrial ecosystems. *Ecology Letters* 11:1351–63.

Vasseur, D. A., and K. S. McCann. 2005. A mechanistic approach for modeling temperature-dependent consumer-resource dynamics. *American Naturalist* 166:184–98.

Visser, M. E. 2008. Keeping up with a warming world: Assessing the rate of adaptation to climate change. *Proceedings of the Royal Society B: Biological Sciences* 275:649–59.

Walters, C. J., and C. S. Holling. 1990. Large-scale management experiments and learning by doing. *Ecology* 71:2060–68.

Wiedermann, M. M., A. Nordin, U. Gunnarsson, M. B. Nilsson, and L. Ericson. 2007. Global change shifts vegetation and plant-parasite interactions in a boreal mire. *Ecology* 88:454–64.

Wilson, K. 2009. Climate change and the spread of infectious ideas. *Ecology* 90:901–2.

7

Rapid Assessment of Plant and Animal Vulnerability to Climate Change

Bruce E. Young, Kimberly R. Hall, Elizabeth Byers, Kelly Gravuer, Geoff Hammerson, Alan Redder, and Kristin Szabo

Although scientists have been concerned about climate change for decades, many policy makers and resource managers have only recently recognized the urgency of the problem. Now resource managers are increasingly asked to identify which of the species on the lands and waters they oversee are most vulnerable to climate change-induced declines. Knowing which species are vulnerable and why is a critical input for developing management strategies to promote persistence of species as climates change. Comparing vulnerabilities across species is difficult, however, because species respond differently to change (Overpeck et al. 1991, Davis and Shaw 2001) and because climate change is likely to impact species both directly and indirectly. Further, the same species may respond differently in different places, due to variations in exposure to climate change or differences in key habitats or species interactions. Also, research on climate change vulnerability is growing rapidly (Brodie et al. this volume) and managers often have little time to keep abreast of new findings (Heller and Zavaleta 2009, Lawler et al. 2009a).

Climate change vulnerability is now on the agenda of international entities such as the European Union and the International Union for the Conservation of Nature (IUCN; CEC 2006, Foden et al. 2008). In the United States, state fish and wildlife agencies increasingly need ways to identify vulnerable species as they begin to revise state wildlife action plans. In the United States, state fish and wildlife agencies increasingly need ways to identify vulnerable species as they begin to revise state wildlife action plans. Wildlife action plans, mandated by the US Congress, require assessments of species and habitats at risk and the development of strategies to prevent species from becoming endangered (AFWA 2009). Revisions of these plans are required every 10 years, but revisions to specifically include climate change are not mandated at this time. Similarly, US federal land managing agencies are seeking ways to address species vulnerability as they begin to modify conservation strategies to account for climate change (Blay and Dombeck, this volume).

Most assessments of vulnerability to climate change tend to focus on single factors, such as changes in distribution (e.g., from bioclimatic models; Peterson et al. 2002, Midgley et al. 2003, Thomas et al. 2004, Lawler et al. 2009b) or changes in phenology and the potential for phenological mismatches (e.g., Bradley et al. 1999, Visser and Both 2005). More recently, scientists have emphasized how key behavioral or demographic characteristics may contribute to vulnerability (e.g., Humphries et al. 2004, Jiguet et al. 2007, Laidre et al. 2008) and to species response patterns at various organizational scales (Parmesan 2007, Willis et al. 2008). Further, several theoretical treatises describe potential frameworks for vulnerability assessments, including evaluations of exposure to climate change, inherent sensitivity, and adaptive capacity (Füssel and Klein 2006, Williams et al. 2008, Austin et al., this volume), as well as guidance on how to incorporate uncertainty and relative risk (Schneider et al. 2007).

Building on these findings, we have developed a "climate change vulnerability index" (hereafter, "index") to serve the needs of wildlife managers for a practical, multifaceted rapid assessment tool. The aim of the index is to provide a means of rapidly distinguishing species likely to be most vulnerable, defined as the degree to which a species is susceptible to detrimental change (Smit et al. 2000). After using the index, managers may wish to perform more in-depth (and resource-intensive) vulnerability analyses of species highlighted by the tool as being particularly vulnerable. The index relies on natural history and distribution factors that are associated with sensitivity to climate change and projections of climatic changes for the assessment area. It does not require advanced technical expertise, so it can be used efficiently by anyone with biological training and access to the relevant natural history and distribution information.

The index is flexible in that it can assess plants and animals from both terrestrial and aquatic habitats, and can handle missing data and uncertainty in species sensitivity measures. It can also handle input from studies that document vulnerability or project future suitable ranges, when available. Its output includes both a vulnerability category for the species of interest and a report on the key factors that have contributed to the ranking, which can help inform conservation actions. Here we discuss the mechanics of the index and report on preliminary results from a case study of vertebrates and mollusks included in Nevada's state wildlife action plan.

Climate Change Vulnerability Index

We divide vulnerability into exposure to changes in climate and species sensitivity (Schneider et al. 2007, Foden et al. 2008, Williams et al.

2008). Exposure is the magnitude of projected climate change across the portion of the range of the focal species that lies within the geographic area considered. Species sensitivity includes intrinsic factors such as natural and life-history traits that promote resilience to change (such as dietary versatility or identification as a habitat generalist), traits that indicate increased risk (such as a strong potential for disruption of key species interactions), and traits that indicate capacity to adapt to change (such as dispersal ability and genetic variation). The index scores a species in relation to multiple intrinsic and extrinsic sensitivity factors and then weights the score depending on the magnitude of climate change projected. Any information available on documented responses of the species to climate change is then combined with the vulnerability score to produce a final index score (figure 7.1).

For simplicity of use, we have developed the index as an MS Excel workbook (available at www.natureserve.org/climatechange) that allows users to enter exposure data and then select categorical answers to questions that assess how the species' natural history may influence its relative vulnerability to climate change. Extensive documentation provides criteria for determining how to "score" sensitivity for each factor, but the user can enter more than one value to indicate uncertainty in species information (Young et al. 2010). The workbook then calculates an index score from the entries on exposure and sensitivity, and converts it to a categorical vulnerability score (extremely vulnerable, highly vulnerable,

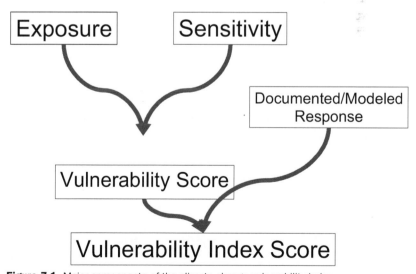

Figure 7.1. Major components of the climate change vulnerability index

moderately vulnerable, not vulnerable/presumed stable, not vulnerable/ increase likely). If a minimum number of factors are not scored or if the exposure data are incomplete, the index reports a value of "insufficient evidence."

Relationship to Existing Conservation Status Assessments

We designed the index to work in concert with, and not duplicate, information contained in standard conservation status assessments such as the IUCN Red List, which is used worldwide, or NatureServe conservation status ranks, which are used extensively in the United States and Canada (Master et al. 2000, Mace et al. 2008). Factors such as population size and range size can influence vulnerability to climate change (Hampe 2004, Aitken et al. 2008, Laidre et al. 2008), but they are also fundamental inputs to assessments of conservation status. To avoid duplication, we have excluded these factors from the index. Because population and range size are major factors in determining conservation status, repeating them in our assessments would cause most threatened species to also be scored as vulnerable to climate change. The purpose of the index is to highlight species with other intrinsic and extrinsic factors that place them at risk.

Indirect Effects

In many cases, climate change impacts species both directly (e.g., by drought-induced declines in reproduction or survival) and indirectly through changes in interspecific interactions (Lawler et al. 2009a). To cite a popular example, the warming experienced in western North America over the past three decades has not directly caused the major declines documented in lodgepole pine (*Pinus contorta*). Instead, warmer winters have allowed mountain pine beetles (*Dendroctonus ponderosae*) to rapidly expand their range northward, leading to the decimation of large stands of pines (Carroll et al. 2004). While we recognize that shifts in competitive, predator-prey, or host-parasite interactions are likely to be very important, we have not attempted to incorporate them into this index. How such interactions change as a result of changes in climate is difficult to predict, even in controlled experiments (Suttle et al. 2007, Spiller and Schoener 2008, Tylianakus et al. 2008). The sheer magnitude of potential biotic and abiotic factors that could contribute to variations in the strength of interactions suggests that grappling with them in a climate-change context will continue to be a major challenge (Tylianakis et al. 2008). However, the index does reflect species' dependence on particular types of interactions (e.g., between plant and pollinator) because these interactions may be uncoupled if the component species respond differently to climate change.

Accounting for Exposure

The index accounts for direct exposure to climate change by integrating the magnitude of predicted change across the range of a species within the geographic area considered. The time horizon is 2050, a date far enough in the future for significant changes to have occured, but before temperature projections from different emissions scenarios and global circulation models diverge substantially (Meehl et al. 2007). Downscaled predictions of climate change are becoming more readily available to facilitate assessment of exposure (e.g., Maurer et al. 2007; data available for viewing and download at www.climatewizard.org).

We considered both the severity and the scope of climate change in our assessment of exposure. The index divides temperature increase and precipitation increase/decrease (severity) into categories and defines the percentage of the species' range within the analysis area that will experience each severity category of temperature and precipitation change (scope). We used multiples of the standard deviation of predicted mid-century change in annual mean temperature and precipitation in the conterminous United States (Maurer et al. 2007; medium [A1B] emission scenario, ensemble average of 16 global circulation models) to delimit categories describing the magnitude of climate exposure. More specific seasonal climatic factors might be more relevant for particular species (e.g., Carroll et al. 2004), but because this information is rarely known, we used the annual data as proxies for severity of climate change.

Indirect Exposure and Species Sensitivity

Next, the index presents four factors to assess extrinsic indirect exposure and 17 factors, each supported in the literature, to evaluate species sensitivity (table 7.1). For each factor, the species is scored according to how much the factor increases or decreases vulnerability to climate change.

Documented or Modeled Vulnerability

For a small but growing number of species, field or modeling studies provide an indication of their vulnerability, as in documenting how their populations have responded to climate change in the recent past. Because these findings are valuable indicators of vulnerability, the index captures them in four factors that are considered separately from exposure and sensitivity (table 7.1).

Table 7.1. Vulnerability factors and exposure weighting used in the index and importance of factors for Nevada test species. See text for descriptions of exposure ratings.

	Factor information			Nevada pilot species		
Factor	Description	References[1]	Exposure weighting	No. of species with adequate knowledge to assess	No. of species for which factor increased vulnerability	No. of species for which factor decreased vulnerability
Indirect exposure factors						
Exposure to sea level rise	Predictions of 0.8–2.0 meter increase in sea level this century suggest that species occurring in coastal zones and low-lying islands will be subject to rapid loss of habitat and vulnerable to associated storm surge.	15, 28	1.0	N.A.	N.A.	N.A.
Distribution relative to natural topographic or geographic habitat barriers	Geographical features of the landscape where a species occurs may naturally restrict it from dispersing to inhabit new areas.	3, 14, 23, 33, 18, 37, 16, 12, 21, 22, 29	Climate stress	216	14	0
Distribution relative to anthropogenic barriers	Dispersal of a species to areas with favorable climates may be hindered by intervening urban or agricultural areas.	25	Climate stress	216	10	0
Impact of land use changes designed to mitigate against climate change by sequestering carbon or reducing dependence on fossil fuels	Strategies designed to mitigate greenhouse gases, such as creating large wind farms, plowing new cropland for biofuel production, or planting trees as carbon sinks have the potential to affect large tracts of land and the species that use those areas in both positive and negative ways.	17, 10, 8	Climate stress	216	96	0

Species sensitivity factors

Dispersal ability	Species with poor dispersal abilities may not be able to track fast-moving, favorable climates.	6, 23, 38, 16	Climate stress	215	78	90
Historical thermal niche	Species that have not experienced much temperature variation in recent historical times (the last 50 years) may not be able to adapt to future change.	31, 35, 37, 9, 5, 12, 20	Temperature change	216	5	74
Physiological thermal niche	Species requiring specific temperature regimes may be less likely to find similar areas as climates change and previously associated temperature and precipitation patterns uncouple.	31, 35, 37, 9, 12, 20	Temperature change	216	14	6
Historical hydrological niche	Species that have not experienced much variation in precipitation during recent historical times (the last 50 years) may not be able to adapt to future change.	31, 35, 37, 9, 12, 20	Precipitation and temperature, weighting precipitation three times as much as temperature	216	159	11
Physiological hydrological niche	Species requiring specific precipitation, hydrological conditions, or moisture regimes may be less likely to find similar areas as climates change and previously associated temperature and precipitation patterns uncouple.	31, 35, 37, 9, 12, 20	Precipitation and temperature, weighting precipitation three times as much as temperature	216	67	1

(Continued)

Table 7.1. (Continued)

	Factor information			Nevada pilot species		
Factor	Description	References[1]	Exposure weighting	No. of species with adequate knowledge to assess	No. of species for which factor increased vulnerability	No. of species for which factor decreased vulnerability
Dependence on ice, ice edge, or snow-cover habitats	The extent of oceanic ice sheets and mountain snow fields is decreasing as temperatures increase, thus imperiling species dependent on these habitats.	34, 15, 20	Climate stress	215	5	N.A.
Physical habitat specificity	Species requiring specific substrates, soils, or physical features such as caves, cliffs, or sand dunes may become vulnerable to climate change if their favored climate conditions shift to areas without these habitat elements.	12	Climate stress	215	20	96
Reliance on interspecific interactions: (a) Dependence on other species to generate habitat	Because species will react idiosyncratically to climate change, those with tight relationships with other species may be threatened.	4, 11, 12	Climate stress	216	21	N.A.
(b) Pollinator versatility (plants)	See above.	12	Climate stress	N./A.	N.A.	N.A.
(c) Dependence on other species for propagule dispersal	See above.	12	Climate stress	216	0	N.A.

(d) Dietary versatility (animals)	See above.	33, 20	Climate stress	216	12	0
(e) Forms part of some other mutualism	See above.	4	Climate stress	212	0	N.A.
Migrations (animals)	Species with very specific migratory destinations are vulnerable, whereas those with broad destinations are less vulnerable to climate change.	19, 16	Climate stress	216	147	60
Measured genetic variation	A species' ability to evolve adaptations to environmental conditions brought about by climate change is largely dependent on its existing genetic variation.	13, 1	Climate stress	0	0	0
Occurrence of bottlenecks in recent evolutionary history	See above.	13, 1	Climate stress	0	0	N.A.
Phenological response to changing seasonal temperature and precipitation regimes	Some species are declining due to their inability to respond to changing annual temperature dynamics (e.g., earlier onset of spring, longer growing season), including European bird species that have not advanced their migration times, and some temperate-zone plants that have not moved their flowering times.	24, 39	Climate stress	0	0	0
Documented or modeled responses to climate change						
Documented change in distribution or abundance attributable to recent climate change	Although conclusively linking species declines to climate change is difficult, convincing evidence relating declines to recent climate patterns has begun to accumulate in a variety of species groups.	25, 26, 30, 7	—	0	0	0

(Continued)

Table 7.1. (*Continued*)

	Factor information			Nevada pilot species		
Factor	Description	References[1]	Exposure weighting	No. of species with adequate knowledge to assess	No. of species for which factor increased vulnerability	No. of species for which factor decreased vulnerability
Modeled future change in range size	Change in the area of the predicted future range relative to the current range is a useful indicator of the species' vulnerability to climate change.	23, 36	—	0	0	0
Overlap of modeled future range with current range	A spatially disjunct predicted future range indicates that the species will need to disperse in order to occupy the newly favored area, and geographical barriers or slow dispersal rates could prevent the species from getting there.	27, 32	—	0	0	N.A.
Occurrence of protected areas in modeled future distribution	If future ranges fall outside of protected areas, long-term viability of populations may be compromised.	38	—	0	0	N.A.

[1] References: 1. Aitken et al. 2008. 2. Archer and Predick 2008. 3. Benito Garzón et al. 2008. 4. Bruno et al. 2003. 5. Calosi et al. 2008. 6. Dyer 1995. 7. Enquist and Gori 2008. 8. Fargione et al. 2009 9. Gran Canaria Declaration 2006. 10. Groom et al. 2008. 11. Hampe 2004. 12. Hawkins et al. 2008. 13. Huntley 2005. 14. IPCC 2002. 15. IPCC 2007. 16. Jiguet et al. 2007. 17. Johnson et al. 2003. 18. Koerner 2005. 19. Laidre and Heide-Jørgensen 2005. 20 Laidre et al. 2008. 21. Lenoir et al. 2008. 22. Loarie et al. 2008. 23. Midgley et al. 2003. 24 Møller et al. 2008. 25. Parmesan 1996. 26. Parmesan and Yohe 2003. 27. Peterson et al. 2002. 28. Pfeffer et al. 2008. 29. Price 2008. 30. Root et al. 2003. 31. Saetersdal and Birks 1997. 32. Schwartz et al. 2006. 33. Simmons et al. 2004. 34. Stirling and Parkinson 2006. 35. Thomas 2005. 36. Thomas et al. 2004. 37. Thuiller et al. 2005. 38. Williams et al. 2005. 39. Willis et al. 2008.

Computing an Index Score

To calculate an overall score, the index first combines information on exposure and sensitivity to produce a numerical sum, calculated by adding subscores for each of the extrinsic and intrinsic species sensitivity factors. Factors receive values (3.0, 2.0, 1.0, 0, –1.0, and –2.0), depending on the degree to which vulnerability is increased or decreased. If a factor is scored in multiple levels, the index uses an average.

The value for each factor is weighted by exposure to calculate a subscore. Climate influences vulnerability factors in different ways. For most factors, the exposure weighting is a climate stress value that combines data on projected change in both temperature and precipitation. In these cases, the weighting factor is the product of weightings for temperature (0.5, 1.0, 1.5, or 2.0, depending on the temperature increase) and precipitation (0.5, 1.0, 1.5, or 2.0, depending on change in precipitation). Table 7.1 summarizes the weighting used for each factor.

The exposure/sensitivity sum is therefore calculated as

$$\Sigma f_i w_i \qquad \text{[eq.1]}$$

where f is the value assigned to each factor according to how it influences sensitivity, and w is the specific exposure weighting for each factor i. The thresholds for the index scores of extremely vulnerable, highly vulnerable, moderately vulnerable, not vulnerable/presumed stable, and not vulnerable/increase likely are 10.0, 7.0, 4.0, and –2.0. The thresholds correspond with possible scenarios of exposure and sensitivity. For example, the "extremely vulnerable" threshold is reached for species with high exposure and at least two indirect exposure/sensitivity factors scored as greatly increase vulnerability, or with high exposure and three factors scored as increase vulnerability.

The documented/modeled response factors are scored identically to the exposure/sensitivity factors and are summed independently with no weighting, because exposure has already been incorporated in the studies upon which the factors are based. The thresholds for the index scores are 6.0, 4.0, 2.0, and –1.0, using the same logic as is used for exposure/sensitivity while accounting for the fewer documented/modeled response factors.

The overall index score is either the exposure/sensitivity score, if there is no documented/modeled response information, or an average of the exposure/sensitivity and documented/modeled response scores. In the case of adjacent scores, such as moderately vulnerable and presumed stable, the average is defined as the score higher on the vulnerability scale. If fewer than 3 indirect exposure or 10 species sensitivity factors are scored, the index score is insufficient evidence.

Uncertainty

Predicting vulnerability to climate change involves uncertainty about future greenhouse gas emissions, how the climate system will respond to these emissions, how species will respond to climate change, and how indirect effects will influence species (Patt et al. 2005, Lawler et al. 2009a). Developing a user-friendly tool requires compromise, and the sheer complexity of exhaustively incorporating uncertainty is beyond the scope of this project. Because our target audience is resource professionals with knowledge of species' natural history, we have allowed users to evaluate the results when more than one level of vulnerability is plausible for one or more factors. The index runs 1,000 Monte Carlo simulations, randomly selecting a single vulnerability level for each factor in which more than one level has been entered. The index calculates a measure of confidence in species information as very high, high, or moderate if more than 90%, 80%, or 60% of the simulation runs, respectively, yield the same score as the original index score. In cases with less than 60% concordance, the confidence is low.

Application of the Climate Vulnerability Index
Nevada Case Study

In 2008, Nevada set out to revise its state wildlife action plan to better address climate change. The Nevada Natural Heritage Program assessed the relative vulnerability of 263 species of "conservation priority," explaining why some species were more vulnerable than others. Although these species are of conservation concern in Nevada, they have range-wide conservation statuses varying from highly threatened to common and secure. Because so many species are involved, Nevada Heritage has used the climate change vulnerability index as a rapid and cost-efficient tool. The project is ongoing, but here we present results for the 216 priority vertebrates and mollusk taxa.

The mid-century climate predictions for Nevada suggest warming of approximately 2.6° C to 3.2° C and variable precipitation scenarios in different parts of the state (figure 7.2). The index sorted taxa into differing levels of vulnerability to climate change (table 7.2, figure 7.3). The majority of taxa fell in the moderately vulnerable and not vulnerable/presumed stable categories. Across taxa, 100% of mollusks, 80% of fish, 38% of amphibians, 30% of reptiles, 35% of mammals, and 4% of birds are at least moderately vulnerable. Natural history and distribution knowledge was

Table 7.2. Taxa scored preliminarily as "extremely vulnerable," "highly vulnerable," and "increase likely" by the climate change vulnerability index applied for distributions within Nevada

Taxon[1]	Group	Conservation status[1]
Extremely vulnerable		
Pygmy rabbit, *Brachylagus idahoensis*	Mammal	G4, S3
Preston White River springfish, *Crenichthys baileyi albivallis*	Fish	T1, S1
Desert dace, *Eremichthys acros*	Fish	G1, S1
Monitor Valley speckled dace, *Rhinichthys osculus* ssp. 5	Fish	T1, S1
Bull trout, *Salvelinus confluentus* pop. 4	Fish	T2, S1
Duckwater springsnail, *Pyrgulopsis aloba*	Snail	G1, S1
Southern Duckwater springsnail, *Pyrgulopsis anatina*	Snail	G1, S1
Elongate Cain Spring springsnail, *Pyrgulopsis augustae*	Snail	G1, S1
Pleasant Valley springsnail, *Pyrgulopsis aurata*	Snail	G1, S1
Fly Ranch springsnail, *Pyrgulopsis bruesi*	Snail	G1, S1
Northern Soldier Meadow pyrg, *Pyrgulopsis militaris*	Snail	G1, S1
Bifid duct springsnail, *Pyrgulopsis peculiaris*	Snail	G2, S1
Antelope Valley springsnail, *Pyrgulopsis pellita*	Snail	G1, S1
Highly vulnerable		
Sierra Nevada mountain beaver, *Aplodontia rufa californica*	Mammal	T3, S1
Sagebrush vole, *Lemmiscus curtatus*	Mammal	G5, S3
Pale kangaroo mouse, *Microdipodops pallidus*	Mammal	G3, S2
Humboldt yellow-pine chipmunk, *Neotamias amoenus celeris*	Mammal	T2, S2
American pika, *Ochotona princeps*	Mammal	G5, S2
California bighorn sheep, *Ovis canadensis californiana*	Mammal	T4, S3
Columbia spotted frog, *Rana luteiventris* (Toiyabe subpopulation)	Amphibian	Not assessed at subpopulation level
Wall Canyon sucker, *Catostomus* sp. 1	Fish	G1, S1
Railroad Valley springfish, *Crenichthys nevadae*	Fish	G2, S2
Fish Lake Valley tui chub, *Gila bicolor* ssp. 4	Fish	T1, S1
Railroad Valley tui chub, *Gila bicolor* ssp. 7	Fish	T1, S1
Big Smoky Valley tui chub, *Gila bicolor* ssp. 8	Fish	T1, S1
Pahranagat roundtail chub, *Gila robusta jordani*	Fish	T1, S1
White River spinedace, *Lepidomeda albivallis*	Fish	G1, S1

(Continued)

Table 7.2. (*Continued*)

Taxon[1]	Group	Conservation status[1]
Lahontan cutthroat trout, *Oncorhynchus clarki henshawi*	Fish	T3, S3
Big Smoky Valley speckled dace, *Rhinichthys osculus lariversi*	Fish	T1, S1
Diamond Valley speckled dace, *Rhinichthys osculus* ssp. 10	Fish	TH, SH
Oasis Valley speckled dace, *Rhinichthys osculus* ssp. 6	Fish	T1, S1
White River speckled dace, *Rhinichthys osculus* ssp. 7	Fish	T2, S2
Steptoe hydrobe, *Eremopyrgus eganensis*	Snail	G1, S1
Turban pebblesnail, *Fluminicola turbiniformis*	Snail	G3, S–
Smooth juga, *Juga interioris*	Snail	G1, S1
Elko pyrg, *Pyrgulopsis leporina*	Snail	G1, S1
Wong's pyrg, *Pyrgulopsis wongi*	Snail	G2, S1
Increase likely		
Clark's grebe, *Aechmophorus clarkii*	Bird	G5, S4
Western grebe, *Aechmophorus occidentalis*	Bird	G5, S4
Cinnamon teal, *Anas cyanoptera*	Bird	G5, S5
Bald eagle, *Haliaeetus leucocephalu*	Bird	G5, S1
Least sandpiper, *Calidris minutilla*	Bird	G5, S4
Short-eared owl, *Asio flammeus*	Bird	G5, S4
Costa's hummingbird, *Calypte costae*	Bird	G5, S3
Lewis's woodpecker, *Melanerpes lewis*	Bird	G4, S3
Olive-sided flycatcher, *Contopus cooperi*	Bird	G4, S2
Mountain willow flycatcher, *Empidonax traillii brewsteri*	Bird	T3, S2
Black phoebe, *Sayornis nigricans*	Bird	G5, S4
Loggerhead shrike, *Lanius ludovicianus*	Bird	G4, S4
Phainopepla, *Phainopepla nitens*	Bird	G5, S2
Virginia's warbler, *Vermivora virginiae*	Bird	G5, S4
Tricolored blackbird, *Agelaius tricolor*	Bird	G2, S1
Hoary bat, *Lasiurus cinereus*	Mammal	G5, S3
Long-eared myotis, *Myotis evotis*	Mammal	G5, S4
Little brown bat, *Myotis lucifugus*	Mammal	G5, S3
Northern river otter, *Lontra canadensis*	Mammal	G5, S2
Brush mouse, *Peromyscus boylii*	Mammal	G5, S3

[1] NatureServe conservation status ranking in which G indicates status for entire global range of a species (T is substituted for G in subspecies), and S indicates status within the state of Nevada. Conservation status scores range from 1 (critically imperiled) to 5 (secure); H indicates species known only from historical records but possibly still extant. A dash (–) indicates that a rank is not applicable. See Master et al. 2000 for more details.

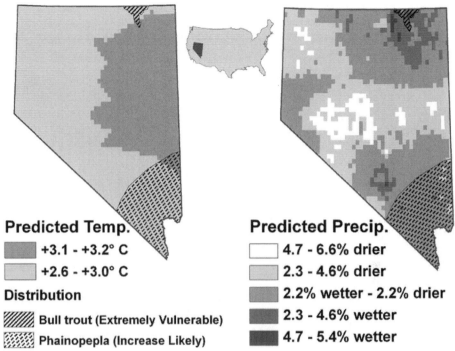

Predicted Temp.

■ +3.1 - +3.2° C

□ +2.6 - +3.0° C

Distribution

▨ Bull trout (Extremely Vulnerable)

▧ Phainopepla (Increase Likely)

Predicted Precip.

□ 4.7 - 6.6% drier

□ 2.3 - 4.6% drier

■ 2.2% wetter - 2.2% drier

■ 2.3 - 4.6% wetter

■ 4.7 - 5.4% wetter

Figure 7.2. Predicted change in temperature and precipitation for Nevada in 2050, under a medium (A1B) emissions scenario

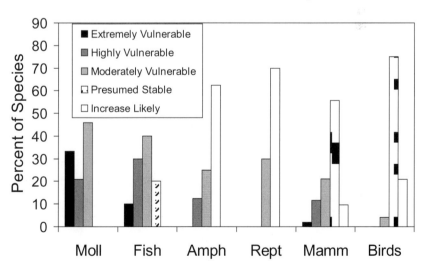

Figure 7.3. Vulnerability to climate change of selected Nevada mollusks (*n* = 24), fish (*n* = 40), amphibians (*n* = 8), reptiles (*n* = 20), mammals (*n* = 52), and birds (*n* = 72).

Table 7.3. Comparison between conservation status at both global (rangewide) and state level and climate change vulnerability. See table 7.2 and Master et al. (2000) for explanation of conservation status rankings.

Global conservation status	Climate change vulnerability				
	Extremely vulnerable	Highly vulnerable	Moderately vulnerable	Presumed stable	Increase likely
Possibly extinct (GH)	0	1	3	1	0
Critically imperiled (G1)	10	11	25	9	0
Imperiled (G2)	2	4	5	3	1
Vulnerable (G3)	0	4	3	11	1
Apparently secure (G4)	1	1	6	24	3
Secure (G5)	0	2	7	61	15
Nevada state conservation status					
Possibly extirpated (SH)	0	1	3	2	0
Critically imperiled (S1)	12	13	28	18	2
Imperiled (S2)	0	5	8	34	4
Vulnerable (S3)	1	3	2	32	5
Apparently secure (S4)	0	0	2	20	8
Secure (S5)	0	0	0	3	1

generally sufficient to allow assessment of all extrinsic factors and 14 of the 17 intrinsic species sensitivity factors (table 7.1).

Vulnerability to climate change was highly correlated with conservation status at both the global (rangewide) and state scale (global, Kendall's $\tau = 0.518$, $p < 0.001$; state, $\tau = 0.465$, $p < 0.001$; table 7.3). Although climate change vulnerability and conservation status are correlated, the relationship is not perfect. Four species ranked as apparently secure or secure (G4 or G5) also scored extremely or highly vulnerable to climate change. For example, the American pika (*Ochotona princeps*) is a widespread mountain inhabitant of western North America, but its dependence on declining snowpack and limited rocky talus slope habitat, together with its difficulty dispersing from one mountaintop to the next, renders it vulnerable to climate change in Nevada. Conversely, 34 (62%) of the 55 globally critically imperiled species examined with the index scored as presumed stable or only moderately vulnerable to climate change. For Nevada, conservation status is therefore an imperfect proxy for vulnerability to climate change.

The Monte Carlo simulations revealed that confidence in the index score was very high or high for 94 (61%) taxa, low for 19 (12%) taxa, and moderate for the rest. In most cases, a low confidence score resulted when the exposure/sensitivity sum was close to the threshold between two index categories.

Limited historical hydrological niches, anticipated impact from mitigation-related land use changes, migration to or through a few potentially vulnerable locations (see also Owen-Smith and Ogutu, this volume), lack of facultative distribution shifts in response to environmental conditions (such as the tracking by seed-eating birds of cone crops of conifers), and dependence on specific vulnerable aquatic/wetland habitats were the factors commonly contributing to vulnerability to climate change (table 7.1). Good dispersal ability, broad physical habitat requirements, migration to broad geographical areas, a tendency to shift distribution in response to environmental conditions, and adaptation to a broad range of temperatures were the factors that most commonly decreased vulnerability (table 7.1).

The climate change vulnerability index enables the state of Nevada to rapidly assess which of the wildlife species deemed of greatest concern are most imperiled by changing climate, and most deserving of more in-depth analysis and management. For each of the six taxonomic groups, the index succeeded in separating taxa into distinct classes of similar vulnerability, thus demonstrating that it is robust to taxonomic affinity for animals. Of course, only time will tell whether its predictions are borne out by range and population contractions or expansions. A more immediate, albeit weak, test of the index would be to compare historical population trends with index scores. Many factors influence population trends, but a preponderance of species scored as vulnerable that began or increased their rate of population decline in the 1970s, when temperatures began to increase sharply, would support the index's ability to identify threatened species.

The index has been a means to identify factors common to many Nevada vertebrates that increase their susceptibility to climate change. A noteworthy finding from this preliminary assessment is that two traits shared by many species in this state—limited historical hydrological niche and dependence on specific vulnerable aquatic/wetland habitats—relate to precipitation. This reflects the aridity of the Nevada climate as well as the dependence of many species of conservation concern on specific hydrological features, such as springs (WAPT 2006). Hence, it would be worthwhile to look more closely at how increasing temperatures will interact with moisture and wildlife habitats. A more surprising result was that anticipated climate-change-mitigation–related land use changes could contribute to several species' vulnerability. In response to the need to reduce emissions, Nevada officials anticipate the construction of solar, wind, and geothermal energy projects that could alter much wildlife habitat. These projects affect habitat used by nearly half of the species assessed, so management actions that mitigate detrimental effects to wildlife should be a priority. On a positive note, the results suggest that 20 species of priority birds and mammals may become more common in response to climate change.

Our results indicate the feasibility of a means to rapidly categorize species by their vulnerability to climate change using readily available natural-history and distribution information. Further testing is warranted. This index should be tested on larger scales to incorporate more spatial variation in climate-change predictions. At small spatial scales, exposure to climate change may be a constant for all species assessed, because they essentially all experience the same climate, so the differences in species' vulnerability would reflect differences in their intrinsic sensitivity. Finally, although we have developed the index using a variety of species as models, testing on larger samples of species that were underrepresented or not included in the Nevada case study, such as insects and plants, would show whether the index is as robust as desired. These results will allow for future refinement of this new resource for land managers.

ACKNOWLEDGMENTS
We thank the Fawcett Family Foundation for generous financial support and encouragement to develop the climate change vulnerability index. A grant from the Nevada Department of Wildlife underwrote the Nevada research. This chapter benefited greatly from comments by J. Brodie, D. Doak, and one anonymous reviewer on a previous draft.

LITERATURE CITED
Aitken, S. N., S. Yeaman, J. A. Holliday, T. Wang, and S. Curtis-McLane. 2008. Adaptation, migration, or extirpation: Climate change outcomes for tree populations. *Evolutionary Applications* 1:95–111.
Archer, S. R., and K. I. Predick. 2008. Climate change and ecosystems of the southwestern United States. *Rangelands* 2008:23–28.
Association of Fish & Wildlife Agencies (AFWA). 2009. Voluntary guidance for states to incorporate climate change into state wildlife action plans and other management plans. AFWA, Washington.
Benito Garzón, M., R. Sánchez de Dios, and H. Sainz Ollero. 2008. Effects of climate change on the distribution of Iberian tree species. *Applied Vegetation Science* 11: 169–78.
Bradley, N. L., A. C. Leopold, J. Ross, and W. Huffaker. 1999. Phenological changes reflect climate change in Wisconsin. *Proceedings of the National Academy of Sciences, USA* 96:9701–4.

Bruno, J. F., J. J. Stachowicz, and M. D. Bertness. 2003. Inclusion of facilitation into ecological theory. *Trends in Ecology and Evolution* 18:119–25.

Calosi, P., D. T. Bilton, and J. I. Spicer. 2008. Thermal tolerance, acclimatory capacity and vulnerability to global climate change. *Biology Letters* 4:99–102.

Carroll, A. L., S. W. Taylor, J. Régnière, and L. Safranyik. 2004. Effects of climate change on range expansion by the mountain pine beetle in British Columbia. Pages 223–232 in In *Mountain Pine Beetle Symposium: Challenges and Solutions*, edited by T. L. Shore, J. E. Brooks, and J. E. Stone. Natural Resources Canada, Canadian Forest Service, Pacific Forestry Centre, Information Report BC-X-399, Victoria, BC, Canada.

Commission of the European Communities (CEC). 2006. Annexes to the communication from the commission, halting the loss of biodiversity by 2010— and beyond. CEC, Brussels, Belgium. Accessed at ec.europa.eu/environment/nature/biodiversity/comm2006/pdf/sec_2006_621.pdf.

Davis, M. B., and R. G. Shaw. 2001. Range shifts and adaptive responses to Quarternary climate change. *Science* 292:673–79.

Dyer, J. M. 1995. Assessment of climatic warming using a model of forest species migration. *Ecological Modeling* 79:199–219.

Enquist, C., and D. Gori. 2008. *A Climate Change Vulnerability Assessment for Biodiversity in New Mexico, Part I: Implications of Recent Climate Change on Conservation Priorities in New Mexico*. Santa Fe: Nature Conservancy.

Fargione, J. E., T. R. Cooper, D. J. Flaspohler, J. Hill, C. Lehman, D. Tilman, T. McCoy, S. McLeod, E. J. Nelson, and K. S. Oberhauser. 2009. Bioenergy and wildlife: Threats and opportunities for grassland conservation. *Bioscience* 59:767–77.

Foden, W., Mace, G., Vié, J.-C., Angulo, A., Butchart, S., DeVantier, L., Dublin, H., Gutsche, A., Stuart, S. and E. Turak. 2008. Species susceptibility to climate change impacts. In *Wildlife in a Changing World: An Analysis of the 2008 IUCN Red List of Threatened Species*, edited by J.-C. Vié, C. Hilton-Taylor and S. N. Stuart, 77–88 IUCN, Gland, Switzerland.

Füssel, H.-M., and R. J. T. Klein. 2006. Climate change vulnerability assessments: An evolution of conceptual thinking. *Climate Change* 75:301–29.

Gran Canaria Declaration II on Climate Change and Plant Conservation. 2006. Published by the Area de Medio Ambiente y Aguas del Cabildo de Gran Canaria Jardín Botánico Canario "Viera y Clavijo" and Botanic Gardens Conservation International. Accessed at www.bgci.org/files/All/Key_Publications/gcdccenglish.pdf.

Groom, M. J., E. M. Gray, and P. A. Townsend. 2008. Biofuels and biodiversity: Principles for creating better policies for biofuel production. *Conservation Biology* 22:602–9.

Hampe, A. 2004. Bioclimatic envelope models: What they detect and what they hide. *Global Ecology and Biogeography* 13:469–71.

Hawkins, B., Sharrock, S. and Havens, K., 2008. *Plants and Climate Change: Which Future?* Richmond, UK: Botanic Gardens Conservation International.

Heller, N. E., and E. S. Zavaleta. 2009. Biodiversity management in the face of climate change: A review of 22 years of recommendations. *Biological Conservation* 142:14–32.

Humphries, M. M., J. Umbanhowar, and K. S. McCann. 2004. Bioenergetic prediction of climate change impacts on northern mammals. *Integrative and Comparative Biology* 44:152–62.

Huntley, B. 2005. North temperate responses. In *Climate Change and Biodiversity*, edited by T. F. Lovejoy and L. Hannah, 109–24. New Haven: Yale University Press.

Intergovernmental Panel on Climate Change (IPCC). 2002. Climate Change and Biodiversity (IPCC Technical Paper V), edited by H. Gitay, A. Suárez, R. T. Watson, and D. Jon Dokken. Accessed at www.grida.no/climate/ipcc_tar/biodiv/pdf/bio_eng.pdf.

———. 2007. *Climate Change 2007: Impacts, Adaptation and Vulnerability. Contribution of Working Group II to the Fourth Assessment Report of the Intergovernmental Panel on Climate Change*, edited by M. L. Parry, O. F. Canziani, J. P. Palutikof, P. J. van der Linden and C. E. Hanson. Cambridge: Cambridge University Press Accessed at www.ipcc.ch/ipccreports/ar4-wg2.htm.

Jiguet, F., A.-S. Gadot, R. Julliard, S. E. Newson, and D. Couvet. 2007. Climate envelope, life history traits and the resilience of birds facing climate change. *Global Change Biology* 13:1672–84.

Johnson, G. D., W. P. Erickson, M. D. Strickland, M. F. Shepherd, D. A. Shepherd, and S. A. Sarappo. 2003. Mortality of bats at a large-scale wind power development at Buffalo Ridge, Minnesota. *American Midland Naturalist* 150:332–42.

Koerner, C. 2005. The green cover of mountains in a changing environment. In *Global Change and Mountain Regions: An Overview of Current Knowledge*, edited by U. M. Huber, M. A. Reasoner, and H. K. M. Bugmann, 367–75. New York: Springer-Verlag.

Laidre, K. L., and M. P. Heide-Jørgensen. 2005. Arctic sea ice trends and narwhal vulnerability. *Biological Conservation* 121:509–17.

Laidre, K. L., I. Stirling, L. F. Lowry, O. Wiig, M. P. Heide-Jørgensen, and S. H. Ferguson. 2008. Quantifying the sensitivity of arctic marine mammals to climate-induced habitat change. *Ecological Applications* 18:S97–125.

Lawler, J. J., S. L. Shafer, D. White, P. Kareiva, E. P. Maurer, A. R. Blaustein, and P. J. Bartlein. 2009b. Projected climate-induced faunal change in the Western Hemisphere. *Ecology* 90:588–97.

Lawler, J. J., T. H. Tear, C. Pyke, M. R. Shaw, P. Gonzalez, P. Kareiva, L. Hansen, L. Hannah, K. Klausmeyer, A. Aldous, C. Benz, and S. Pearsall. 2009a. Resource management in a changing and uncertain climate. *Frontiers in Ecology and the Environment* 7:10.1890/070146.

Lenoir, J., J. C. Gégout, P. A. Marquet, P. de Ruffray, and H. Brisse. 2008. A significant upward shift in plant species optimum elevation during the 20th century. *Science* 320:1768–71.

Loarie, S. R., B. E. Carter, K. Hayhoe, S. McMahon, R. Moe, C. A. Knight, and D. D. Ackerly. 2008. Climate change and the future of California's endemic flora. *PLoS ONE* 3:e2502.

Mace, G. M., N. J. Collar, K. J. Gaston, C. Hilton-Tailor, H. R. Akçakaya, N. Leader-Williams, E. J. Milner-Gulland, and S. N. Stuart. 2008. Quantification of extinction risk: IUCN's system for classifying threatened species. *Conservation Biology* 22:1424–42.

Master, L. L., B. A. Stein, L. S. Kutner, and G. A. Hammerson. 2000. Vanishing assets: Conservation status of US species. In *Precious Heritage: The Status of Biodiversity in the United States*, edited by B. A. Stein, L. S. Kutner, and J. S. Adams, 93–118. New York: Oxford University Press.

Maurer, E. P., L. Brekke, T. Pruitt, and P. B. Duffy. 2007. Fine-resolution climate projections enhance regional climate change impact studies. *Eos Trans. AGU* 88:504.

Meehl, G. A., T. F. Stocker, W. D. Collins, P. Friedlingstein, A. T. Gaye, J. M. Gregory, A. Kitoh, R. Knutti, J. M. Murphy, A. Noda, S. C. B. Raper, I. G. Watterson, A. J. Weaver, and Z.-C. Zhao. 2007. Global climate projections. In *Climate Change 2007: The Physical Science Basis. Contribution of Working Group I to the Fourth Assessment Report of the Intergovernmental Panel on Climate Change*, edited by S. Solomon, D. Qin, M. Manning, Z. Chen, M. Marquis, K. B. Averyt, M. Tignor, and H. L. Miller, 747–845. New York: Cambridge University Press.

Midgley, G. F., L. Hannah, D. Millar, W. Thuiller, and A. Booth. 2003. Developing regional and species-level assessments of climate change impacts on biodiversity in the Cape Floristic Region. *Biological Conservation* 112:87–97.

Møller, A. P., D. Rubolini, and E. Lehikoinen. 2008. Populations of migratory bird species that did not show a phenological response to climate change are declining. *Proceedings of the National Academy of Sciences (US)* 105:16195–200.

Overpeck, J. T., P. J. Bartlein, and T. Webb III. 1991. Potential magnitude of future vegetation change in eastern North America: Comparisons with the past. *Science* 254:692–95.

Parmesan, C. 1996. Climate change and species' range. *Nature* 382:765–66.

———. 2007. Influences of species, latitudes and methodologies on estimates of phenological response to global warming. *Global Change Biology* 13:1860–72.

Parmesan, C. and G. Yohe. 2003. A globally coherent fingerprint of climate change impacts across natural systems. *Nature* 421:37–42.

Patt, A., R. J. T. Klein, and A. Vega-Leinert. 2005. Taking the uncertainty in climate-change vulnerability assessment seriously. *Geoscience* 337:411–24.

Peterson, A. T., M. A. Ortega-Huerta, J. Bartley, V. Sanchez-Cordero, J. Soberon, R. H. Buddemeier, and D. R. B. Stockwell. 2002. Future projections for Mexican faunas under global climate change scenarios. *Nature* 416:626–29.

Pfeffer, W. T., J. T. Harper, and S. O'Neil. 2008. Kinematic constraints on glacier contributions to 21st-century sea-level rise. *Science* 321:1350–43.

Price, M. F. 2008. Maintaining mountain biodiversity in an era of climate change. In *Managing Alpine Future: Proceedings of International Conference October 15–17, 2007*, edited by A. Borsdorf, J. Stötter, and E. Veulliet, 17–33. Vienna: Austrian Academy of Sciences Press.

Root, T. L., J. T. Price, K. R. Hall, S. H. Schneider, C. Rosenzweig, and J. A. Pounds. 2003. Fingerprints of global warming on wild animals and plants. *Nature* 421: 57–60.

Saetersdal, M., and H. J. B. Birks. 1997. A comparative ecological study of Norwegian mountain plants in relation to possible future climatic change. *Journal of Biogeography* 24:127–52.

Schneider, S. H., S. Semenov, A. Patwardhan, I. Burton, C. H. D. Magadza, M. Oppenheimer, A. B. Pittock, A. Rahman, J. B. Smith, A. Suarez, and F. Yamin. 2007. Assessing key vulnerabilities and the risk from climate change. In *Climate Change 2007: Impacts, Adaptation and Vulnerability: Contribution of Working Group II to the Fourth Assessment Report of the Intergovernmental Panel on Climate*

Change, edited by M. L. Parry, O. F. Canziani, J. P. Palutikof, P. J. van der Linden, and C. E. Hanson, 779–810. Cambridge: Cambridge University Press.

Schwartz, M. W., L. R. Iverson, A. M. Prasad, S. N. Mathews, and R. J. O'Conner. 2006. Predicting extinctions as a result of climate change. *Ecology* 87:1611–15.

Simmons, R. E., P. Barnard, W. R. J. Dean, G. F. Midgley, W. Thuiller, and G. Hughes. 2004. Climate change and birds: Perspectives and prospects from southern Africa. *Ostrich* 75:295–308.

Smit, B., I. Burton, R. J. T. Klein, and J. Wandel. 2000. An anatomy of adaptation to climate change and variability. *Climate Change* 45:223–51.

Spiller, D. A., and T. W. Schoener. 2008. Climatic control of trophic interaction strength: The effect of lizards on spiders. *Oecologia* 154:763–71.

Stirling, I., and C. L. Parkinson. 2006. Possible effects of climate warming on selected populations of polar bears (*Ursus maritimus*) in the Canadian Arctic. *Arctic* 59: 261–75.

Suttle, K. B., M. A. Thomsen, and M. E. Power. 2007. Species interactions reverse grassland responses to changing climate. *Science* 315:640–42.

Thomas, C. D. 2005. Recent evolutionary effects of climate change. In *Climate Change and Biodiversity*, edited by T. E. Lovejoy and L. Hannah, 75–88. New Haven: Yale University Press.

Thomas, C. D., A. Cameron, R. E. Green, et al. 2004. Extinction risk from climate change. *Nature* 427:145–48.

Thuiller, W., S. Lavorel, M. B. Araújo, M. T. Sykes, and I. C. Prentice. 2005. Climate change threats to plant diversity in Europe. *Proceedings of the National Academy of Sciences, USA* 102:8245–50.

Tylianakis, J. M., R. K. Didham, J. Bascompte, and D. A. Wardle. 2008. Global change and species interactions in terrestrial ecosystems. *Ecology Letters* 11:1351–63.

Visser, M. E., and C. Both. 2005. Shifts in phenology due to global climate change: The need for a yardstick. *Proceedings of the Royal Society* B 272:2561–69.

Wildlife Action Plan Team (WAPT). 2006. Nevada wildlife action plan. Reno: Nevada Department of WildlifeAccessed at www.ndow.org/wild/conservation/cwcs/index.shtm#plan.

Williams, P., L. Hannah, S. Andelman, G. Midgley, M. Araújo, G. Hughes, L Manne, E. Martinez-Meyer, and R. Pearson. 2005. Planning for climate change: Identifying minimum dispersal corridors for the Cape Proteaceae. *Conservation Biology* 19:1063–74.

Williams, S. E., L. P. Shoo, J. L. Isaac, A. A. Hoffmann, and G. Langham. 2008. Towards an integrated framework for assessing the vulnerability of species to climate change. *PLoS Biology* 6:10.1371/journal.pbio.0060325.

Willis, C. G., B. Ruhfel, R. B. Primack, A. J. Miller-Rushing, and C. C. Davis. 2008. Phylogenetic patterns of species loss in Thoreau's woods are driven by climate change. *Proceedings of the National Academy of Sciences, USA* 105:17029–33.

Young, B., E. Byers, K. Gravuer, K. Hall, G. Hammerson, and A. Redder. 2010. *Guidelines for Using the NatureServe Climate Change Vulnerability Index*. NatureServe, Arlington, VA. Accessed at www.natureserve.org/climatechange.

PART

2

Case Studies
of Climatic Effects
on Wildlife
Conservation

Precipitation   A1B: 2080-2099   DJF

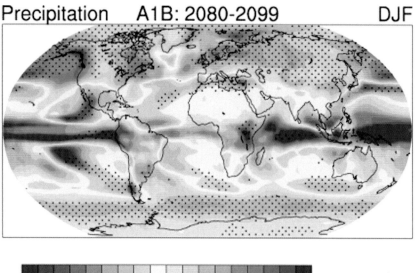

(mm day⁻¹)

-0.8 -0.6 -0.4 -0.2   0   0.2   0.4   0.6   0.8

Precipitation   A1B: 2080-2099   JJA

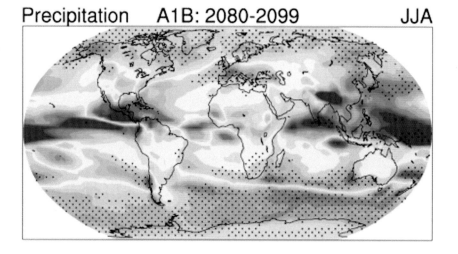

(mm day⁻¹)

-0.8 -0.6 -0.4 -0.2   0   0.2   0.4   0.6   0.8

Plate 1. Relative changes in precipitation (in percent) for the period 2090–99, relative to 1980–99. Values are multimodel averages based on the SRES A1B scenario for December to February (*left*) and June to August (*right*). White areas indicate where fewer than 66% of the models agree in the sign of the change, and stippled areas indicate where more than 90% of the models agree in the sign of the change.

Plate 2. Large-scale relative changes in annual runoff for the period 2090–99, relative to 1980–99. White areas indicate where less than 66% of the ensemble of 12 models agrees on the sign of change, and hatched areas indicate where more than 90% of models agree on the sign of change. Source: Milly et al., 2005.

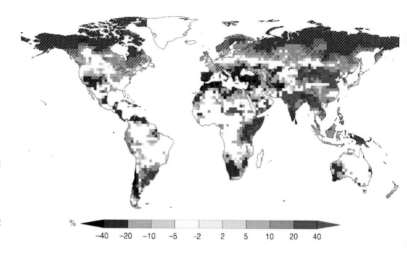

March — April Snow Departure
(1988 - 2004) minus (1967 - 1987)

Plate 3. Differences between Northern Hemisphere distribution of March-April average snow cover in earlier (1967–87) and later (1988–2004) parts of the satellite era (expressed in percentage of coverage). Negative values indicate greater extent in the earlier portion of the record. Extents are derived from NOAA/NESDIS snow maps. Red curves show the 0°C and 5°C isotherms averaged for March and April 1967 to 2004, from the Climatic Research Unit (CRU) gridded land surface temperature version 2 (CRUTEM2v) data. Source: IPCC 2007a.

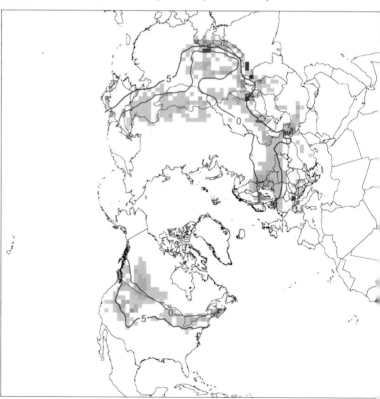

-36 to -26　-25 to -16　-15 to -6　-5 to 5　6 to 15　16 to 25　26 to 38

8

Changing Rainfall and Obstructed Movements: Impact on African Ungulates

Norman Owen-Smith and Joseph O. Ogutu

Declining ungulate populations within national parks and wildlife reserves have become an object of growing concern with regard to the preservation of Africa's rich large mammal diversity (Caro and Scholte 2007), especially in the case of migratory populations (Harris et al. 2009). A collapse of large mammal diversity within protected areas was forecast by Soulé et al. (1979) simply from principles of island biography, as these areas became increasingly insularized remnants of formerly vast ecosystems. The actual threats are now more tangible, however, in the form of expanding human settlements, cultivation, fencing, and other land transformations in surrounding regions, together with incursions of human hunters and domestic livestock into supposedly protected areas. Fences both along boundaries of protected areas and within the adjoining regions block the formerly wider-ranging movements of wildlife, livestock, and people. They consolidate the fragmentation and isolation of large mammal populations within protected areas (Newmark 2008).

A new threat is global climate change, due largely to human activities in regions remote from Africa—especially with regard to its influence on rainfall patterns. Rainfall governs vegetation growth and hence food production for herbivores within the savanna ecosystems where the greatest richness of African large mammals occurs. Because of the erratic nature of rainfall, plant growth is less predictable in both time and space within these ecosystems than in north temperate latitudes, where temperature exerts the main control over the alternation between summer and winter. To counteract this unpredictability, large herbivores rely on movement towards more favorable, or at least less adverse, localities, sometimes in the form of mass migrations. Opportunities for such movement are becoming restricted at a time when they may be crucial to cope with the wider climatic variability that is expected as a consequence of global warming. Besides food production, rainfall affects the retention of surface water for water-dependent grazers. Conflict with humans and their livestock may be especially acute around the few water sources remaining during the dry season. These worsening circumstances present a challenge for efforts to conserve ungulates within and beyond the bounds of protected areas. What responses would best ensure the resilience of these populations within the fragmented remnants that remain of vaster ecosystems?

In addressing this challenge, we will first describe the rainfall patterns typical of African savanna ecosystems, along with associated vegetation responses. We then outline how rainfall variability affects the population dynamics and movement patterns of various ungulate species. This leads us to consider the consequences of obstructions to movement, as well as other influences contributing to downward population trends. From this perspective, we consider what can be done to counteract or mitigate the compounded effects of climatic variation and range compression on African ungulate populations. It would be ideal to expand protected areas to fully encompass seasonal ranges and the migratory routes that connect them, but this option is restricted by human settlements and activities pressing against the boundaries of those areas (unpublished observations by N. O.-S.). An alternative would be to establish forms of land use in the surrounding region that would be compatible with the presence of large wildlife. If these options are impractical, the fallback option would be to promote the functional components of ecosystem heterogeneity that confer population resilience within the protected areas as much as possible. We offer suggestions for how this might be achieved.

Climatic Variation and Vegetation Responses

An alternation between wet and dry seasons is a defining feature of the savanna ecosystems prevalent through much of eastern and southern Africa. Within the tropics, the timing of these seasons is controlled by latitudinal shifts in the intertropical convergence zone. Near the equator, the wet season is subdivided into short rains through November and December, and long rains from March into May, with a relatively dry interlude through January and February. Towards the south, a single wet season extends from October or November through March or April, with the dry season encompassing cool months from May to July and hot months from August to October. In subtropical South Africa, the wet season occupies the summer months from October to March, while the dry season spans the winter months of April through September. Typically about 80% of annual rainfall falls during the summer months and only 20% during winter, although in parts of equatorial East Africa the dry season may be less intense. The range of variation in the mean annual rainfall for savannas extends from about 250 mm at the arid end towards 1200 mm at the moist end, grading into a forest-grassland mosaic. However, the annual rainfall actually received can vary from more than twice to less than half of the long-term mean.

Shifts in atmospheric pressure cells driven by the El Niño Southern Oscillation (ENSO) have a strong influence on seasonal patterns as well as the amount of rain received, but a modifying influence from the Indian Ocean dipole has recently become recognized (Webster et al. 1999, Marchant et al. 2007). In southeastern and south-central Africa, decades with mean annual rainfall above or below the long-term mean alternated through much of the 20th century (Tyson and Gatebe 2001; see figure 8.1). In East Africa, rainfall patterns have shown a five-year quasiperiodicity associated with fluctuations in ENSO (Ogutu et al. 2008a, Ritchie 2008). The El Niño phase associated with warm sea surface temperatures in the eastern Pacific Ocean promotes midsummer droughts in southern Africa, while the opposite (La Niña) phase generates elevated rainfall. Contrasting conditions tend to prevail in eastern Africa, with El Niño conditions affecting primarily the early or short rains (Nicholson 2000, Tyson et al. 2002). The intensification and persistence of El Niño conditions through 1981–98 (Fedorov and Philander 2000) was associated with a prolonged dry period in southern Africa with extreme droughts in 1983 and 1992. Rainfall received during the dry season months was also exceptionally low through much of this period (Ogutu and Owen-Smith 2003). The intense El Niño phase experienced in 1997–98, compounded by an additional influence from the Indian Ocean dipole, led to extreme floods in East Africa but had surprisingly little effect on rainfall in southern Africa. In East Africa, an extreme drought was experienced in 1999 and 2000 during a La Niña phase (Ritchie 2008, Ogutu et al. 2008a), while parts of southern Africa had record floods. A drought in East Africa during 2005 and 2006 was due solely to the Indian Ocean dipole (Hastenrath et al. 2007).

Global warming is evident in both central and southern Africa, with a rise in mean daily temperature by almost 0.5° C experienced during the course of the 20th century (Tyson and Gatebe 2001, Ogutu et al. 2008a). Minimum daily temperatures were most strongly affected, rising by as much as 1.4 to 3.6° C in parts of Kenya (Ogutu et al. 2008a). Rainfall records for Kruger National Park show the greater variation in annual rainfall that is expected (Easterling et al. 2000) to be a result; the coefficient of variation has risen from 20% or less prior to 1963 to 40% or more over the most recent decades (figure 8.1; the high variance during the early 1920s was due to exceptional floods in 1925). In contrast, in the Serengeti region of Tanzania both the mean and the variance of annual rainfall have

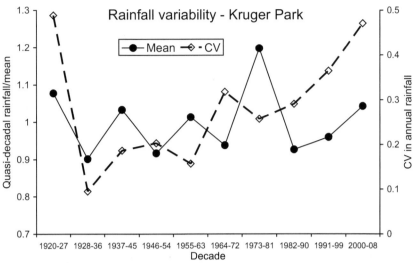

Figure 8.1. Changes in mean annual rainfall total and its coefficient of variation (CV) between nine-year blocks approximating opposite phases of the 16- to 20-year oscillation, averaged across eight rainfall recording stations in the Kruger National Park

decreased locally, along with an increase in the amount received during the dry season (Ritchie 2008).

In savanna climates, both the annual cycle of growth and dieback in above-ground grass parts and the foliage retention on deciduous trees and shrubs respond to seasonal variation in rainfall (Owen-Smith 2002, 2008). Peak herbaceous biomass is almost linearly related to the annual or wet season rainfall total (Rutherford 1980, Deshmukh 1984), and the coefficient of variation in grass biomass can exceed that of the rainfall (O'Connor et al. 2001). Grasses dry out rapidly after rain showers cease, and most of the forage available by the late dry season consists of brown, metabolically inactive leaves and stems. Crude protein levels in this grass tend to fall below the minimum maintenance requirements for medium-sized ungulates, especially in regions of low soil fertility (Owen-Smith 1982). The seasonal growth of woody plants is less affected by rainfall than is grass production (Rutherford 1984). For many trees, annual foliage production takes place before the main rains and depends on rainfall received during the previous wet season, while the amount of time that leaves remain on deciduous trees and shrubs depends on the amount of rain falling during the later part of the wet season. Rainfall also varies spatially, depending on where local storms drop their moisture, with the effect on vegetation affected additionally by the redistribution of runoff within landscapes. Bottomlands and swamps tend to retain green vegetation longer than uplands.

Diminishing rainfall coupled with rising temperatures through the 1990s and early 2000s appeared responsible for progressive habitat desiccation and reduced vegetation production in East Africa (Ogutu et al. 2008b). Low rainfall may lead to changes in plant species composition towards more pioneer or even annual grasses retaining less forage through the dry season (O'Connor 1985).

Herbivore Population Dynamics

Because of the influence of rainfall on forage production, trends in herbivore populations are expected to be positively related to the preceding rainfall. However, not all herbivore populations show this pattern (Ogutu and Owen-Smith 2003, Ogutu et al. 2008b). In Kruger Park, the ungulate species showing the strongest positive responses to rainfall variation were buffalo (*Syncerus caffer*), waterbuck (*Kobus ellipsiprymnus*), and greater kudu (*Tragelaphus strepsiceros*; Mills et al. 1995). The former two species are grazers favoring fairly tall grass, while kudu as browsers respond to forb production in the herbaceous layer as well as to leaf retention on trees and shrubs. Most resistant to annual variation in rainfall were grazing zebra (*Equus burchelli*) and wildebeest (*Connochaetes taurinus*), as well as browsing giraffe (*Giraffa camelopardalis*). The former two species may be favored by the shorter and thus less fibrous grass that is prevalent when rainfall is lower. Taller grass also provides more concealment for stalking lions (*Panthera leo*; Smuts 1978, Hopcraft et al. 2005). In the Laikipia region of northern Kenya, where predation by lions is less rife, zebra responded positively to rainfall variation in their population trend (Georgiadis et al. 2003). The dry season component of the annual rainfall exerted a positive influence on the dynamics of wildebeest in both Kruger Park (Ogutu and Owen-Smith 2003) and Serengeti (Mduma et al. 1999), and affected a wider range of species than the annual rainfall total in Kruger Park, presumably by improving green leaf retention and hence forage quality during this adverse period.

Extreme droughts can lead to severe herbivore mortality through starvation (Hillman and Hillman 1977), although the magnitude of the population reduction rarely exceeds 50% in extensive ecosystems. In Kruger Park, ungulate population declines over the 1983 drought ranged from 20% for zebra to 40% for buffalo, with wildebeest almost unaffected (figure 8.2). In Serengeti, the 1993 drought brought about a 25% reduction in the migratory wildebeest population (Mduma et al. 1999), while the

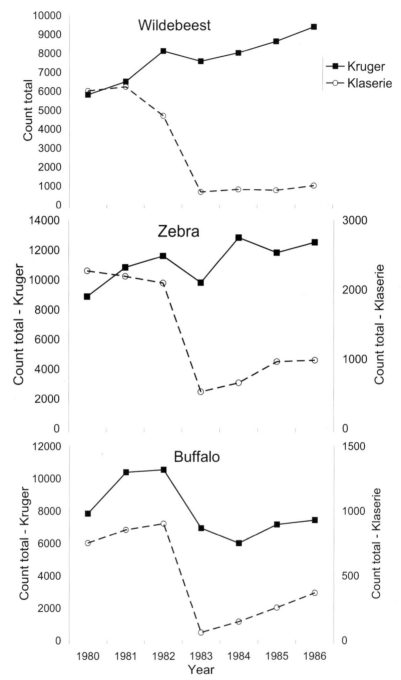

Figure 8.2. Changes in wildebeest, zebra, and buffalo numbers counted in the Klaserie Private Nature Reserve and the adjoining Central Region of Kruger National Park through the 1982–83 drought period

buffalo population within Serengeti fell by more than 40% (Metzger et al. 2010), and within the Mara region by more than 70% (Sinclair et al. 2007), exacerbated by competition from the cattle driven into the reserve (Butt et al. 2009). Competition from cattle for both food and water led to a 44% reduction in the total biomass of wild ungulates in the Athi-Kaputiei Plains of Kenya during the severe drought experienced in 1961 (Stewart and Zaphiro 1963). Large-scale mortality among cattle and wildlife was reported widely through Kenya when severe drought conditions took hold again in 2009. The nutritional stress associated with droughts can result in reduced calf production by the surviving females in the following year, and can affect the synchrony between births and seasonal rainfall patterns (Estes et al. 2006, Ogutu et al. 2010a).

Megaherbivores (species weighing more than 1,000 kg) seem less vulnerable to annual droughts than smaller herbivores. In Kruger Park, the persistent upward trend in the elephant (*Loxodonta africana*) population has shown little response to annual rainfall variation (Mills et al. 1996). In Tsavo East National Park in Kenya, almost 7000 elephants died during exceptionally severe drought conditions experienced over 1970–71, but this constituted less than 20% of the regional elephant population (Corfield 1973, Phillipson 1975). Mortality was exacerbated by the compression of elephants into the park due to hunting and settlements in the surrounding area (Parker 1983), and most of the deaths occurred in the most arid north-central region, where the annual rainfall remained under 200 mm through two successive years (Myers 1973, Phillipson 1975). The extreme drought conditions experienced in Kenya in 2009 led to severe mortality among elephants in Amboseli National Park, which was accentuated by illegal hunting in its environs, reducing the population there by perhaps 25%. In Hwange National Park, Zimbabwe, drought-related mortality amounted to only a few percent of the elephant population, despite a regional density exceeding two elephants per km^2 (Dudley et al. 2001). No drought-related mortality of elephants has been recorded in northern Botswana, despite dry-season concentrations near rivers amounting to 6 to 20 elephants per km^2. The expanding white rhino (*Ceratotherium simum*) population in the Hluhluwe-iMfolozi Park was unaffected by the drought conditions prevailing between 1968 and 1971, and apparently favored by the shorter grass more prevalent in dry years (Owen-Smith 1988).

In wetter savannas, flooding of swampy areas following exceptionally high rainfall can be adverse for herbivore populations, restricting their accessible foraging area and promoting rank grass growth. Accordingly, trends in several herbivore populations in the Masai Mara Reserve showed a humped relationship with rainfall (Ogutu et al. 2008b). Flooding due to exceptionally high rainfall over 1997–98 caused several ungulate species to move out of Nairobi National Park into the Athi-Kaputiei Plains, and few wildebeest entered the park in the subsequent dry season (Ogutu and Owen-Smith, submitted a). In Lake Nakuru National Park, populations of several ungulate species showed a negative response to higher rainfall, because the associated rise in lake level contracted the area of high-quality grazing available around the lake's margin (Ogutu and Owen-Smith, in press 2012). Both droughts and floods can also promote outbreaks of infectious diseases—especially anthrax after low rainfall, and rift valley fever along with other tick-born diseases after high rainfall (Bengis et al. 2003).

Herbivore Movement

The movements of large herbivores are governed mainly by temporal and spatial variation in the relative availability of green foliage and drinking water. Within a local landscape the seasonal relocation sometimes extends no further than between hillcrests, which offer the most nutritious grass during the wet season, and valley bottoms, which retain most forage through the dry season (Bell 1970, 1971). Bottomlands that retain forage during the late dry season have been regarded as "key resource areas" because they contribute disproportionately to the support of larger herbivore populations (Scoones 1995, Illius and O'Connor 2000). Wider animal movements may be forced by seasonal restrictions in surface water distribution, leading to dry season concentrations near rivers or other sources of drinking water, and to wet season dispersal over a wider area, especially among grazers (Western 1975). The distance between the seasonally distinct ranges may be as much as 50 to 100 km (Whyte and Joubert 1988, Joos-Vandewalle 2000). Outside protected areas, the concentration of humans and their livestock around water points restricts access by wildlife, at least during the day (Ogutu et al. 2010b). The continued flow of rivers traversing protected areas is threatened by the abstraction of water for irrigation and other uses upstream as well as by the destruction of forests in catchments, a situation becoming especially acute in parts of Kenya (e.g., Mati et al. 2008).

Migrations may also be made to exploit food resources that are especially favorable in their nutritional value during the wet season. This is best exemplified by the migrations of wildebeest, zebra, and Thomson's gazelles (*Gazella thomsoni*) in the Serengeti ecosystem. The short-grass plains in the southeast, underlain by volcanic ash, offer exceptionally high-quality forage during the rains but dry out rapidly in the dry season. Both wildebeest (Talbot and Talbot 1963, Wilmshurst et al. 1999) and Thomson's gazelles (Fryxell et al. 2004, 2005) move widely during the wet season in response to locally erratic rain showers and consequent flushes of new grass. Modeling suggests that the gazelles need access to an area larger than 1600 km^2 to secure the green vegetation they need (Fryxell et al. 2005).When the dry season sets in, the wildebeest and zebra migrate almost 200 km towards the Masai Mara Reserve in Kenya, where a mean annual rainfall exceeding 1,000 mm ensures the retention of adequate green forage during the dry season (Pennycuick 1975, Maddock 1979, Thirgood et al. 2004). During their northward migration, most of the wildebeest move beyond the park boundaries through communal settlements to the west. Although some 50,000 animals are killed illegally during this time (Mduma et al. 1998), this amounts to only 4% of the population. While migrating back toward the plains, the wildebeest may take advantage of areas where fire has removed accumulated dead grass, promoting green regrowth.

Migrations by wildebeest and zebra documented elsewhere are similarly related to seasonal variations in food and water availability. Current or former migrations by one or both species entailed seasonal shifts between the Masai Mara Reserve and the Loita Plains in Kenya (Stellfox et al. 1986, Ottichilo et al. 2001), between Nairobi National Park and the Athi-Kaputiei Plains (Foster and Coe 1968, Gichohi 2000, Reid et al. 2007), between the Amboseli Basin and surrounding group ranches in Kenya (Western 1975, 1982), between Tarangire National Park and the Simanjiro plains in Tanzania (Kahurananga and Silkiluwasha 1997, Bolger et al. 2008), between the Liuwa plains in western Zambia and a region of Angola (Daly 2005), between Etosha Pan and parts of Ovamboland in Namibia (Berry 1997), between the Linyanti River and Savuti marsh in Botswana (Joose-Vandewalle 2000), and in Kruger Park between the Sabie River and the Lindanda-Sweni region, as well as between the Sand River and the Orpen region (Whyte and Joubert 1988). Vast herds of Thomson's gazelles and zebras reportedly migrated between Lake Baringo and the Nakuru-Elementaita region of Kenya until the early part of the 20th century (Percival, 1928, Kutilek, 1974). In Boma National Park in southeastern Sudan, the mass migration by white-eared kob (*Kobus kob leucotis*) occurs in a somewhat different context (Fryxell and Sinclair 1988). During the dry season, these animals congregate in wet meadows near rivers that support continuing grass growth, but are forced to shift toward higher terrain after floodwaters inundate much of the lower-lying area. African elephants may also move over vast distances between dry season ranges near rivers or other perennial water sources, and localities where rainfall has promoted the active growth of vegetation during the wet season (Rasmussen et al. 2006).

In relatively arid savanna regions with less predictable rainfall, ungulates may show wide but less regular movements in response to local rainfall conditions. Long journeys to seek remaining surface water during severe droughts have been documented for wildebeest and eland (*Taurotragus oryx*) in the southern Kgalagadi region of Botswana (Knight 1995), and by wildebeest in the central Kgalagadi (Williamson et al. 1988). Spectacular "treks" by springbok (*Antidorcas marsupialis*) occurred in Northern Cape during the 19th century (Skinner 1993), seemingly representing a mass exodus in search of better conditions during severe droughts (Roche 2004). Even relatively sedentary ungulates may undertake excursions outside their usual home ranges during adverse periods (personal observations on kudu, buffalo, and sable antelope [*Hippotragus niger*]). Individual buffalo may also move between the spatially distinct ranges of different herds, perhaps in search of better foraging conditions (Halley et al. 2002, Cross et al. 2005).

Consequences of Restricting Migrations

Migratory populations achieve much higher abundance levels than resident populations of the same species (Fryxell et al. 1988). A substantial advantage comes through partial escape from sedentary predators. Predator numbers in both the wet season and dry season ranges of the Serengeti wildebeest are limited by the abundance of nonmigratory ungulates remaining in these regions year-round (Scheel and Packer 1995).

When migrations have become blocked or restricted, populations especially of wildebeest have collapsed to much lower numbers. In the Athi-Kaputiei region that includes Nairobi National Park, human settlements, associated fencing, competition for water, mining, and other developments have led to a reduction in wildebeest numbers from around 30,000 to about 5,000, and few wildebeest currently enter the park during the dry season (Reid et al. 2007, Imbahale et al. 2008; figure 8.3). Zebra numbers have shown little change, perhaps because zebra are somewhat more tolerant than wildebeest of the tall grass conditions that have become prevalent within the park. Furthermore, a smaller proportion of the zebra population was dependent on the park as a dry-season refuge (Gichohi 2000). The wildebeest that migrate northward from the Masai Mara Reserve toward the Loita Plains were reduced from 119,000

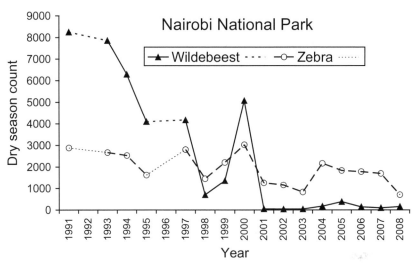

Figure 8.3. Changes in wildebeest and zebra totals counted in Nairobi National Park, averaged over the dry season months of August and October

to 22,000 animals following expansion of wheat farming in their wet-season range (Ottichilo et al. 2001, Serneels and Lambin 2001). In the Maasai steppe region of northern Tanzania, the wildebeest migrating seasonally from Tarangire National Park to the Simanjiro Plains have declined from about 43,000 in 1988 to a mere 5,000 in 2001, following the blocking of migratory corridors by land cultivation and expanding settlements (Bolger et al. 2008, Msoffe et al. 2011). In the Etosha Pan region of Namibia, fencing resulted in the collapse of the wildebeest population from 30,000 to around 2,500, while zebra numbers declined less drastically from 23,000 to 4,000 (Gasaway et al. 1996, Berry 1997). In central Kruger Park, the wildebeest that formerly migrated toward the Sand River became reduced from 6,000 animals to 1,500 animals following completion of the boundary fence confining them within their wet-season range (Whyte and Joubert 1988). In the central Kgalagadi region of Botswana, fences blocking movement towards the only remaining surface water during the 1983 drought led to a collapse by the wildebeest population from more than 100,000 animals to around 10% of that number (Williamson and Mbane 1988, Spinage and Matlhare 1992). Other ungulate species showed less extreme population reductions in this region: about 70% for hartebeest (*Alcelaphus buselaphus*), but only 30% to 35% for arid-adapted gemsbok (*Oryx gazella*) and springbok (*Antidorcas marsupialis*). These populations have not regained their former abundance.

Other Substantial Ungulate Declines

Substantial declines by ungulates within protected areas have not been restricted to migratory populations. In the Klaserie Private Nature Reserve adjoining Kruger National Park, population crashes by 80% to 90% were documented for several grazers, including wildebeest and zebra, over the 1983 drought (Walker et al. 1987; figure 8.3). A predisposing factor was excessive water provision via dams built by each landowner, which had promoted local increases in resident numbers of these species. Forage became eliminated over an extensive area when the zones of degradation around these water sources coalesced. In the adjoining section of Kruger Park, a wider spacing between water points meant that animals were able to find forage still remaining far from water when rains eventually broke the drought, alleviating the food deficit at this critical time. Within the Masai Mara Reserve, resident populations of buffalo, topi (*Damaliscus lunatus korrigum*), hartebeest, waterbuck, warthogs (*Phacochoerus africanus*) and giraffes all decreased to a quarter or less of their former numbers between 1977 and 1997, with land use developments outside and cattle influxes inside exacerbating the effects of recurrent droughts (Ottichilo et al. 2000, Ogutu et al. 2009). The shrinkage of these herbivore populations in this protected area has continued, fundamentally because of spreading human influences (Ogutu et al. 2011).

Within Kruger Park, where human intrusions are not a factor, 450 roan antelope (*Hippotragus equinus*) counted in 1986 had decreased to around 30 by 1996 (Harrington et al. 1999). Over the same period, tsessebe (*Damaliscus lunatus*) decreased from 1,200 to less than 200 animals, and sable antelope from more than 2,000 to around 550 animals. Populations of kudu, waterbuck, eland, and warthog fell by a similar proportion, but with less threat to their viability. These population declines occurred four years after the 1982–83 drought, during and after a less severe drought (Ogutu and Owen-Smith 2003). Expanded surface water distribution contributed to these population declines, but by a mechanism different from the one identified at Klaserie. The dams and drinking troughs provided by park managers, meant to reduce water restrictions on forage access during droughts (Pienaar 1970), attracted more zebra into the distribution ranges of the rarer antelope species, thus supporting more lions (Owen-Smith et al. 2005, Owen-Smith and Mills 2006). Lions shifted their hunting towards alternative prey species when their principal prey—wildebeest, zebra and buffalo—became less available (Owen-Smith and Mills 2008). A decline in the abundance of hartebeest in the Laikipia region of northern Kenya likewise appeared to be related to increased predator abundance, following a shift in animals stocked toward more wildlife (Georgiadis et al. 2007a).

Functional Components of Ecosystem Heterogeneity

Animals move to seek the resources and conditions they need, and to avoid local attenuation in those resources, during distinct stages of the seasonal cycle. Hence the key resources concept needs to be widened to encompass (1) ephemeral "hot spots," which offer especially high-quality forage in the wet season, thus promoting reproductive success; (2) staple forage, which provides principal support at other times during the wet season; (3) reserve forage, which retains sufficient quality in the dry season; (4) bridging forage, which provides temporary subsistence during periods when other resources are only sparsely available; and (5) buffer resources, which retard starvation during critical times when little other food remains (Owen-Smith and Cooper 1989, Owen-Smith 2002). Each of these resource components needs to be available within the ambit of movement of animals for their populations to remain viable and retain the resilience to recover from extreme conditions. Modeling has suggested how confining herbivores year-round to the hot spot of the Serengeti Plains, which becomes a dust bowl during the dry season, would lead to a somewhat lower and more variable population, while restricting them to the more abundant but generally lower-quality resources in Masai Mara would result in a much lower, though relatively constant, population (Owen-Smith 2004).

Conserving a diverse assemblage of large herbivores, especially given the additional threat posed by greater seasonal variability in rainfall, means ensuring that these animals retain access to the functional components of ecosystem heterogeneity that promote population resilience (Hobbs et al. 2008, 2010). The extent of the area needed depends on the spatial scale of landscape and rainfall gradients. In the Serengeti region, most of the wildebeest must travel over a distance of almost 200 km twice annually to span the annual rainfall gradient of 500 mm between the wet-season hot spot constituted by the short grass plains and the dry-season forage reserve persisting in the north (Holdo et al. 2009a). The sedentary wildebeest subpopulations exploiting local landscape gradients in the western corridor are vastly less abundant than the migrants. Large herds of grazers more tolerant of mature grass, such as buffalo, move over home ranges covering 200 to 500 km² to secure their seasonal food requirements in drier savanna regions (Ryan et al. 2006; unpublished observations by N.O-S.). White rhino occupy much smaller home ranges of 10 to 20 km² by cultivating local mosaics of nutritious grazing lawns interspersed with taller grass reserves (Owen-Smith 1988). Browsers exploit the more fine-scale heterogeneity of woody plant resources within local landscapes (Owen-Smith 2008), apart from eland, which may be more or less regularly migratory (Hillman 1988).

The population declines of rarer antelope recorded within Kruger Park, described above, expose a generally overlooked aspect of functional ecosystem heterogeneity: spatial variation in the risk of predation. These antelope species occurred mainly in the north of Kruger Park, where limited surface water availability had restricted the abundance of zebra and wildebeest, the main prey species for lions. The partial refuge this region had provided against predation became occluded when water sources were augmented.

Counteracting the Effects of Climatic Variability

For African savanna ungulates, the major threat from global warming is the increased frequency and severity of droughts and floods that is likely to result. Barriers to movement restrict these herbivores' ability to respond to such extreme conditions by seeking places where things are less adverse. If access to those localities is blocked, populations may crash by 50% or more, with consequently retarded recovery. While numbers remain low, recovery may be hindered by so-called Allee effects, such as reduced herd sizes exposing survivors to greater risk of predation, and perhaps even inbreeding.

Mawdsley et al. (2009) listed 16 adaptation strategies to alleviate the threats posed by climate change to wildlife and biodiversity conservation (table 8.1). Those relevant to our assessment of in situ conservation of African large mammals can be grouped into three categories: (1) expanding protected areas, (2) making surrounding land uses compatible with the presence of wildlife, and (3) retaining, restoring, or enhancing functional ecosystem heterogeneity within protected areas.

Increasing the extent of the formally protected area will be of most benefit when it can encompass regional rainfall or landscape gradients and migration routes. Opportunities for such conservation planning are extremely limited in Africa because of human pressures on land, especially as wetter areas are most likely to be given over to cultivation. Nevertheless, adding more of the same habitat at least enlarges the populations that can be supported, and increases opportunities for animals to exploit local spatial variation in rainfall. In South Africa, an ambitious plan is being implemented to expand the Addo Elephant National Park from its original 103 km² to an area of 3,410 km², encompassing a gradient from

Table 8.1. Climate-change adaptation strategies for wildlife management and biodiversity conservation relevant to in situ conservation of African large mammals listed by Mawdsley et al. (2009)

	Strategy	Type
1	Increase extent of protected areas	Expansion
2.	Improve representation and replication of protected area networks	Expansion
3.	Establish new protected areas designed to improve regional resilience	External
4.	Increase permeability of surrounding landscape to animal movements	External
5.	Develop dynamic conservation plans for regional landscape	External
6.	Protect movement corridors, stepping stones, and refugia	External
7.	Ensure that wildlife conservation needs are considered as part of broader societal adaption	External
8.	Improve management and restoration of existing protected areas to facilitate resilience	Internal
9.	Manage and restore ecosystem function rather than focusing on specific components	Internal
10.	Improve monitoring programs for wildlife and ecosystems	Internal
11.	Focus resources on species that are most threatened	Internal
12.	Reduce pressures on species from other sources	Internal
13.	Incorporate predicted climate-change impacts into all management plans, programs, and activities	Cross-cutting
14.	Review and modify existing laws and policies relating to wildlife and land use	Cross-cutting
15.	Translocate species at risk of extinction	Ex situ
16.	Establish captive populations of species that would otherwise go extinct	Ex situ

escarpment foothills to the coast (Kerley and Boshoff 1997). This entails purchasing land currently under private ownership. The Kruger National Park has had its effective area enlarged by 25% through the proclamation of the Limpopo National Park in an adjoining section of Mozambique, the two areas together forming the Greater Limpopo Transfrontier Park. This action was possible due to low human density in the Mozambique section (from which people have yet to be relocated). Further expansion is planned to link this area with Gonarezhou National Park in Zimbabwe and perhaps other regions of Mozambique, so that the broadened protected area would potentially cover 100,000 km². An even more grandiose scheme entails the development of the Kavango Zambezi Transfrontier Conservation Area extending through Botswana, Zimbabwe,

Zambia, Namibia, and Angola. This would include 36 national parks, game reserves, community conservancies, and game management areas, which together with intervening areas of human settlement potentially encompass 287,000 km².

In most situations the only practical way to expand or at least retain the scope for wildlife movement is through contractual agreements with adjoining areas under private or communal tenure. A notable example led to removal of the veterinary fence separating Kruger Park from the neighboring Sabi-Sand, Timbavati, Klaserie, and Umbabat private nature reserves, and from Manyaleti provincial reserve. In Kenya, land within group ranches bordering Masai Mara Reserve, recently transferred to private tenure, is being leased to private investors for wildlife tourism (Ogutu et al. 2009). In the Laikipia district, land use agreements are being negotiated to allow elephants and other wildlife to travel between community wildlife reserves in the Samburu and Meru districts and privately owned ranches through intervening group ranches (Georgiadis et al. 2007b). Leases with landowners in the Athi-Kaputiei Plains have been arranged to allow wildlife movement between this area and Nairobi National Park to continue (Reid et al. 2007). Carnivore protection programs pay consolation fees to owners who lose livestock to lions and hyenas dispersing from the park. In Tanzania, the Grumeti Reserve west of Serengeti National Park has come under effective protection from a concession holder who switched the land use from hunting to photographic safaris. In northern Botswana the area available to wildlife, including an elephant population approaching 150,000, encompasses more than 100,000 km² through the leasing of blocks surrounding Chobe and Moremi National Parks to private safari operators. In Namibia, community-based conservancies have been established whereby legal rights over wildlife and tourism have been conceded to representative councils, who have responded by setting aside some of their land for wildlife reserves. In Zimbabwe, wildlife conservancies have been established by removing fences between privately owned ranches, permitting animals to roam more widely.

However, many parks have become surrounded by expanding human settlements and associated cultivation, creating sharp boundaries (Lamprey and Reid 2004, Wittemeyer et al. 2008, Ogutu et al. 2009). Within Kenya, examples include Nairobi National Park adjoining the city of Nairobi (Imbehale et al. 2008), Lake Nakuru National Park (Ogutu and Owen-Smith in press 2012), and Aberdares National Park (as shown by Naughton-Treves et al. 2005). In South Africa, similarly abrupt land-use transitions surround the Hluhluwe-iMfolozi Park and also exist along sections of the southern and western boundaries of Kruger Park. In these situations, fences become essential to restrict damage to livestock and crops from wild animals, particularly elephants and lions, dispersing out of the park, and to prevent incursions by game poachers or livestock into the park.

Public roads flanked by fences can prevent the wider movements of wildlife through privately owned or leased land. Major highways that traverse protected areas bring a cost in terms of animals killed by speeding vehicles (Newmark et al. 1996), although this loss may be a small price to pay in exchange for the benefits of the enlarged area. The tarred road that traverses Chobe National Park in Botswana between Kasane and the Ngoma border post has incurred few problems, because of tight control of vehicle travel between the entry and exit points. The practicality of constructing underpasses for wildlife remains to be investigated.

Where the potential for expanding the area under effective protection does not exist, attention needs to be directed towards retaining or promoting the ecosystem heterogeneity needed to support diverse large mammal species within the confines of the protected area. The Hluhluwe-iMfolozi Park in the KwaZulu-Natal province of South Africa is a striking example of success in achieving this within a tightly fenced area of 900 km^2 (Brooks and Macdonald 1983). It contains both white and black rhino (*Diceros bicornis*), 19 other ungulate species, five large predators, and has recently reintroduced elephants. This protected area has high topographic diversity and relatively fertile soils, plus a rainfall gradient from 650 to 960 mm over a distance of 30 km. White rhinos enhance the diversity of grazing conditions through their cultivation of grazing lawn mosaics (Waldram et al. 2008). Earlier management of this park entailed culling herbivore populations to restrict population crashes during drought years, and was achieved largely through live capture and sales (Walker et al. 1987). Since the restoration of a full complement of large predators, however, these animal removals have been curtailed. Management is currently targeted at replicating ecological processes that once operated over larger areas. Because opportunities for wider dispersal are prevented, the white rhino population is managed by designating sections of the protected area as dispersal sinks (Owen-Smith 1983). White rhinos settling within these regions are captured and trucked elsewhere. Culling takes place in this way without killing, and without imposing any arbitrary ceiling on the white rhino population in the core section of the park. How to deal with the rapidly growing elephant population still has to be resolved, because these animals traverse the entire park area.

Large parks may contain insufficient spatial heterogeneity to buffer herbivore populations against widely varying climatic conditions, especially if they are situated in interior plateau regions where landscape gradients are wide, as in the Serengeti region. The temptation is to provide more water to expand herbivore populations and improve game viewing for tourists. As a consequence, the extent of the area not heavily grazed is reduced (Owen-Smith 1996), or, as in Klaserie Private Reserve, virtually eliminated, further constricting the spatial aspect of ecosystem heterogeneity. Retaining buffer forage contributes more to the animal numbers and diversity supported in the long term than does the temporary benefit provided by additional surface water. Furthermore, sedentary herbivore populations become subjected to higher rates of predation, because predators can find prey nearby year-round. More than half of the artificial water points established in Kruger Park have recently been closed, following recognition of the homogenizing effect of water points on animal distributions (Gaylard et al. 2003, Smit et al. 2007). Nevertheless, artificial water points may be necessary in parks established in regions with few or no perennial water sources, Hwange National Park in Zimbabwe being a notable example (Owen-Smith 1996). The wildlife wealth in Hwange is almost entirely dependent on 50 pumped pans, which led to a crisis when many of the diesel engines supplying these pans failed. A farsighted policy would be to concentrate the water sources in particular regions, thereby alleviating the impact of elephants on vegetation and other species elsewhere (Chamaille-Jammes et al. 2008).

Fire can also be used to promote heterogeneity by means of point ignitions at different stages of the dry season (Brockett et al. 2001). Fire management in Kruger Park has been changed accordingly (van Wilgen et al. 2004), although this has not suppressed the progressive expansion of woody shrubs across much of the park (Eckhardt et al. 2000). In Serengeti, the densification of trees has been ascribed to reduction in fires caused by the huge grazing impact of the wildebeest (Sinclair et al. 2007, Holdo et al. 2009b). However, there may be an insidious climatic threat underlying the expansion by woody plants here and elsewhere in savanna regions: the effect of increased atmospheric carbon dioxide concentrations on the growth rates of woody plants relative to those of the C4 grasses that are adapted to low carbon dioxide availability in hot and dry regions (Bond and Midgley 2000). It may seem desirable to have more elephants to reverse this trend by uprooting shrubs, but instead elephants have promoted the expansion of shrubs by felling or debarking canopy trees, thereby killing them. Effective management of elephant populations and their vegetation impact may require mega-parks covering 100,000 km^2 or more (van Aarde and Jackson 2007). Within protected areas smaller than this, the best response would be to limit surface water distribution in order to restrict access by elephants to sensitive regions of the landscape during stages of the dry season when their impact on trees is most severe (Owen-Smith et al. 2006).

Conclusions

By amplifying seasonal and annual variability in rainfall, climate change poses a threat to the abundance and diversity of African large mammals that are supposedly protected in parks and reserves. Large herbivores respond to the consequent temporal variability in their food resources by moving to exploit spatial heterogeneity in resources and conditions. The components of resource heterogeneity act functionally as hot spots and staple, reserve, bridging, and buffer resources. Refugia from high predation also contribute by assisting survival through adverse conditions. Surface water distribution provides a seasonal constraint on forage access. The resilience of herbivore populations in coping with widely variable rainfall is being reduced by expanding human settlements, fences, and cultivation that crowd the boundaries of protected areas. The most far-reaching response would be to extend the area over which large mammals can move. This can be accomplished through enlarged parks, transfrontier conservation areas, contractual agreements, conservancies and leases of land. In the absence of such measures, the aim should be to retain, restore, or enhance within protected areas the components of ecosystem heterogeneity that help to support a high diversity of large mammals.

ACKNOWLEDGMENTS
We acknowledge helpful comments from Jedediah F. Brodie, Eric S. Post, Daniel F. Doak, and John Wilmshurst.

LITERATURE CITED
Bell, R. H. V. 1970. The use of the herb layer by grazing ungulates in the Serengeti. In *Animal Populations in Relation to Their Food Resources*, edited by A. Watson, 111–24. Oxford: Blackwell.
———. 1971. A grazing ecosystem in the Serengeti. *Scientific American* 225:86–93.
Bengis, R. G., R. Grant and V. de Vos. 2003. Wildlife diseases and veterinary controls: A savanna ecosystem perspective. In *The Kruger Experience*, edited by J. T. du Toit, K. H. Rogers, and H. C. Biggs, 349–69. Washington: Island Press.
Berry, H. H. 1997. The wildebeest problem in the Etosha National Park: A synthesis. *Madoqua* 13:151–57.
Bolger, D. T., W. D. Newmark, T. A. Morrison, and D. F. Doak. 2008. The need for integrative approaches to understand and conserve migratory ungulates. *Ecology Letters* 11:63–77.
Bond, W. J., and G. F. Midgley. 2000. A proposed CO_2 mechanism of woody plant invasion in grasslands and savannas. *Global Change Biology* 6:1–5.

Brockett, B. H., H. C. Biggs, and B. W. van Wilgen. 2001. A patch mosaic burning system for conservation areas in southern African savannas. *International Journal of Wildland Fire* 10:169–83.

Brooks, P. M., and I. A. W. Macdonald. 1983. The Hluhluwe-Umfolozi Reserve: An ecological case history. In *Management of Large Mammals in African Conservation Areas*, edited by R. N. Owen-Smith, 51–78. Pretoria: Haum.

Butt, B., A. Shortridge, and A. M. G. A. WinklerPrins. 2009. Pastoral herd management, drought coping strategies, and cattle mobility in southern Kenya. *Annals of the Association of American Geographers* 99:309–34.

Caro, T., and Scholte, P., 2007. When protection falters. *African Journal of Ecology* 45:233–35.

Chamaille-Jammes, S., H. Fritz, M. Valeix, F. Murindagomo, and J. Clobert. 2008. Resource variability, aggregation and direct density dependence in an open context: The local regulation of an African elephant population. *Journal of Animal Ecology* 77:135–44.

Corfield, T. F. 1973. Elephant mortality in Tsavo National Park, Kenya. *East African Wildlife Journal* 11:339–68.

Cross, P. C., J. I. Lloyd-Smith, and W. M. Getz. 2005. Disentangling association patterns in fission-fusion societies using African buffalo as an example. *Animal Behaviour* 69:499–506.

Daly, R. 2005. Good gnus, bad gnus. *Getaway Magazine*, April, 38–43.

Deshmukh, I. K. 1984. A common relationship between precipitation and grassland peak biomass for east and southern Africa. *African Journal of Ecology* 22:181–86.

Dudley, J. P., G. C. Craig, D. S. C.Gibson, G. Haynes and J. Klimowicz. 2001. Drought mortality of bush elephants in Hwange National Park, Zimbabwe. *African Journal of Ecology* 39:187–94.

Easterling, D. R., G. A. Meehl, C. Parmesan, S. A. Changnon, T. R. Karl, and L. O. Mearns.2000. Climate extremes: Observations, modeling, and impacts. *Science* 289:2068–74.

Eckhardt, H. C., B. W. van Wilgen, and H. C. Biggs. 2000. Trends in the woody vegetation cover in the Kruger National Park, South Africa, between 1940 and 1998. *African Journal of Ecology* 38:108–15.

Estes, R. D., J. L. Atwood, and A. B. Estes. 2006. Downward trends in Ngorongoro Crater ungulate populations 1986–2005: Conservation concerns and the need for ecological research. *Biological Conservation* 131:106–20.

Federov, A. V., and S. G. Philander. 2000. Is El Nino changing? *Science* 288:1997–2002.

Foster, J. B., and M. J. Coe. 1968. The biomass of game animals in Nairobi National Park 1960–1966. *Journal of Zoology London* 155:413–25.

Fryxell, J. M., J. Greever, and A. R. E. Sinclair. 1988. Why are migratory ungulates so abundant? *American Naturalist* 131:781–98.

Fryxell, J. M. and A. R. E. Sinclair. 1988. Seasonal migration by white-eared kob in relation to resources. *African Journal of Ecology* 26:17–31.

Fryxell, J. M., J. F. Wilmshurst, and A. R. E. Sinclair. 2004. Predictive models of movement by Serengeti grazers. *Ecology* 85:2429–35.

Fryxell, J. M., J. F. Wilmshurst, A. R. E. Sinclair, D. T. Haydon, R. D. Holt, and P. A. Abrams. 2005. Landscape scale, heterogeneity, and the viability of Serengeti grazers. *Ecology Letters* 8:328–35.

Gasaway, W. C., K. T. Gasaway, and H. H. Berry. 1996. Persistent low densities of plains ungulates in Etosha National Park, Namibia: Testing the food regulating hypothesis. *Canadian Journal of Zoology* 74:1556–72.

Gaylard, A., N. Owen-Smith, and J. Redfern. 2003. Surface water availability: Implications for heterogeneity and ecosystem processes. In *The Kruger Experience*, edited by J. T. du Toit, K. H. Rogers and H. C. Biggs, 171–88. Washington: Island Press.

Georgiadis, N., M. Hack, and K. Turpin. 2003. The influence of rainfall on zebra population dynamics: Implications for management. *Journal of Applied Ecology* 40:125–36.

Georgiadis, N. J., F. Ihiwagi, J. G. N. Olwero, and S. S. Romanasch. 2007a. Savanna herbivore dynamics in a livestock-dominated landscape. II. Ecological, conservation and management implications of predator restoration. *Biological Conservation* 137:473–83.

Georgiadis, N. J., J. G. N. Olivero, G. Ojwang, and S. S. Romanasch. 2007b. Savanna herbivore dynamics in a livestock-dominated landscape. I. Dependence on land use, rainfall, density, and time. *Biological Conservation* 137:461–72.

Gichohi, H. 2000. Functional relationships between parks and agricultural areas in East Africa: The case of Nairobi National Park. In *Wildlife Conservation by Sustainable Use*, edited by H. H. T. Prins, J. G. Grootenhuis and T. T. Dolan, 141–68. Dordrecht: Kluwer.

Halley, D. J., M. E. J. Vandewalle, M. Mari, and C. Taolo. 2002. Herd-switching and long-distance dispersal in female African buffalo. *African Journal of Ecology* 40:97–99.

Harrington, R., N. Owen-Smith, P. Viljoen, H. Biggs, and D. Mason. 1999. Establishing the causes of the roan antelope decline in the Kruger National Park, South Africa. *Biological Conservation* 90:69–78.

Harris, G., S. Thirgood, J. G. C. Hopcraft, J. P. G. M. Cromsigt, and J. Berger. 2009. Global decline in aggregated migrations of large terrestrial mammals. *Endangered Species Research* 7:55–76.

Hastenrath, S., D. Polzin, and C. Mutai. 2007. Diagnosing the 2005 drought in equatorial East Africa. *Journal of Climate* 20: 4628–37.

Hillman, J. C. 1988. Home range and movement of the common eland in Kenya. *African Journal of Ecology* 26:135–48.

Hillman, J. C., and A. K. K. Hillman. 1977. Mortality of wildlife in Nairobi National Park during the drought of 1973–1974. *African Journal of Ecology* 15:1–18.

Hobbs, N. T., R. S. Reid, K. A. Galvin, and J. E. Ellis. 2008. Fragmentation of arid and semi-arid ecosystems: implications for people and animals. In *Fragmentation of Semi-arid and Arid Landscapes: Consequences for Human and Natural Systems*, edited by K. A. Galvin, R. S. Reid, R. H. Behnke, and N. T. Hobbs, 25–44. Dordrecht: Springer.

Hobbs, N. T., and I. J. Gordon. 2010. How does landscape heterogeneity shape population dynamics? In *Dynamics of Large Herbivore Populations in Changing Environments: Towards Appropriate Models*, edited by N. Owen-Smith, 141–64. Oxford: Wiley-Blackwell.

Holdo, R. M., R. D. Holt, and J. M. Fryxell. 2009a. Opposing rainfall and plant nutritional gradients best explain the wildlife migration in the Serengeti. *American Naturalist* 173:431–45.

Holdo, R. M., A. R. E. Sinclair, A. P. Dobson, K. L. Metzger, B. M. Bolker, M. E. Ritchie, and R. D. Holt. 2009b. A disease-mediated trophic cascade in the Serengeti and it implications for ecosystem C. *PLoS Biology* 7:1–12.

Hopcraft, G. C., A. R. E. Sinclair, and C. Packer. 2005. Planning for success: Serengeti lions seek prey accessibility rather than abundance. *Journal of Animal Ecology* 74:559–66.

Illius, A. W. and T. G. O'Connor. 2000. Resource heterogeneity and ungulate population dynamics. *Oikos* 89:283–94.

Imbahale, S. S., J. M. Githaiga, R. M. Chira, and M. Y. Said. 2008. Resource utilization bylarge migratory herbivores of the Athi-Kapiti ecosystem. *African Journal of Ecology* 46 (Suppl. 1): 43–51.

Joos-Vandewalle, M. E. 2000. Movement of migratory zebra and wildebeest in northern Botswana. PhD thesis, University of the Witwatersrand.

Kahurananga, J., and F. Silkiluwasha. 1997. The migration of zebra and wildebeest between Tarangire National Park and Simanjiro Plains, northern Tanzania, in 1972 and recent trends. *African Journal of Ecology* 35:179–85.

Kerley, G. I. H., and A. F. Boshoff. 1997. A proposal for a Greater Addo National Park: A regional and national conservation and development opportunity. Terrestrial Ecology Research Unit Report 17, 62 pp. University of Port Elizabeth.

Knight, M. H. 1995. Drought-related mortality of wildlife in southern Kalahari and the role of man. *African Journal of Ecology* 33:377–94.

Kutilek, M.J. 1974. The density and biomass of large mammals in Lake Nakuru National Park. *East African Wildlife Journal* 12:201–12.

Lamprey, R. H., and R. S. Reid. 2004. Expansion of human settlement in Kenya's Maasai Mara: What future for pastoralism and wildlife? *Journal of Biogeography* 31:997–1032.

Maddock, L. 1979. The "migration" and grazing succession. In *Serengeti: Dynamics of an Ecosystem*, edited by A. R. E. Sinclair and M. Norton-Griffiths, 104–29. Chicago: University of Chicago Press.

Marchant, R., C. Mumbi, S. Behera and T. Yamagata. 2007. The Indian Ocean dipole: The unsung driver of climatic variability in East Africa. *African Journal of Ecology* 45:4–16.

Mati, B. M., S. Mutie, H. Gadain, P.Home, and F. Mtalo,. 2008. Impacts of use/cover changes on the hydrology of the transboundary Mara River, Kenya/Tanzania. *Lakes and Reservoirs: Research and Management* 13:169–77.

Mawdsley, J. R., R. O'Malley, and D. S. Ojima. 2009. A review of climate-change adaptation strategies for wildlife management and biodiversity conservation. *Conservation Biology* 23:1080–89.

Mduma, S., R. Hilborn and A. R. E. Sinclair. 1998. Limits to exploitation of Serengeti wildebeest and implications for its management. In *Dynamics of Tropical Communities*, edited by D. M. Newbery, H. H. T. Prins, and N. D. Brown, 243–65. Oxford: Blackwell.

Mduma, S. A. R., A. R. E. Sinclair, and R. Hilborn. 1999. Food regulates the Serengeti wildebeest: A 40-year record. *Journal of Animal Ecology* 68:1101–22.

Metzger, K. L., A. R. E. Sinclair, R. Hilborn, J. G. C Hopcraft, and S. A. R. Mduma. 2010. Evaluating the protection of wildlife in parks: The case of African buffalo in Serengeti. *Biodiversity Conservation* 10.1007/s10531-010-9904-z.

Mills, M. G. L., H. C. Biggs and I. J. Whyte. 1995. The relationship between rainfall, lion predation and population trends in African herbivores. *Wildlife Research* 22:75–88.

Mills, M. G. L., I. J. Whyte, A. J. Viljoen, N. Zambatis, and A. L. F. Potgieter. 1996. Background information for the National Parks Board's review of the Kruger National Park's elephant management policy. Scientific Report no. 1/96. Scientific Services Division, Kruger Natonal Park, Skukuza.

Msoffe, F. U., S. K. Kifugo, M. Y. Said, M. O. Neselle, P. van Gardingen, R. S. Reid, J. O. Ogutu, M. Herero, and J. de Leeuw. 2011. Drivers and impacts of land-use change in the Maasai Steppe of northern Tanzania: An ecological-social-political analysis. *Journal of Land Use Science* 6:261–81. DOI: 10.1080/1747423X.2010.511682.

Myers, N. 1973. Tsavo National Park, Kenya, and its elephants: An interim appraisal. *Biological Conservation* 5:123–32.

Naughton-Treves, L., M. B. Holland, and K. Brandon. 2005. The role of protected areas in conserving biodiversity and sustaining local livelihoods. *Annual Review of Environment and Resources* 30:219–52.

Newmark, W. D. 2008. Isolation of African protected areas. *Frontiers in Ecology and the Environment* 6:321–28.

Newmark, W. D., J. I. Boshe, H. I. Sariko, and G. K. Makumbule. 1996. Effects of a highway on large mammals in Mikumi National Park, Tanzania. *African Journal of Ecology* 34:15–31.

Nicholson, S. E. 2000. The nature of rainfall variability over Africa on time scales of decades to millennia. *Global and Planetary Change* 26:137–58.

O'Connor, T. G. 1985. A synthesis of field experiments concerning the grass layer in the savanna regions of southern Africa. South African National Scientific Programmes Report no. 114. Foundation for Research Development, Pretoria.

O'Connor, T. G., L. M. Haines, and H. A. Snyman. 2001. Influence of precipitation and species composition on phytomass of a semi-arid grassland. *Journal of Ecology* 89:850–60.

Ogutu, J. O., and N. Owen-Smith. 2003. ENSO, rainfall and temperature influences on extreme population declines among African savanna ungulates. *Ecology Letters* 6:412–19.

Ogutu, J. O., H.-P. Piepho, H. T. Dublin, N. Bhola, and R. S. Reid. 2008a. El Nino–Southern Oscillation, rainfall, temperature and Normalized Vegetation Difference Index fluctuations in the Mara-Serengeti ecosystem. *African Journal of Ecology* 46:132–43.

———. 2008b. Rainfall influences on ungulate population abundance in the Mara-Serengeti ecosystem. *Journal of Animal Ecology* 77:814–29.

———. 2009. Dynamics of Mara-Serengeti ungulates in relation to land use changes. *Journal of Zoology* 277:1–14.

———. 2010a. Rainfall extremes explain interannual shifts in timing and synchrony of breeding in topi and warthog. *Population Ecology* 52:89–92.

Ogutu, J. O., H.-P. Piepho, R. S. Reid. M. E. Rainy, R. L. Kruska, J. S. Worden, M. Nyabenge, and N. T. Hobbs. 2010b. Large herbivore responses to water and settlements in savannas. *Ecological Monographs* 80:241–66.

Ogutu, J. O., N. Owen-Smith, H.-P. Piepho, and M. Y. Said. 2011. Continuing wildlife population declines and range contraction in the Mara region of Kenya during 1977–2009. *Journal of Zoology* 285:99–109.

Ogutu, J. O., N. Owen-Smith, H.-P. Piepho, B. Kuloba, and J. Edebe. In press 2012. Dynamics of ungulates in relation to climatic and land use changes in an insularized African savanna ecosystem. *Biodiversity Conservation*.

Ottichilo, W. K., J. De Leeuw, A. K. Skidmore, H. H. T. Prins, and M. Y. Said. 2000. Population trends of large non-migratory wild herbivores and livestock in the Masai-Mara ecosystem, Kenya, between 1977 and 1997. *African Journal of Ecology* 38:202–16.

Ottichilo, W. K., J. De Leeuw, and H. H. T. Prins. 2001. Population trends of resident wildebeest and factors influencing them in Masai Mara ecosystem, Kenya. *Biological Conservation* 97:271–82.

Owen-Smith, N. 1982. Factors influencing the consumption of plant products by large herbivores. In *The Ecology of Tropical Savannas*, edited by B. J. Huntley and B. H. Walker, 359–404. Berlin and Hamburg: Springer-Verlag.

———. 1996. Ecological guidelines for waterpoints in extensive protected areas. *South African Journal of Wildlife Research* 26:107–12.

———. 2002. *Adaptive Herbivore Ecology: From Resources to Populations in Variable Environments*. Cambridge: Cambridge University Press.

———. 2004. Functional heterogeneity within landscapes and herbivore population dynamics. *Landscape Ecology* 19:761–71.

———. 2008. The comparative population dynamics of browsing and grazing ungulates. In *The Ecology of Grazing and Browsing in Mammalian Herbivores*, edited by I. Gordon & H. H. T. Prins, 149–78. Berlin: Springer Verlag.

Owen-Smith, N., and S. M. Cooper. 1989. Nutritional ecology of a browsing ruminant, the kudu, through the seasonal cycle. *Journal of Zoology* 219:29–43.

Owen-Smith, N., G. I. H. Kerley, B. Page, R. Slotow & R. J. Van Aarde. 2006. A scientific perspective on the management of elephants in the Kruger National Park and elsewhere. *South African Journal of Science* 102:389–94.

Owen-Smith, N. D. R. Mason, and J. O. Ogutu. 2005. Correlates of survival rates for 10 African ungulate populations: Density, rainfall and predation. *Journal of Animal Ecology* 74:774–88.

Owen-Smith, N., and M. G. L. Mills. 2006. Manifold interactive influences on the population dynamics of a multi-species ungulate assemblage. *Ecological Monographs* 76:93–109.

———. 2008. Shifting prey selection generates contrasting herbivore dynamics within a large-mammal predator-prey web. *Ecology* 89:1120–33.

Owen-Smith, R. N. 1983. Dispersal and the dynamics of large herbivores in enclosed areas: Implications for management. In *Management of Large Mammals in African Conservation Areas*, edited by R. N. Owen-Smith, 127–43. Pretoria: Haum.

———. 1988. *Megaherbivores: The Influence of Very Large Body Size on Ecology*. Cambridge Studies in Ecology. Cambridge: Cambridge University Press,.

Parker, I. S. C. 1983. The Tsavo story: An ecological case history. In *Management of Large Mammals in African Conservation Areas*, edited by R. N. Owen-Smith, 37–50. Pretoria: Haum.

Pennycuick, L. 1975. Movements of the migratory wildebeest population in the Serengeti area between 1960 and 1973. *East African Wildlife Journal* 13:65–87.

Percival, A. B. 1928. *A Game Ranger on Safari*. London: Nisket and Company.

Pienaar, U. deV. 1970. Water resources of the Kruger Park. *African Wild Life* 24: 180–91.

Phillipson, J. 1975. Rainfall, primary production and carrying capacity of Tsavo National Park (East), Kenya. *East African Wildlife Journal* 13:171–201.

Rasmussen, H. B., G. Wittemyer, and I. Douglas-Hamilton. 2006. Predicting time-specific changes in demographic processes using remote-sensing data. *Journal of Applied Ecology* 43:366–76.

Reid, R. S., H. Gichohi, M. Y. Said et al. 2008. Fragmentation of a peri-urban savanna in the Athi-Kaputei Plains, Kenya. In *Fragmentation of Semi-Arid and Arid Landscapes: Consequences for Human and Natural Systems*, edited by K. A. Galvin, R. S. Reid, R. H. Behnke, and N. T. Hobbs, 195–224. Dordrecht: Springer.

Ritchie, M. E. 2008. Global environmental changes and their impacts on the Serengeti. In *Serengeti III*, edited by A. R. E. Sinclair, C. Packer, S. A. R. Mduma, and J. M. Fryxell, 183–208. Chicago: University of Chicago Press.

Roche, C. 2004. Ornaments of the desert: Springbok treks in the Cape Colony, 1774–1908. M.Sc. thesis, University of Cape Town.

Rutherford, M. C. 1980. Annual plant production: Precipitation relations in arid and semi-arid regions. *South African Journal of Science* 76:53–56.

———. 1984. Relative allocation and seasonal phasing of growth of woody plant components in a South African savanna. *Progress in Biometeorology* 3:200–221.

Ryan, S. J., C. U. Knechtel, and W. M. Getz. 2006. Range and habitat selection of African buffalo in south Africa. *Journal of Wildlife Management* 70:764–76.

Scheel, D., and C. Packer. 1995. Variation in predation by lions: Tracking a moveable feast. In *Serengeti II*, edited by A. R. E. Sinclair and P. Arcese, 299–314. Chicago: University of Chicago Press.

Scoones, I. 1995. Exploiting heterogeneity: Habitat use by cattle in dryland Zimbabwe. *Journal of Arid Environments* 29:221–37.

Serneels, S., and E. F. Lambin. 2001. Impact of land-use change in the wildebeest migration in the northern part of the Serengeti-Mara ecosystem. *Journal of Biogeography* 28:391–407.

Sinclair, A. R. E., S. A. R. Mduma, J. G. C. Hopcraft, J. M. Fryxell, R. Hilborn, and S. Thirgood. 2007. Long-term ecosystem dynamics in the Serengeti: Lessons for conservation. *Conservation Biology* 21:580–90.

Skinner, J. D. 1993. Springbok treks. *Transactions of the Royal Society of South Africa* 48:291–305.

Smit, I. P. J., C. C. Grant, and B. J. Devereux. 2007. Do artificial waterholes influence the way herbivores use the landscape? Herbivore distribution patterns around rivers and artificial water sources in a large African savanna park. *Biological Conservation* 136:85–99.

Smuts, G. L. 1978. Interrelations between predators, prey and their environment. *BioScience* 28:316–20.

Soule, M. E., B. A.Wilcox and C. Holtby. 1979. Benign neglect: Model of faunal collapse in the game reserves of East Africa. *Biological Conservation* 15:259–72.

Spinage, C. A., and J. M. Matlhare. 1992. Is the Kalahari cornucopia fact or fiction? A predictive model. *Journal of Applied Ecology* 29:605–10.

Stellfox, J. G., D. G. Peden, H. Epp, R. J. Hudson, S. W. Mbugua, and J. L Agatsiva. 1986. Herbivore dynamics in Southern Narok, Kenya. *Journal of Wildlife Management* 50:339–47.

Stewart, D. R. M., and D. R. P. Zaphiro. 1963. Biomass and density of wild herbivores in different East African habitats. *Mammalia* 27:483–96.

Talbot, L. M., and M. H. Talbot. 1963. The wildebeest in western Masailand. *Wildlife Monographs* 12:88.

Thirgood, S., A. Mosser, S. Tham, G. Hopcraft, E. Mwangomo, T. Mlengeya, M. Kilewo, J. Fryxell, A. R. E. Sinclair, and M. Borner. 2004. Can parks protect migratory ungulates? The case of the Serengeti wildebeest. *Animal Conservation* 7:113–20.

Tyson, P. D. and C. K. Gatebe. 2001. The atmosphere, aerosols, trace gases and biogeochemical change in southern Africa: A regional integration. *South African Journal of Science* 97:106–18.

Tyson, P. D., J. Lee-Thorpe, K. Holmgren, and J. F. Thackeray. 2002. Changing gradients of climate change in southern Africa during the past millennium: Implications for population movemens. *Climate Change* 52:129–35.

Van Aarde, R. J., and T. Jackson. 2007. Megaparks for metapopulations: Addressing the causes of locally high elephant numbers in South Africa. *Biological Conservation* 134:289–97.

Van Wilgen, B. W., N. Govender, H. C. Biggs, D. Ntsala, and X. N. Funda. 2004. Response of savanna fire regimes to changing fire-management policies in a large African national park. *Conservation Biology* 18:1533–40.

Waldram, M. S., W. J. Bond, and W. D. Stock. 2008. Ecological engineering by a mega-grazer: White rhino impacts on a South African savanna. *Ecosystems* 11:101–12.

Walker, B. H., R. H. Emslie, R. N. Owen-Smith and R. J. Scholes. 1987. To cull or not to cull: Lessons from a southern African drought. *Journal of Applied Ecology* 24:381–401.

Webster, P. J., A. M. Moore, J. P. Loschnigg, and R. R. Leben. 1999. Coupled ocean-atmosphere dynamics in the Indian Ocean during 1997–1998. *Nature* 401:356–60.

Western, D. 1975. Water availability and its influence on the structure and dynamics of a savannah large mammal community. *East African Wildlife Journal* 13:265–86.

———. 1982. Amboseli National Park: Enlisting landowners to conserve migratory wildlife. *Ambio* 11:302–8.

Whyte, I. J., and S. C. J. Joubert. 1988. Blue wildebeest population trends in the Kruger National Park and the effects of fencing. *South African Journal of Wildlife Research* 18:78–87.

Williamson, D., and B. Mbane. 1988. Wildebeest mortality during 1983 at Lake Xau, Botswana. *African Journal of Ecology* 26:341–44.

Wilmshurst, J. F., J. M. Fryxell, B. P. Farm, A. R. E. Sinclair, and C. P. Henschel. 1999. Spatial distribution of Serengeti wildebeest in relation to resources. *Canadian Journal of Zoology* 77:1223–32.

Wittemyer, G., P. Elsen, W. T. Bean, A. Coleman O. Burton, and J. S. Brashares. 2008. Accelerated human population growth at protected area edges. *Science* 321:123–26.

9

Ecological Effects of Climate Change on European Reptiles

Jean François Le Galliard, Manuel Massot,
Jean-Pierre Baron, and Jean Clobert

The destruction and fragmentation of habitats, overexploitation of natural resources, and invasion of pest species have resulted in worldwide declines of many animal and plant species (Morris and Doak 2002). It is increasingly apparent that the effects of this deadly cocktail will be reinforced by global warming, which might disrupt and eventually push towards extinction entire ecosystems and an increasing number of species (e.g., Walther et al. 2002, Thomas et al. 2004, Parmesan 2006). As a consequence, there is a pressing need for data and predictions on wildlife responses to global warming, and for an agenda of conservation strategies to mitigate its deleterious impacts (Hannah et al. 2005).

Responses to climate change have been summarized for plants, passerine birds, butterflies, mammals, and some amphibians (Parmesan 2006, Rosenzweig et al. 2007, and references therein). Until now, however, they have not been reviewed for reptiles, despite the fact that these animals are often at the forefront of conservation priorities (Pounds et al. 1999, Gibbons et al. 2000, Thomas et al. 2004). Reptiles constitute an assemblage of terrestrial and aquatic vertebrates broadly defined as a paraphyletic group (i.e., excluding birds; Pough et al. 2001, Pianka and Vitt 2003). Squamates, or "scaled reptiles," include snakes and lizards and make up the most diverse order of reptiles, distributed almost worldwide (e.g., Tinkle and Ballinger 1972, Huey et al. 1983, Pianka and Vitt 2003). Key characteristics of squamates include ectothermy (reliance on external means to control body temperature); continuous growth through repeated molting, resulting in great size variation within a single population; intermediate levels of mobility; and the presence of two major reproductive modes, viviparity and oviparity, associated with an absence of parental care in most species (Pough et al. 2001). Ectothermy is a fundamental characteristic of squamates, and it has generated the evolution of physiological and behavioral tactics that enable partial control of body temperatures (Huey 1982, Adolph and Porter 1993, Kearney et al. 2009). The demographic effects of climate change in reptiles should be mediated by interactions between the direct effects of warming on behavior and physiology and the indirect ecological effects that result from trophic interactions and habitat changes (Matthews et al. this volume).

Climate warming has begun to be considered as a major extinction threat for these animals, especially in tropical environments, in species with restricted spatial ranges, and in species with temperature-dependent sex determination (Janzen 1994, Gibbons et al. 2000, Cox and Temple 2009, Sinervo et al. 2010). In this chapter we assess ecological consequences and challenges posed by global warming in Europe for conservation biology of reptiles. We focus our attention on field data, model predictions, and conservation problems in the European geographic area, defined to include all species north of the Mediterranean Sea and west of Russia (see figure 9.1 and Gasc et al. 1997). We discuss only data for squamate reptiles (lizards and snakes) for which long-term demographic studies are available.

Historical reconstructions of species distributions in Europe have improved our understanding of the factors that shaped their spatial ranges during the Quaternary ages. These studies all point out that major range shifts were driven by change in climatic conditions and were constrained by geographic barriers and life-history adaptations (e.g., Surget-Groba et al. 2001, Ursenbacher et al. 2006, Pinho et al. 2007, Ursenbacher et al. 2008). Areas of endemism and high species richness occur now in southern refugia to which reptiles retreated during the ice ages. Range expansions to the north during interglacial ages have been more complex than a simple uniform expansion wave. Different species have followed distinct expansion routes constrained by the presence of mountainous barriers (Pyrenees, Alps and Balkan, and Carpathian) and also by characteristics that provide the physiological capacity to live in colder climates, such as viviparity or the ability to perform critical performances at low temperatures. Alotgether, these data suggest that some species of reptiles may be able to expand their range into northern habitats with future warming.

During the last decades, however, European reptiles have suffered great numeric declines (Corbett 1989, Gasc et al. 1997, Hickling et al. 2006, IUCN 2008, Cox and Temple 2009). Major factors of these declines include habitat loss and fragmentation, change in land use, and invasion by pest species (Cox and Temple 2009). Some species also suffer from persecution and can be threatened by illegal captures for the trade market . Major threats to the 31 species listed in table 9.1, obtained from IUCN 2008 (Cox and Temple 2009), were considered to be changes in habitat, collection of live specimens, persecution, and negative effects of invasive species. Climate change was also considered as a potential threat to a significant number of species. Unfortunately, quantitative data for

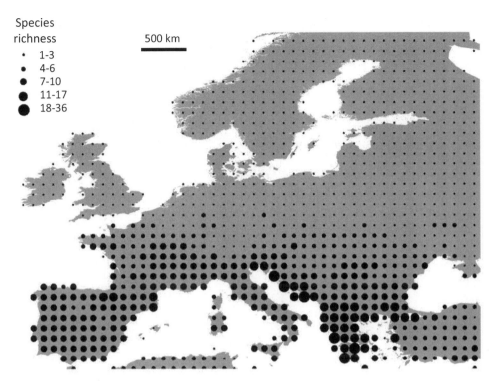

Species richness
- • 1-3
- • 4-6
- ● 7-10
- ● 11-17
- ● 18-36

500 km

Figure 9.1. Spatial distribution of species richness for European squamate reptiles (snakes and lizards). A map of species richness (number of species per half-degree of latitude indicated by diameter of circles around points spaced one degree apart) was calculated from spatial data downloaded in August 2010 from the IUCN website (www.iucnredlist.org/technical-documents/spatial-data). Spatial data were corrected for numerous taxonomic and biogeographic errors. The diversity of European squamate amounts to approximately 90 species of lizards belonging to 6 families, 1 amphisbaenian species, and about 40 species of snakes belonging to 4 families (Arnold et al. 2002, 2007). Figure kindly drawn by B. Decencière Ferrandière.

climate change effects are missing in these evaluations (IUCN 2008, Cox and Temple 2009). For most reptiles, a link between demography and climatic conditions therefore remains to be established. Below we use field data and predictive models to describe the effects of climate change for European squamates.

The Significance of Climate Change for Reptiles in Europe

In Europe, the average climate warming was on the order of more than 0.4 °C per decade during the last century, with stronger warming at high altitudes and in central and northeastern Europe, stronger average

Table 9.1. List of critically endangered (CR), endangered (EN), vulnerable (VU), and near threatened (NT) squamates in Europe according to the IUCN 2008 Red List, downloaded on August 12, 2009 (IUCN 2008, Cox and Temple 2009). The total number of native squamate reptiles amounts to approximately 90 species of lizards from five families (Agamidae, Anguidae, Chamaeleonidae, Gekkonidae, Lacertidae, Scincidae), as well as one amphisbenian and approximately 40 species of snakes from four families (Boidae, Colubridae, Typhlopidae, Viperidae). Species from the Canary Islands and from northern Africa were excluded from the list. Assessment criteria: (1) restricted and fragmented range, (2) decline in size and quality of habitat, and (3) decline in abundance.

Assessment	Family	Species	Geographic distribution	Major threats
CR (1, 3)	Lacertidae	Iberolacerta martinezricai*	Sierra de Francia (Spain)	Collection, habitat loss, and future climate change
CR (1, 3)	Lacertidae	Podarcis raffonei	Aeolian Islands (Italy)	Collection and invasive species, intrinsic factors
EN (1, 2)	Lacertidae	Iberolacerta aranica*	Pyrenees Mountains (France and Spain)	Land use, habitat loss, collection, and future climate change
EN (1, 2)	Lacertidae	Iberolacerta aurelioi*	Pyrenees Mountains (France and Spain)	Land use, habitat loss, collection, and future climate change
EN (1, 2)	Lacertidae	Iberolacerta cyreni	Central Spain	Habitat loss and future climate change
EN (1, 2)	Lacertidae	Podarcis cretensis*	Crete and satellite islands	Habitat loss and degradation
EN (1, 2, 3)	Lacertidae	Algyroides marchi	Southeast Spain	Habitat loss, land use, and invasive species
EN (1, 2, 3)	Viperidae	Macrovipera schweizeri	Western Cyclade Islands (Greece)	Collection, persecution, and habitat loss
EN (1, 2, 3)	Lacertidae	Podarcis carbonelli*	Western and central Portugal	Habitat loss and future climate change
VU (1)	Lacertidae	Podarcis gaigeae*	Aegean Islands (Greece)	No major threat at present
VU (1)	Lacertidae	Podarcis levendis	Greekislets	No major threat at present
VU (1)	Lacertidae	Podarcis milensis	Aegean Islands (Greece)	No major threat at present
VU (1, 2)	Lacertidae	Dinarolacerta mosorensis†	Balkans (Bosnia and Montenegro)	Habitat loss, land use, and collection

Status	Family	Species	Distribution	Threats
VU (1, 2)	Lacertidae	*Iberolacerta monticola*	Northern Spain and Portugal	Habitat loss and land use
VU (1, 2)	Viperidae	*Vipera ursinii*	Southeast and central Europe	Habitat loss, land use, collection, and climate change
VU (2)	Viperidae	*Vipera latastei*	Portugal and Spain	Persecution, habitat loss, and natural mortality
NT (1)	Anguidae	*Anguis cephalonnica*	Greece	Persecution
NT (1)	Gekkonidae	*Euleptes europaea*	Corsica and Sardinia (France and Italy)	Habitat loss and invasive species
NT (1)	Lacertidae	*Iberolacerta bonnali**	Pyrenees Mountains (France and Spain)	Land use, habitat loss, collection, and future climate change
NT (1)	Lacertidae	*Iberolacerta galani**	Montes de León (Spain)	Habitat loss and land use
NT (1)	Lacertidae	*Iberolacerta horvathi*	Austria, Italy, and Balkans	Intrinsic factors but no major threat at present
NT (1)	Lacertidae	*Podarcis pityusensis*	Balearic Islands (Spain)	No major threat at present
NT (1, 2)	Lacertidae	*Algyroides moreoticus*	Greece	Habitat loss, land use, and persecution
NT (1, 2)	Lacertidae	*Archaeolacerta bedriagae*	Corsica and Sardinia (France and Italy)	Habitat loss, land use, and intrinsic factors
NT (1, 2)	Lacertidae	*Hellenolacerta graeca*	Southern Greece	Land use
NT (2)	Scincidae	*Chalcides bebriagai*	Spain and Portugal	Habitat loss and natural predation
NT (2)	Lacertidae	*Darevskia praticola**	Eastern Europe (Serbia, Romania, Bulgaria)	Habitat loss and land use
NT (2)	Colubridae	*Elaphe quatuorlineata*	Italy, Balkans, and Greece	Habitat loss and persecution
NT (2)	Lacertidae	*Lacerta schreiberi*	Spain and Portugal	Habitat loss and land use
NT (2)	Colubridae	*Macroprotodon brevis**	Southern Spain	Habitat loss, land use, and natural mortality
NT (2)	Lacertidae	*Timon lepidus*	Portugal, Spain, France, and Italy	Habitat loss, pollution, and land use

* Recently described species. † Recently proposed species.

warming during the winter, and an increase in the occurrence of warm extremes (IPCC 2007). Less obvious and spatially variable trends in precipitation were also observed with a pattern of wetter climates in oceanic and northern Europe, and drier climates around the Mediterranean basin. Climate scientists now forecast further changes in climatic conditions for the next century, even under the most optimistic scenarios of reduced emission of greenhouse gases (Alcamo et al. 2007). Annual mean temperatures in Europe should increase by 0.2 to 0.5° C per decade, with mean summer levels as high as 0.6° C per decade in parts of France or the Iberian Peninsula. Precipitation should also decrease in the south, but will most likely increase as much as 15% to 30% in the north and west, with significant seasonal and regional contrasts. Summer precipitation may strongly decrease by as much as 70% in southern and central Europe, as well as by as much as 30% in western Europe. Daily precipitation is likely to become increasingly variable, and dry summer spells may increase by as much as 50% in southern Europe, France, and central Europe.

These changes in climate conditions are significant for squamates because heat and water availability are direct determinants of their whole-organism performances (e.g., Huey 1982, Nagy et al. 1991, Angilletta 2001). Thermal performance curves for locomotor or behavioral activities typically exhibit a bell shape: optimal performance values are observed within a mode or narrow range of temperatures, called the thermal mode (Huey 1982, Huey and Kingsolver 1989). Reptiles can control their body temperatures by means of behavioral tactics, and preferred body temperatures usually match the thermal mode. The thermal plasticity of life-history strategies suggests several potential effects of climate changes on the population demography of lizards and snakes. In thermally constrained environments (upland or northern habitats), warming should increase growth rates and result in earlier maturation because of direct, positive effects on body temperatures, because warmer environments enable longer activity periods, and because of potential indirect effects of warmer environments on prey abundance (Sinervo and Adolph 1989, Adolph and Porter 1993). Growth and reproduction are tightly linked since body size has a positive effect on fecundity (Clobert et al. 1998). Thus, warming may also have positive effects on reproductive effort. Also, thermal conditions are known to be important for the development of embryos in utero or ex utero. Embryonic development usually requires heat but temperature extremes may cause deaths or injury among embryos. In addition, abiotic conditions during embryonic development can influence the physiology, morphology, and behavior of neonate reptiles (Shine and Harlow 1993, Shine and Downes 1999, Massot et al. 2002, Brown and Shine 2005).

Although less often studied than thermal conditions, water availability may constrain reptiles' physiology even more severely through water balance (e.g. Nagy et al. 1991, Peterson 1996). Reptiles can control their evaporative water balance by reductions in water loss from activity, exposure to wind, and temperature control, or by increases in water gain from habitat and food selection, but water limitation during embryonic development may have lethal or deleterious effects. Variation in rainfall may induce further direct change in growth rates (Lorenzon et al. 1999), and it is also likely to have indirect effects through habitat quality and prey abundance (Marquis et al. 2008). In reality, temperature and humidity conditions may have interactive effects on reptiles' life history, resulting in a subtle balance between positive effects of warming and negative effects of drought.

Modeling Reptile Responses to Changing European Climates

As described above, future climate changes in Europe should directly affect the conditions for physiological processes and therefore the distribution of physiological niches. Physiological data can be used to forecast the dynamics and spatial range of each species in response to global warming (Grant and Porter 1992). A typical physiological niche model predicts mass-energy balance, water budget, and body growth and translates these predictions into demographic data in a specific area. These models have been used to map the niche of various physiological processes (Kearney and Porter 2004), to predict the spatial range and abundance of reptiles (Buckley 2008, Deutsch et al. 2008), or to test the effects of behavioral flexibility on responses to climate warming (Kearney et al. 2009).

In Europe, a physiological niche model based on thermal performance curves for sprint speed suggests positive effects of warming on lizards (Deutsch et al. 2008). Another model, which includes more complex assumptions, predicts further that species around the Mediterranean basin will be at greater risk of heat stress (Kearney et al. 2009). These predictions must, however, be interpreted with caution. Indeed, forecasts from Deutsch et al. (2008) depend on the questionable assumption that sprint speed is tightly linked to fitness. Also, the physiological model by Kearney et al. (2009) was built up to predict the physiology and behavior of a "standard" small ectotherm. In reality, species have physiological and behavioral adaptations to abiotic conditions, such as summer quiescence in some Mediterranean species (Arnold et al. 2002), or low-temperature digestion in cold-temperate vipers (Naulleau 1983). Future models should try to account explicitly for these differences, because they may represent adaptive buffers against the ecological consequences of warming (Grant and Porter 1992).

Climate niche models offer an alternative approach to forecasting a species' response to global warming. These models rely on distribution data and statistical tools to predict the range of a species. Araùjo et al. (2006) used this approach to forecast potential range shifts of European reptiles in response to global warming and predicted dramatic changes. Range contractions for 5% to 35% of species predominated in southern areas subject to important reductions in annual precipitation (the Iberian, Italian, and Anatolian peninsula). Many species (44% to 65%) should expand in cooler areas if they have unlimited dispersal capacity. However, the ranges of most species (97% to 100%) should contract if the species lack dispersal capacity. Species whose ranges are predicted to consistently retract are endemic lizards with restricted ranges (*Iberolacerta spp, Podarcis tiligertus*), species distributed in humid and cold habitats (the common lizard *Zootoca vivipara*, the adder *Vipera berus*, and the sand lizard *Lacerta agilis*), the Seoane's viper (*Vipera seoanei*) from temperate humid habitats in southwestern Europe, and the Schreibers' green lizard (*Lacerta schreiberi*) in Spain. Again, these predictions must be interpreted with caution, since we lack information on dispersal capacity (see below). Furthermore, these models ignore the ecological diversity of reptiles. For example, the northern or upper altitudinal ranges of several squamate species may be subject to clear thermal constraints, but the southern and lower altitudinal ranges of those same species and the distribution limits of Mediterranean species are much more complex and could involve constraints due not only to temperature but also to water limitations, interspecific interactions, and historical factors (e.g., Gasc et al. 1997).

Observed Ecological Effects of Climate Change

Niche models indicate that the effects of global warming may be substantial, but these predictions may suffer from not considering species-specific demographic and evolutionary mechanisms. Here, we review the links between climatic variation and phenology, life history, spatial dynamics, and evolutionary dynamics.

Phenological Changes

Changes in the annual timing of life-history events have been linked with climate warming in many species (Parmesan 2006) and are also expected for squamates (Saint Girons 1985, Adolph and Porter 1993, and references therein). Timing of the breeding cycles is controlled by external cues, such as light and temperature, and by internal cues, such as age, sex, and body condition (reviewed in Licht 1972). All else being

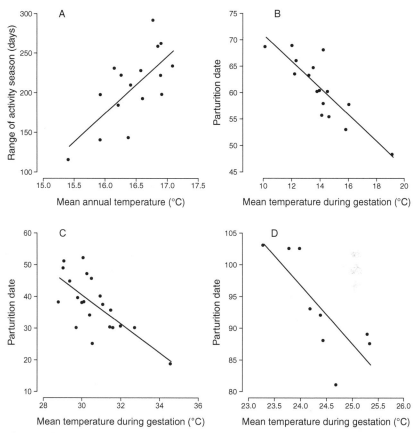

Figure 9.2. Phenological changes associated with climate warming in squamate reptiles from Europe. In response to local warming, increase in the length of the activity season was observed in (a) the Montpellier snake (after Moreno-Rueda et al. 2009), and advancement of breeding dates was found in (b) the common lizard (after Le Galliard et al. 2010), in (c) the meadow viper (after Baron and Le Galliard, unpub. data), and in (d) the asp viper (after Lourdais et al. 2002). Parturition date is calculated from June 1 in b, from August 1 in c, and from June 10 in d.

equal, warmer climates should result in longer activity periods and earlier breeding times. Some of the best evidence available so far comes from a study of the Montpellier snake (*Malpolon monspessulanus*), in which the annual activity period has increased in concert with climate warming (Moreno-Rueda et al. 2009; see figure 9.2A). Anecdotal observations also suggest that warmer climates are associated with earlier winter emergence in northern species, prolonged activity into the autumn and winter, and increased winter activity (Henle et al. 2008). In most species investigated so far, during the activity season the gestation length is shorter in a warmer climate (figure 9.2B to 9.2D). For example, a detailed investigation of thermal constraints on reproductive timing in the asp viper concluded that temperature was most critical in mid-gestation, and this explained up to half of the temporal variation in gestation length (Lourdais et al. 2004).

The match-mismatch hypothesis for consumer-resource systems predicts that the ecological consequences of global warming depend critically on the temporal concordance between the phenology of a species and the availability of the resources it needs (Post et al. 2001, Stenseth and Mysterud 2002). Unfortunately, it remains unclear whether this hypothesis provides a useful scenario for ectothermic consumers such as reptiles. We suspect that major differences in demographic response to phenological changes may exist between insectivorous species (most lizards and some snakes) and other carnivorous species (most snakes), as well as between short- and long-lived species. Insectivorous species prey on a highly sensitive resource that responds quickly to climate variation, while the availability of prey for carnivorous species (e.g., mammals, lizards, or amphibians) may be less sensitive to climate. Also, earlier breeding may be beneficial in short-lived, fast-growing species because it creates an extended growth period for the offspring prior to wintering, or because offspring born earlier suffer less from competition with their congeners (Olsson and Shine 1997, Sinervo 1999). In long-lived, slow-growing species, the fitness effects of an earlier breeding may be weaker or even potentially negative (Baron et al. 2010a).

Sex Ratio at Birth

Temperature-dependent sex determination (TSD) implies that changes in incubation temperatures could lead to imbalanced sex ratios at birth (Janzen 1994). However, TSD systems based on heteromorphic sex chromosomes are absent in snakes, and reports of TSD in lizards from Europe are equivocal (Viets et al. 1994, Ciofi and Swingland 1997, Andrews 2005, Janzen and Phillips 2006). More work is needed to clarify the phylogenetic distribution of TSD systems in squamates, but we suspect that climate-induced change in sex ratio at birth is unlikely to be a major threat to these species in Europe (Baron et al. 2010b).

Growth, Survival, and Reproduction

Results on life-history sensitivity to climatic conditions in European reptiles come primarily from long-term studies of the common lizard (*Zootoca vivipara*) and the meadow viper (*Vipera ursinii ursinii*) in France, as well from two field studies of the asp viper (*Vipera aspis*). Common lizard populations have been monitored in mountains from central France (ca. 1400 m above sea level) since the mid-1980s. This species is very widespread and is not considered at risk in Europe, though its populations are located in endangered habitats such as peat bogs and wetlands (Clobert et al. 1994). The asp viper has been monitored in a high-density population located around sea level in western France (Lourdais et al. 2004), as well as in marginal low-density populations in Switzerland (ca. 400–1,100 m above sea level) where the species is critically endangered (Altwegg et al. 2005). The meadow viper has been monitored intensively in one study population located in southeast France (ca. 1400 m above sea level) at the most western part of its distribution (Baron et al. 1996). This snake is listed as vulnerable across Europe and as critically endangered in France.

In general, climate warming has positive effects on most life-history traits (see table 9.2). In the common lizard this includes a clear trend towards larger body sizes and litters (figure 9.3A; Chamaillé-Jammes et al. 2006), faster growth at the yearling stage (Le Galliard et al. 2010), weak positive effects on adult survival (Chamaillé-Jammes et al. 2006), and strong positive effects on maternal body condition after parturition (Le Galliard et al. 2010). The increased body size and fecundity at the population level come primarily from increased growth opportunities due to earlier birth dates (see figure 9.2B) and longer activity seasons (Le Galliard et al. 2010). In vipers, warmer environments are associated with faster growth during nonreproductive years (Forsman 1993, Lourdais et al. 2002), increased maternal condition after parturition (Baron and Le Galliard, unpub. data), and fewer late embryonic failures (Lourdais et al. 2004, Baron and Le Galliard, unpub. data). However, temperatures negatively affect the mass gain during reproduction in the asp viper, thus suggesting metabolic costs of gestation at high temperatures in this species (Lourdais et al. 2002, Ladyman et al. 2003). In general, thermal effects are important during gestation and the juvenile stage (table 9.2). However, Altwegg et al. (2005) also discovered negative effects of winter freezing on juvenile survival, which points to the importance of thermal conditions during the inactivity season (figure 9.3B).

Limitation of growth and reproduction by availability of water has been quantified in the common lizard, in which complex patterns of positive and negative effects have been revealed (Marquis et al. 2008, Le Galliard et al. 2010; also see table 9.2 and figure 9.3C). This complexity could be explained by a combination of positive, indirect effects of rainfall on the summer availability of food and direct, negative effects of rainfall on insolation and therefore on growth opportunities (Marquis et al. 2008, Le

Table 9.2. Summary of observed climate effects on life-history traits (growth, survival, and reproduction) in natural populations of squamate reptiles subject to long-term demographic studies in Europe. Source references are provided in the main text.

Climate factor	Season	Study species	Life stage	Demographic trait	Climate effect
Temperature	Summer	Vipera aspis	Adult, gestation	Litter success	Positive
			Adult, gestation	Offspring size at birth	Positive
			Adult, gestation	Maternal condition	Negative
			Mean annual conditions	Nonreproductive adult body growth	Positive
Temperature	Summer	Vipera ursinii	Adult, gestation	Litter success	Positive
			Adult, gestation	Maternal condition	Positive
Temperature and cloud cover	Summer	Vipera berus	Mean annual conditions	Adult growth	Positive
Number of freezing days	Winter	Vipera aspis	Juvenile	Survival	Negative
Temperature	Spring	Zootoca vivipara	Adult, emergence	Survival	Positive
Temperature	Summer	Zootoca vivipara	Mean annual conditions	Yearling and adult size	Positive
			Mean annual conditions	Fecundity	Positive
			Juvenile	Yearling body growth	Positive
			Adult, gestation	Maternal condition	Positive
Rainfall	Summer	Zootoca vivipara	Adult, gestation	Litter success	Negative
			Adult, gestation	Offspring growth and survival	Positive
			Adult, gestation	Fecundity at adulthood	Negative
			Adult, gestation	Size of F1 offspring at adulthood	Positive
			Juvenile	Juvenile growth	Negative
			Juvenile	Juvenile survival	Positive

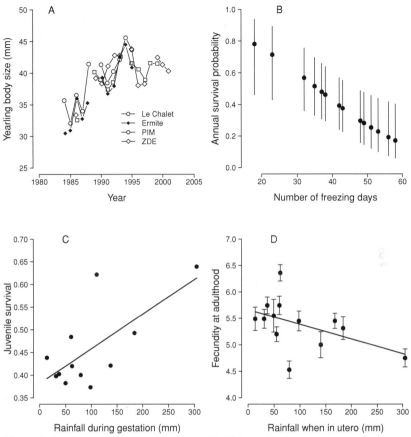

Figure 9.3. Effects of climatic conditions on life-history traits in squamate reptiles from Europe. In the common lizard, long-term field studies have shown positive effects of global warming on (*a*) body size (after Chamaillé-Jammes et al. 2006), positive effects of rainfall on (*c*) juvenile survival (after Le Galliard et al. 2010), and negative delayed effects of rainfall on (*d*) female fecundity (after Marquis et al. 2008). Juvenile survival shrinks during the coldest winters in (*b*) the asp viper (after Altwegg et al. 2005).

Galliard et al. 2010). Altogether, these studies suggest that the total demographic effects of climate change will depend on a balance between positive thermal effects and the potentially negative effects of water limitation. For example, in the common lizard, warming has little effect on juvenile survival and positive effects on size and fecundity, but drier environments are associated with a lower juvenile survival (figure 9.3D). A matrix population model that includes temporal variation in all demographic traits shows that this species is most sensitive to changes in juvenile survival, and that the observed variation in juvenile survival contributes most to the variation in population growth (Le Galliard et al. 2010). Thus, even though positive fitness effects of local warming have been found (Chamaillé-Jammes et al. 2006), a trend towards lower juvenile survival in drier environments could contribute to population decrease in this species. In sharp contrast, the population dynamics of vipers are most sensitive to variation in adult survival, and then to variations in juvenile survival and reproduction (Altwegg et al. 2005). Thus, the positive effects of warming on reproduction are unlikely to cause major changes in the dynamics of viper populations, while warming during the winter may be beneficial for the persistence of endangered populations of asp vipers from Switzerland.

Range Shifts and Dispersal Behavior

Recent latitudinal or elevational expansions have been observed in mobile animals and in plants (Parmesan 2006). Range expansion in European squamates has been described for two southern species, *Vipera aspis* and *Hierophis viridiflavus*, that have shifted 60 km north in 40 years (Naulleau 2003). Such limited evidence in reptiles may be due to their intermediate mobility, a lack of data, or confounding effects of habitat loss and shifting land use (Hickling et al. 2006). Many reptiles are also predicted to retreat from their southern or lowland margins (Sinervo et al. 2010). For example, the adder (*Vipera berus*) has retreated from southern habitats in some parts of France (Naulleau 2003). Field studies on numerous populations across the distribution area of the common lizard illustrate another case of range retraction. Monitoring the southern margin of the species range has shown that several lowland populations went extinct in 10 years' time, or that their density was reduced by more than 50% after a warm spell (Sinervo, Massot. and Clobert, unpub. data).

The ability of a species to expand its range or to persist after range contraction is critically dependent on its dispersal potential, which is a key parameter in the processes of colonization and invasion (Thomas et al. 2001, Clobert et al. 2004, Parmesan 2006). Unfortunately, relatively little is known about dispersal in reptiles. For example, Bowne and Bowers (2004) only found three suitable studies from two reptiles in their review of interpatch movements. Also, rates of colonization are poorly known, with the best data coming from two European species introduced in North America. The Mediterranean gecko *Hemidactylus turcicus*, introduced some decades ago in Texas, increased its range at a rate of about 20m per year (Locry and Stone 2008). The wall lizard *Podarcis mura-*

lis, introduced in Canada, expanded its range more rapidly (400 m per year, after Bertram 2004). These examples illustrate colonization ability in introduced populations, but say little about responses in dispersal behavior to climate change in the native range. Future studies should focus on understanding the effects of global warming on dispersal, and on distinguishing processes at work at the southern and northern margins of a species' range. At the northern margins, new habitats become available, and the major process driving range shift is colonization. It is now recognized that colonizers are not a random sample of individuals (Clobert et al. 2009). For example, in the common lizard, individuals that disperse to avoid mother-offspring competition are asocial, are prepared to take important risks, and prefer empty habitats where they achieve higher fitness (Le Galliard et al. 2003, Cote and Clobert 2007, Cote et al. 2007, Cote and Clobert 2010). Thus, avoidance of kin competition may be the main process by which colonization is achieved at the northern margin (Clobert et al. 2009). In contrast, processes at work at the southern margins are characterized mainly by the loss of suitable habitats (Thomas et al. 2004, Massot et al. 2008). As suitable habitats deteriorate, the distance between hospitable patches increases, and therefore the costs of dispersal should increase. Moreover, in a southern mountainous population of the common lizard, ongoing climate warming clearly inhibits dispersal (Massot et al. 2008, Lepetz et al. 2009a; also see figure 9.4). This plastic response of dispersal should further increase the speed of extinction at the southern margin by decreasing colonization and rescue effects.

A species that cannot track environmental changes in space could still adapt to new environmental conditions. Adaptation may involve phenotypic plasticity or evolutionary changes driven by selection. The dominant view is that life-history adaptations to climate change in animals are mainly caused by phenotypic plasticity (Parmesan 2006, Charmantier et al. 2008). Unfortunately, there are very few examples of adaptive phenotypic plasticity in response to global warming in squamates (Telemeco et al. 2009). Selection on preexisting genetic polymorphism has also been implicated in birds and mammals (Réale et al. 2003, Nussey et al. 2005). In squamates, substantial genetic variation exists for traits ranging from morphology to physiology and life history (reviewed in Shine 2005), but no study has yet documented evolutionary responses to warming. Color morphs in lizards can be determined by genetic and environmental cues (Sinervo et al. 2001, Cote et al. 2008), and may be good markers for studying the interplay between genetic and plastic responses to global warming. In Europe, Lepetz et al. (2009a) reported on

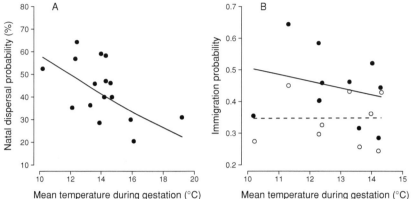

Figure 9.4. Dispersal inhibition in response to climate warming in a natural population of the common lizard (*Zootoca vivipara*). Natal dispersal inside a natural population (movement probability out of the natal home range) and immigration into this natural population (proportion of immigrants) decreased concurrently with local warming. Data are (*a*) dispersal frequencies from Massot et al. (2008), and (*b*) immigration frequencies for linear morphs (filled circles) and reticulated morphs (empty circles) from Lepetz et al. (2009a).

a correlation between mean temperature and the population frequency of dorsal, melanin-based coloration in the common lizard. As expected, it was observed that common lizards with less melanin on their backs were more numerous in more recent and warmer years. However, this change was due not to phenotypic plasticity or to differential selection, but to a temporal trend towards a lower immigration rate of lizards with more melanin. This result provides the first evidence of the impact of climate warming on population composition due to migration.

Interspecific Interactions

The effects of climate change may involve interspecific interactions (Walther et al. 2002). For example, studies of squamates in deserts from North America indicate that rainfall determines the abundance of plants, insects, and small mammals that are prey for a whole community of lizards and snakes (e.g., Tinkle et al. 1993). In Europe, similar trophic interactions have yet to be quantified but are expected in water-limited environments and for squamates specialized for fluctuating resources (e.g., the asp viper; Lourdais et al. 2002). In addition, warming may lead to community reassembly and may promote coexistence with novel competitors, predators, or invasive species. For example, ophiophagous snakes have started to colonize habitats occupied by meadow vipers (Baron, personal observation) and the green lizard has been recently seen in high-altitude habitats occupied by the common lizard (Clobert, personal observation). However, the link between these extinction factors and climate change remains to be more thoroughly examined.

Promoting Climate Change Resilience in European Reptiles
Synthesis and Knowledge Gaps

Our review shows clearly that some climate change effects are already occurring. Some of the observed demographic effects are positive (enhanced growth and fecundity, earlier maturation and higher reproductive success, enhanced survival) but others are negative (lower survival, increased water and heat stress, reduced dispersal). Demographic growth or declines caused by climate change are not yet apparent, except in data collected in southern populations of the common lizard. On the other hand, niche models predict that future projected changes in climate should result in substantial range shifts for most reptiles in Europe. Several reptiles may benefit from climate warming if they can expand their range northward or upward, but cool-adapted, water-limited endemic species around the Mediterranean basin are predicted to decline. The relative effects of temperature and water availability are difficult to tease apart in most studies, and it is premature to compare the results of field studies with those of niche models.

Our ability to understand the ecological consequences of global warming is still limited by gaps in our knowledge of European squamates. Four major shortcomings can be recognized. First, current evidence for the impact of climate change comes only from three model species out of about 130. Demographic data are surprisingly rare for the most threatened species that attract considerable attention and funding (Cox and Temple 2009). Second, even the best studies are focused on only a few target populations, rather than across the entire geographic range, with the notable exception of data collected along the southern limit of the common lizard's range (Sinervo, Massot and Clobert, unpub. data). Third, even the best available studies do not allow us to distinguish between the direct and indirect effects of climatic conditions. Fourth, basic information on dispersal is lacking for most species. Only the common lizard has been investigated in sufficient detail for us to make predictions concerning spatial population dynamics (Massot et al. 2008). This critical gap is not specific to squamates, since we lack information on the dispersal potential of other animals (Kokko and Lopez-Sepulcre 2006).

Implications for Conservation

Environmental agencies and stakeholders point out the need to conserve European biodiversity, including squamates, in the face of climate change (see table 9.1), but no clear recommendation has been made for the management of squamates potentially threatened by climate change. Extinction risk from climate change has been investigated across a wide range of species on the basis of climate niche models (Araùjo et al. 2006), and in the common lizard on the basis of field census (Sinervo, Massot, and Clobert, unpub. data). These studies suggest that several species will be threatened in the future, especially common and cold-adapted species at their southern margins, water-limited species, and range-restricted endemics of the Mediterranean basin. Yet it remains difficult to quantify the risks posed by global warming relative to other extinction threats. Monitoring programs aimed at tracking the spatial and temporal changes of reptiles are rare (see eumon.ckff.si), and funding such programs should be considered a research priority (Lengyel et al. 2008, Lepetz et al. 2009b). Collaboration and consensus over monitoring schemes and monitored species will be needed to achieve these aims. Compared to birds and mammals, many species of reptiles are easily amenable to experiments at the individual or population level. Experiments on climate change effects should also be conducted to gain insights into mechanisms, causal pathways involved, and nonlinear responses to warmer climates expected in the future. These new data should stimulate the development of predictive models, such as population viability analyses, that would help in predicting the future of reptile populations in Europe.

Firm measures to promote climate change resilience in European reptiles will also be needed (Brodie et al., this volume; Popescu and Hunter, this volume). First, we must promote action against existing threats to improve conditions so that species can respond to future climate changes. Squamates are highly sensitive to their landscape, and securing the size, quality, and connectivity of their natural habitat should be treated as a top conservation priority (Cox and Temple 2009, Silva et al. 2009). In the most common species potentially threatened by climate change (e.g., widespread, cool-adapted species), the preservation of natural habitat should help maintain large populations within a connected network of habitat patches, which in turn will facilitate ecological and evolutionary responses to global warming. As a general rule, we recommend three major conservation priorities for these species: (1) restraints on existing protected areas to secure the size and evolutionary potential of existing populations, even if they are not currently threatened; (2) definition and protection of suitable habitats in the future, such as those at the northern margin and in higher altitudinal ranges; and (3) definition and protection of denser networks of protected areas, including habitat corridors, to enable dispersal movement.

In addition, species at risk of extinction from climate change (e.g., island endemics or range-restricted species) and species that are already threatened (see table 9.1) should be the focus of further studies to test on-the-ground management strategies to mitigate the effects of climate in local populations. For example, Kearney et al. (2009) have noted that the availability of shade and moisture will be crucial for the persistence of small lizards in semiarid regions. In these areas, conservation action to promote shade and moisture and facilitate movement may include changes in land use practices and effective restoration of natural corridors (i.e., maintenance of mixed structures in open and shaded areas). Similarly, information on the reproductive biology and its sensitivity to climate conditions can be used for field management (Shine and Bonnet 2009). For snakes, ponds and artificial shelters will probably prove useful to create sites for resting, basking, and nesting and to mitigate some impacts of climate change. Hence, knowledge of the natural history of squamates will provide "best of the bad job" management options. Because climate change effects on ectothermic vertebrates are unavoidable, and because these animals' dispersal capacity may be limited, proactive measures to assist their movement might be needed. Unfortunately, we lack information on the success of ex situ conservation and assisted dispersal (Silva et al. 2009). Furthermore, assisted migration of a given species may interfere with evolutionary responses or unassisted range shifts in other species. More studies are needed to evaluate the feasibility and potential negative impacts of assisted colonization strategies in European reptiles before we decide to implement such measures.

ACKNOWLEDGMENTS
We thank Jedediah F. Brodie, Daniel F. Doak, Eric S. Post, and two anonymous reviewers for comments on an earlier version of this chapter. We especially thank the editors for their patience and hard work. Gregorio-Moreno Rueda, Res Altwegg, and Chris Thomas kindly shared published data and Barry Sinervo provided us with unpublished results. This research program is supported by the Centre National de la Recherche Scientifique (CNRS). Work on this book chapter was facilitated by an Agence Nationale de la Recherche grant (07-JCJC-0120) and a Région Ile-de-France grant R2DS (2007–06) to Jean-François Le Galliard, as well as by an Agence Nationale de la Recherche grant DIAME and project BIODIVERSITAS Tenlamas to Jean Clobert.

LITERATURE CITED

Adolph, S. C., and W. P. Porter. 1993. Temperature, activity, and lizard life-histories. *American Naturalist* 142:273–95.

Alcamo, J., J. M. Moreno, B. Nováky, M. Bindi, R. Corobov, R. J. N. Devoy, C. Giannakopoulos, E. Martin, J. E. Olesen, and A. Shvidenko. 2007. Europe. Climate change 2007: Impacts, adaptation and vulnerability. In *Contribution of Working Group II to the Fourth Assessment Report of the Intergovernmental Panel on Climate Change*, edited by M. L. Parry, O. F. Canziani, J. P. Palutikof, P. J. van der Linden, and C. E. Hanson, 541–80. Cambridge: Cambridge University Press.

Altwegg, R., S. Dummermuth, B. R. Anholt, and T. Flatt. 2005. Winter weather affects asp viper *Vipera aspis* population dynamics through susceptible juveniles. *Oikos* 110:55–66.

Andrews, R. M. 2005. Incubation temperature and sex ratio of the veiled chameleon (*Chamaeleo calyptratus*). *Journal of Herpetology* 39:515–18.

Angilletta, M. J. J. 2001. Thermal and physiological constraints on energy assimilation in a widespread lizard (*Sceloporus undulatus*). *Ecology* 82:3044–56.

Araùjo, M. B., W. Thuiller, and R. G. Pearson. 2006. Climate warming and the decline of amphibians and reptiles in Europe. *Journal of Biogeography* 33:1712–28.

Arnold, E. N., O. Arribas, and S. Carranza. 2007. Systematics of the Palaearctic and Oriental lizard tribe Lacertini (Squamata : Lacertidae : Lacertinae), with descriptions of eight new genera. *Zootaxa* 1430: 1–86.

Arnold, E. N., J. A. Burton, and D. W. Ovenden. 2002. *Field Guide to Reptiles and Amphibians of Britain and Europe*. London: Harper Collins.

Baron, J. P., R. Ferrière, J. Clobert, and H. Saint Girons. 1996. Stratégie démographique de *Vipera ursinii* au Mont-Ventoux. *Comptes Rendus de l'Académie des Sciences de Paris* 319: 57–69.

Baron, J.-P., J.-F. Le Galliard, T. Tully, and R. Ferrière. 2010a. Cohort variation in offspring growth and survival: prenatal and postnatal factors in a late-maturing viviparous snake. *Journal of Animal Ecology* 79:640–49.

Baron, J.-P., T. Tully, and J.-F. Le Galliard. 2010b. Sex-specific fitness returns are too weak to select for nonrandom patterns of sex allocation in a viviparous snake. *Oecologia* 164:369–78.

Bertram, N. A. 2004. Ecology of the introduced European wall lizard, *Podarcis muralis*, near Victoria, British Columbia. MS thesis, University of Victoria.

Bowne, D. R., and M. A. Bowers. 2004. Interpatch movements in spatially structured populations: a literature review. *Landscape Ecology* 19:1–20.

Brown, G. P., and R. Shine. 2005. Female phenotype, life history, and reproductive success in free-ranging snakes (*Tropidonophis mairii*). *Ecology* 86:2763–70.

Buckley, L. B. 2008. Linking traits to energetics and population dynamics to predict lizard ranges in changing environments. *American Naturalist* 171:E1–19.

Chamaillé-Jammes, S., M. Massot, P. Aragon, and J. Clobert. 2006. Global warming and positive fitness response in mountain populations of common lizards *Lacerta vivipara*. *Global Change Biology* 12:392–402.

Charmantier, A., R. H. McCleery, L. R. Cole, C. Perrins, L. E. B. Kruuk, and B. C. Sheldon. 2008. Adaptive phenotypic plasticity in response to climate change in a wild bird population. *Science* 320:800–803.

Ciofi, C., and I. R. Swingland. 1997. *Environmental Sex Determination in Reptiles*. Pp. 251–65. Elsevier Science Bv.

Clobert, J., T. J. Garland, and R. Barbault. 1998. The evolution of demographic tactics in lizards. *Journal of Evolutionary Biology* 11:329–64.

Clobert, J., R. A. Ims, and F. Rousset. 2004. Causes, mechanisms and consequences of dispersal. In *Ecology, Genetics and Evolution of Metapopulations*, edited by I. Hanski and O. Gagiotti, 307–35. London: Elsevier, Academic Press.

Clobert, J., J.-F. Le Galliard, J. Cote, M. Massot, and S. Meylan. 2009. Informed dispersal, heterogeneity in animal dispersal syndromes and the dynamics of spatially structured populations. *Ecology Letters* 12:197–209.

Clobert, J., M. Massot, J. Lecomte, G. Sorci, M. de Fraipont, and R. Barbault. 1994. Determinants of dispersal behavior: The common lizard as a case study. In *Lizard Ecology. Historical and Experimental Perspectives*, edited by L. J. Vitt and E. R. Pianka, 183–206. Princeton, NJ: Princeton University Press.

Corbett, C. K. 1989. *Conservation of European Reptiles and Amphibians*. London: C. Helm.

Cote, J., and J. Clobert. 2007. Social personalities influence natal dispersal in a lizard. *Proceedings of the Royal Society of London B* 274:383–90.

———. 2010. Risky dispersal: Avoiding kin competition despite uncertainty. *Ecology* 91:1485–93.

Cote, J., J. Clobert, and P. S. Fitze. 2007. Mother-offspring competition promotes colonization success. *Proceedings of the National Academy of Sciences USA* 104:703–8.

Cote, J., J.-F. Le Galliard, J. M. Rossi, and P. S. Fitze. 2008. Environmentally induced changes in carotenoid-based coloration of female lizards: A comment on Vercken et al. *Journal of Evolutionary Biology* 21:1165–72.

Cox, N. A., and H. J. Temple. 2009. *European Red List of Reptiles*. Office for Official Publications of the European Communities, Luxembourg.

Deutsch, C. A., J. J. Tewksbury, R. B. Huey, K. S. Sheldon, C. K. Ghalambor, D. C. Haak, and P. R. Martin. 2008. Impacts of climate warming on terrestrial ectotherms across latitude. *Proceedings of the National Academy of Sciences of the United States of America* 105:6668–72.

Forsman, A. 1993. Growth rate in different colour morphs of the adder, *Vipera berus*, in relation to yearly weather variation. *Oikos* 66:279–85.

Gasc, J.-P., A. Cabela, J. Crnobrnja-Isailovic, D. Dolmen, K. Grossenbacher, P. Haffner, J. Lescure, H. Martens, J. P. Martínez Rica, H. Maurin, M. E. Oliveira, T. S. Sofianidou, M. Veith, and A. Zuiderwijk, eds. 1997. *Atlas of Amphibians and Reptiles in Europe*. Paris: National Natural History Museum of Paris.

Gibbons, J. W., D. E. Scott, T. J. Ryan, K. A. Buhlmann, T. D. Tuberville, B. S. Metts, J. L. Greene, T. Mills, Y. Leiden, S. Poppy, and C. T. Winne. 2000. The global decline of reptiles, Deja Vu amphibians. *Bioscience* 50:653–66.

Grant, B. W., and W. P. Porter. 1992. Modeling global macroclimatic constraints on ectotherm energy budgets. *American Zoologist* 32:154–78.

Hannah, L., T. E. Lovejoy, and S. H. Schneider. 2005. Biodiversity and climate change in context. In *Climate Change and Biodiversity*, edited by T. E. Lovejoy and L. Hannah, 3–14. New Haven: Yale University Press.

Henle, K., D. Dick, A. Harple, I. Kühn, O. Schweiger, and J. Settele. 2008. *Climate Change Impacts on European Amphibians and Reptiles*. Convention on the Conservation of European Wildlife and Natural Habitats, Strasbourg, France.

Hickling, R., D. B. Roy, J. K. Hill, R. Fox, and C. D. Thomas. 2006. The distributions of a wide range of taxonomic groups are expanding polewards. *Global Change Biology* 12:450–55.

Huey, R. B. 1982. Temperature, physiology, and the ecology of reptiles. In *Biology of the Reptilia*, edited by C. Gans and F. H. Pough, 25–91. New York: Academic Press.

Huey, R. B., and J. G. Kingsolver. 1989. Evolution of thermal sensitivity of ectotherm performance. *Trends in Ecology and Evolution* 4:131–35.

Huey, R. B., E. R. Pianka, and T. W. Schoener, eds. 1983. *Lizard Ecology: Studies of Model Organism*. Cambridge, MA: Harvard University Press.

IPCC (International Panel for Climate Change). 2007. *Climate Change 2007: Synthesis Report. Summary for Policymakers*. Geneva.

IUCN (International Union for the Conservation of Nature). 2008. *Red List of Threatened Species*.

Janzen, F. J. 1994. Climate-change and temperature-dependent sex determination in reptiles. *Proceedings of the National Academy of Sciences of the United States of America* 91:7487–90.

Janzen, F. J., and P. C. Phillips. 2006. Exploring the evolution of environmental sex determination, especially in reptiles. *Journal of Evolutionary Biology* 19:1775–84.

Kearney, M., and W. P. Porter. 2004. Mapping the fundamental niche: Physiology, climate, and the distribution of a nocturnal lizard. *Ecology* 85:3119–31.

Kearney, M., R. Shine, and W. P. Porter. 2009. The potential for behavioral thermoregulation to buffer "cold-blooded" animals against climate warming. *Proceedings of the National Academy of Sciences of the United States of America* 106:3835–40.

Kokko, H., and A. Lopez-Sepulcre. 2006. From individual dispersal to species range: perspectives for a changing world. *Science* 313:789–90.

Ladyman, M., X. Bonnet, O. Lourdais, D. Bradshaw, and G. Naulleau. 2003. Gestation, thermoregulation, and metabolism in a viviparous snake, *Vipera aspis*: Evidence for fecundity-independent costs. *Physiological and Biochemical Zoology* 76:497–510.

Le Galliard, J.-F., R. Ferrière, and J. Clobert. 2003. Mother-offspring interactions affect natal dispersal in a lizard. *Proceedings of the Royal Society of London B* 270:1163–69.

Le Galliard, J.-F., O. Marquis, and M. Massot. 2010. Cohort variation, climate effects and population dynamics in a short-lived lizard. *Journal of Animal Ecology*. 79:1296–1307.

Lengyel, S., A. Kobler, L. Kutnar, E. Framstad, P. Y. Henry, V. Babij, B. Gruber, D. Schmeller, and K. Henle. 2008. A review and a framework for the integration of biodiversity monitoring at the habitat level. *Biodiversity and Conservation* 17:3341–56.

Lepetz, V., M. Massot, A. Chaine, and J. Clobert. 2009a. Climate warming and the evolution of morphotypes in a reptile. *Global Change Biology* 15:454–66.

Lepetz, V., M. Massot, D. Schmeller, and J. Clobert. 2009b. Biodiversity monitoring: some proposals to adequately study species' responses to climate change. *Biodiversity and Conservation* 18:3185–3203.

Licht, P. 1972. Environmental physiology of reptilian breeding cycles: Role of temperature. *General and Comparative Endocrinology Supplement* 3:477–88.

Locry, K. J., and P. A. Stone. 2008. Ontogenetic factors affecting diffusion dispersal in the introduced Mediterranean gecko, *Hemidactylus turcicus*. *Journal of Herpetology* 42:593–99.

Lorenzon, P., J. Clobert, A. Oppliger, and H. B. John-Alder. 1999. Effect of water constraint on growth rate, activity and body temperature of yearling common lizard (*Lacerta vivipara*). *Oecologia* 118:423–30.

Lourdais, O., X. Bonnet, M. Guillon, and G. Naulleau. 2004. Climate affects embryonic development in a viviparous snake, *Vipera aspis*. *Oikos* 104:551–60.

Lourdais, O., X. Bonnet, R. Shine, D. Denardo, G. Naulleau, and M. Guillon. 2002. Capital-breeding and reproductive effort in a variable environment: a longitudinal study of a viviparous snake. *Journal of Animal Ecology* 71:470–79.

Marquis, O., M. Massot, and J.-F. Le Galliard. 2008. Intergenerational effects of climate generate cohort variation in lizard reproductive performance. *Ecology* 89:2575–83.

Massot, M., J. Clobert, and R. Ferrière. 2008. Climate warming, dispersal inhibition and extinction risk. *Global Change Biology* 14:1–9.

Massot, M., J. Clobert, P. Lorenzon, and J.-M. Rossi. 2002. Condition-dependent dispersal and ontogeny of the dispersal behaviour: an experimental approach. *Journal of Animal Ecology* 71:253–61.

Moreno-Rueda, G., J. M. Pleguezuelos, and E. Alaminos. 2009. Climate warming and activity period extension in the Mediterranean snake *Malpolon monspessulanus*. *Climatic Change* 92:235–42.

Morris, W. F., and D. F. Doak. 2002. *Quantitative Conservation Biology*. Sunderland, MA: Sinauer Associates.

Nagy, K. A., B. C. Clarke, M. K. Seely, D. Mitchell, and J. R. B. Lighton. 1991. Water and energy balance in Namibian desert sand-dune lizards *Angolosorus skoogi* (Andersson, 1916). *Functional Ecology* 5:731–39.

Naulleau, G. 1983. The effects of temperature on digestion in 5 species of European vipers of genus *Vipera*. *Bulletin de la Societe Zoologique de France: Evolution et Zoologie* 108:47–58.

Naulleau, G. 2003. Evolution de l'aire de répartition en France, en particulier au Centre Ouest, chez trois serpents: Extension vers le nord (la couleuvre verte et jaune, *Coluber viridiflavus* Lacépède et la Vipère aspic, *Vipera aspis* Linné) et régression vers le nord (la Vipère péliade, *Vipera berus* Linné). *Biogeographica* 79:59–69.

Nussey, D. H., E. Postma, P. Gienapp, and M. E. Visser. 2005. Selection on heritable phenotypic plasticity in a wild bird population. *Science* 310:304–6.

Olsson, M., and R. Shine. 1997. The seasonal timing of oviposition in sand lizards (*Lacerta agilis*) : Why early clutches are better? *Journal of Evolutionary Biology* 10:369–81.

Parmesan, C. 2006. Ecological and evolutionary responses to recent climate change. *Annual Review of Ecology and Systematics* 37:637–69.

Peterson, C. C. 1996. Ecological energetics of the desert tortoise (*Gopherus agassizii*): Effects of rainfall and drought. *Ecology* 77:1831–44.

Pianka, E. R., and L. J. Vitt. 2003. *Lizards: Windows to the Evolution of Diversity*. Berekeley: University of California Press.

Pinho, C., D. J. Harris, and N. Ferrand. 2007. Contrasting patterns of population subdivision and historical demography in three western Mediterranean lizard species inferred from mitochondrial DNA variation. *Molecular Ecology* 16: 1191–1205.

Post, E., M. C. Forchhammer, N. C. Stenseth, and T. V. Callaghan. 2001. The timing of life-history events in a changing climate. *Proceedings of the Royal Society of London Series B* 268:15–23.

Pough, F. H., R. M. Andrews, J. E. Cadle, M. L. Crump, A. H. Savitsky, and K. D. Wells. 2001. *Herpetology*. Upper Saddle River, NJ: Prentice-Hall.

Pounds, J. A., M. P. L. Fogden, and J. H. Campbell. 1999. Biological response to climate change on a tropical mountain. *Nature* 398:611–15.

Réale, D., A. G. McAdam, S. Boutin, and D. Berteaux. 2003. Genetic and plastic responses of a northern mammal to climate change. *Proceedings of the Royal Society of London B* 270:591–96.

Rosenzweig, C., G. Casassa, D. J. Karoly, A. Imeson, C. Liu, A. Menzel, S. Rawlins, T. L. Root, B. Seguin, and P. Tryjanowski. 2007. Assessment of observed changes and responses in natural and managed systems. In *Climate Change 2007: Impacts, Adaptation and Vulnerability. Contribution of Working Group II to the Fourth Assessment Report of the Intergovernmental Panel on Climate Change*, edited by M. L. Parry, O. F. Canziani, J. P. Palutikof, P. J. van der Linden, and C. E. Hanson, 79–131. Cambridge: Cambridge University Press.

Saint Girons, H. 1985. Influence des facteurs de l'environnement sur les cycles annuels et reproducteurs des reptiles. *Bulletin de la Société Zoologique de France* 110:307–19.

Shine, R. 2005. Life-history evolution in reptiles. *Annual Review of Ecology and Systematics* 36:23–46.

Shine, R., and X. Bonnet. 2009. Reproductive biology, population viability, and options for field management. In *Snakes: Applied Ecology and Conservation*, edited by S. J. Mullin and R. A. Seigel, 172–200. Ithaca, NY: Cornell University Press.

Shine, R., and S. J. Downes. 1999. Can pregnant lizards adjust their offspring phenotypes to environmental conditions? *Oecologia* 119:1–8.

Shine, R., and P. Harlow. 1993. Maternal thermoregulation influences offspring viability in a viviparous lizard. *Oecologia* 96:122–27.

Silva, J. P., J. Toland, W. Jones, J. Eldridge, T. Hudson, and E. O'Hara. 2009. *LIFE and Europe's Reptiles and Amphibians*. Luxembourg: European Commission.

Sinervo, B. 1999. Mechanistic analysis of natural selection and a refinement of Lack's and William's principles. *American Naturalist* 154:S26–42.

Sinervo, B., and S. C. Adolph. 1989. Thermal sensitivity of growth rate in hatchling *Sceloporus* lizards: Environmental, behavioral and genetic aspects. *Oecologia* 78:411–19.

Sinervo, B., C. Bleay, and C. Adamopoulou. 2001. Social causes of correlational selection and the resolution of a heritable throat color polymorphism in a lizard. *Evolution* 55:2040–52.

Sinervo, B., F. Mendez-de-la-Cruz, D. B. Miles, B. Heulin, E. Bastiaans, M. V. S. Cruz, R. Lara-Resendiz, N. Martinez-Mendez, M. L. Calderon-Espinosa, R. N. Meza-Lazaro, H. Gadsden, L. J. Avila, M. Morando, I. J. De la Riva, P. V. Sepulveda, C. F. D. Rocha, N. Ibarguengoytia, C. A. Puntriano, M. Massot, V. Lepetz, T. A. Oksanen, D. G. Chapple, A. M. Bauer, W. R. Branch, J. Clobert, and J. W. Sites. 2010. Erosion of lizard diversity by climate change and altered thermal niches. *Science* 328:894–99.

Stenseth, N. C., and A. Mysterud. 2002. Climate, changing phenology, and other life history traits: Non-linearity and match-mismatch to the environment. *Proceedings of the National Academy of Sciences of the United States of America* 99:13379–81.

Surget-Groba, Y., B. Heulin, C.-P. Guillaume, R. S. Thorpe, L. Kupriyanova, N. Vogrin, R. Maslak, S. Mazzotti, M. Venczel, I. Ghira, G. Odierna, O. Leontyeva, J. C. Monney, and N. Smith. 2001. Intraspecific phylogeography of *Lacerta vivipara* and the evolution of viviparity. *Molecular Phylogenetics and Evolution* 18:449–59.

Telemeco, R. S., M. J. Elphick, and R. Shine. 2009. Nesting lizards (*Bassiana duperreyi*) compensate partly, but not completely, for climate change. *Ecology* 90:17–22.

Thomas, C. D., E. J. Bodsworth, R. J. Wilson, A. D. Simmons, Z. G. Davies, M. Musche, and L. Conradt. 2001. Ecological and evolutionary processes at expanding range margins. *Nature* 411:577–81.

Thomas, C. D., A. Cameron, R. E. Green, M. Bakkene, L. J. Beaumont, Y. C. Collingham, B. F. N. Erasmus, M. Ferreira de Siqueira, A. Grainger, L. Hannah, L. Hughes, B. Huntley, A. S. van Jaarsveld, G. F. Midley, L. Miles, M. A. Ortega-Huerta, A. Towsend Petersen, O. L. Phillips, and S. E. Williams. 2004. Extinction risks from climate change. *Nature* 427:145–48.

Tinkle, D. W., and R. E. Ballinger. 1972. *Sceloporus undulatus*: A study of the intraspecific comparative demography of a lizard. *Ecology* 53:570–84.

Tinkle, D. W., A. E. Dunham, and J. D. Congdon. 1993. Life history and demographic variation in the lizard *Sceloporus graciosus*: a long-term study. *Ecology* 74:2413–29.

Ursenbacher, S., M. Carlsson, V. Helfer, H. Tegelstrom, and L. Fumagalli. 2006. Phylogeography and Pleistocene refugia of the adder (*Vipera berus*) as inferred from mitochondrial DNA sequence data. *Molecular Ecology* 15:3425–37.

Ursenbacher, S., S. Schweiger, L. Tomovic, J. Crnobrnja-Isailovic, L. Fumagalli, and W. Mayer. 2008. Molecular phylogeography of the nose-horned viper (*Vipera ammodytes*, Linnaeus [1758]): Evidence for high genetic diversity and multiple refugia in the Balkan peninsula. *Molecular Phylogenetics and Evolution* 46:1116–28.

Viets, B. E., M. A. Ewert, L. G. Talent, and C. E. Nelson. 1994. Sex-determining mechanisms in squamate reptiles. *Journal of Experimental Zoology* 270:45–56.

Walther, G.-R., E. Post, P. Convey, A. Menzel, C. Parmesan, T. J. C. Beebee, J.-M. Fromentin, O. Hoegh-Guldberg, and F. Bairlein. 2002. Ecological responses to recent climate change. *Nature* 416:389–95.

Arctic Shorebirds:
Conservation of a Moving Target
in Changing Times

Steve Zack and Joe Liebezeit

Shorebirds are among the most long-distant migratory wildlife in the world, and for the majority of migrant species in the families Charadriidae (plovers) and Scolopacidae (shorebirds or waders; order Charadriiformes) the Arctic is the destination breeding ground. Unlike that of most other avian taxa, shorebird diversity increases poleward. At least 50 species of shorebirds breed across the circumpolar Arctic (Piersma and Wiersma 1996) and are the most diverse and most abundant wildlife in the region (see figure 10.1).

In the Arctic, climate change has occurred at twice the average global rate over the past 100 years (IPCC 2007). Changes to arctic land- and seascapes are occurring at an alarming pace, and climate models project a very different and transformed Arctic in the near future (ACIA 2004, Anisimov et al. 2007, Lenton et al. 2008). Transformative processes now underway include recession of arctic sea ice and increasing dominance of open water, melting of permafrost triggering numerous changes including widespread thermokarst (changes in surface topology due to melting permafrost), shoreline erosion, and encroachment of woody vegetation into sedge-tundra–dominated systems (Sturm et al. 2005, Smol et al. 2005, Tape et al. 2006). Although it seems inevitable that there will be dramatic changes for wildlife in the changing Arctic, there is little consensus on what those changes might be because of the current paucity of information on real and potential climate change impacts, and because of ignorance about how such species might respond to projected changes. Because of this, conservation planning for wildlife with respect to climate change in the region is still in its infancy.

Yet it is clear that the Arctic is changing more quickly due to climate change than any other region on earth, with changes so dramatic that it is expected to undergo a transformation into a different ecosystem altogether. The changing Arctic means new challenges to the conservation of migratory shorebirds that already face threats elsewhere in their migratory world. Because they are highly vagile, shorebirds depend on disparate habitats across their breeding, stopover, and wintering grounds. Thus,

Buff-breasted Sandpiper, a uniparental breeder

Whimbrel may benefit from a drier Arctic

Red-necked Phalarope, vulnerable to predators in oil fields

Dunlin migrate through Asia, including Saemangeum

American Golden Plover is a species of Special Concern

Long-billed Dowitcher breed in wet tundra habitat

Figure 10.1. Shorebirds in arctic Alaska facing climate change. Captions refer to issues discussed in the text. All images by Steve Zack, Wildlife Conservation Society.

changes in seasonal patterns of food availability and timing of critical life-cycle phenomena for shorebirds (e.g., insect emergence and hatching of young) could be decoupled by forces associated with the changing climate (Coppack and Both 2002, Tulp and Schekkerman 2008) and could negatively impact populations. Shorebirds also depend disproportionately on littoral habitats, and thus are susceptible to potential changes in sea level rise brought on by climate change. Shorebirds are also particularly vulnerable, because they have less genetic variation than other bird groups (Baker and Strauch 1988, Wennerberg et al. 2002), perhaps due to population bottlenecks caused by previous climate changes. Indeed, shorebirds may be considered sentinels of the effects of climate change (Piersma and Lindstrom 2004).

The End of Remote: Development in the Arctic and Its Effects on Shorebirds

Arctic Alaska (figure 10.2) is the only arctic landscape of the United States. To date, the only region with large-scale development is in the state-owned land encompassing Prudhoe Bay. The Prudhoe Bay Oilfield, the adjacent Kuparuk Oilfield, and smaller satellite oilfields make up the largest oil development complex in North America (NAS 2003), covering an area as large as the state of Rhode Island (1,545 mi^2). Although in the late 1980s expansion of oil development slowed down, since the late 1990s there has been renewed interest in tapping the vast National Petroleum Reserve–Alaska (NPR-A). At 23.5 million acres, nearly the size of the state of Indiana, the NPR-A is the largest piece of public land in the United States, encompassing virtually all of western arctic Alaska. From 1998 to 2010, the Bureau of Land Management (BLM), administrators of the NPR-A, sold leases for oil exploration on roughly three million acres. To the east of Prudhoe Bay there is continuing interest in expanding oil development into the 1002 area of the Arctic National Wildlife Refuge, the coastal plain region. In addition, offshore extraction of oil is planned for both the Beaufort and Chukchi Seas, with exploration activities currently underway. All new oil development in arctic Alaska must inevitably connect to the existing and developing pipeline system that feeds into the Trans-Alaska Pipeline System, running from Prudhoe Bay through the heart of Alaska to the port of Valdez in the southeast.

In addition to the expansion of oil development, there are plans to extract natural gas from areas within the current oilfields and also in other parts of arctic Alaska, most notably in the Brooks Range foothills. Gas extraction will require completely new infrastructure, and thus will substantially increase human activity in the region. Vast coal resources are believed to lie in the western Arctic under the Utukok Uplands (BLM/MMS 1998). Although there are no immediate plans for coal extraction, there is tremendous international interest in eventually mining that region. Thus, human activity in the Alaskan Arctic will increase significantly in this region in decades to come.

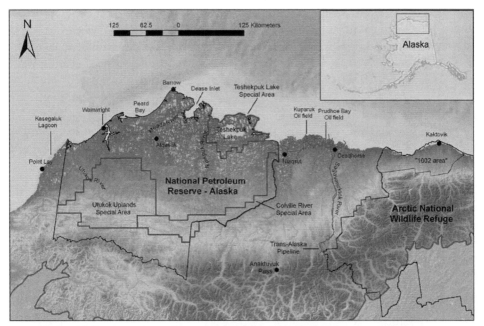

Figure 10.2. Map of arctic Alaska, including the prominent federal lands and the special areas of the National Petroleum Reserve–Alaska. The immense wetland complex of western arctic Alaska covers the region from the Prudhoe Bay Oilfield west to Wainwright.

What does all this development and infrastructure mean for arctic wildlife? For shorebirds, it has meant loss of breeding habitat where structures are placed. Currently this habitat loss is negligible, since the total amount of loss is quite minimal ($< 1\%$) with respect to the total amount of habitat available on the arctic coastal plain. Degradation of available habitat near the oilfields is a concern. Dust accumulation from vehicle traffic on tundra adjacent to oilfield roadways limits nesting opportunities for shorebirds. Early in the season the dust may actually attract usage by foraging shorebirds because it accelerates snowmelt, but nesting birds avoid dust-affected areas (Troy 2000). The main impact on nesting birds, however, is likely an indirect effect of the structures and edible refuse that arise from such development, as generalist species that also act as nest predators have increased with the infrastructure in the Prudhoe Bay region (NAS 2003). Arctic fox have increased in association with structures like elevated gravel pads and culverts where they can den their young, and they access edible wastes (Eberhardt et al. 1982, Ballard et al. 2000). Glaucous gull populations are believed to have increased, again in association with access to edible refuse and road kills (Day 1998). Finally,

Christmas Bird Count data indicates that common ravens have increased in the Prudhoe Bay oilfields over the past 20 years (Wildlife Research Center, Laurel, MD data available online) and they increasingly use oilfield infrastructure for nesting. All three of these species are nest predators on arctic nesting birds, including shorebirds. There is recent evidence that nest survivorship of some bird species is lower in areas closer to oilfield infrastructure (within 5 km; Liebezeit et al. 2009).

The very remoteness of the Arctic, and particularly its remoteness from nest predators, is a key factor in understanding its importance to nesting shorebirds. A recent study demonstrated that arctic nesting birds had higher nesting success with increasing latitude (McKinnon et al. 2010), thus supporting the assertion that the Arctic's remoteness may afford reduced predation risk. Development in the Arctic is bringing heightened nest predation pressure on shorebirds and other migratory birds that have historically depended on this region for nesting.

To examine the effect of increased nest predators, we collaboratively engaged several partners in a large-scale, multiyear examination of nest productivity of arctic-breeding birds, primarily shorebirds and songbirds, near and distant from the Prudhoe Bay oilfields (Liebezeit et al. 2009). Our analyses indicated a strong effect of the proximity of infrastructure on the nesting success of Lapland longspurs, the most common songbird in the Arctic. Our efforts to assess the overall effect of nearby infrastructure on shorebirds' nest productivity were frustrated by the very high natural variation in nest productivity among sites and between years. Overall, proximity of infrastructure did not add to the model effects on nest variation. A subsequent ad hoc analysis restricted to the Prudhoe Bay oilfields did detect an effect on the two phalarope species, but not on other shorebird species.

In a separate study, arctic fox were the most common nest predators identified by remote cameras at shorebird nests (Liebezeit and Zack 2008). They were also the least commonly detected of the potential nest predators in our surveys. This result is interesting in that arctic fox are among the species that are increasing with human development, and relative to other predators they have an apparently disproportionate effect on nest predation.

Development in the Arctic is bringing heightened nest predation pressure on shorebirds and other migratory birds that historically have depended on this region for nesting. As more development ensues (including natural gas extraction and mining), it seems clear that the Arctic is becoming less remote and less a breeding haven for shorebirds.

Arctic Shorebirds in a Changing Arctic

Arctic shorebirds face a "double whammy" with climate change: both their high-latitude breeding grounds in the arctic north and their (often) high-latitude wintering habitat in the south are being disrupted and are projected to be dramatically affected by climate change, as high-latitudinal environments are more effected (IPCC 2007). Their long-distance migrations will not allow them to escape these changes; rather, they will make them confront quite different and separate effects at both ends of their journeys, and changes in their stopover habitats as well.

It is very hard to predict exactly how changes associated with climate change will affect shorebirds, because of all the uncertainties, indirect effects, and potential interactions (Mustin et al., 2007; Martin et al. 2009). However, changes brought by global warming are considered to be happening too rapidly for most vertebrate species to respond to in an evolutionary sense (Berteaux et al. 2004; also see Austin et al. this volume). Thus, any adjustments shorebirds and other species make with the changing climate are either phenotypic adjustments or reflections of evolutionary abilities that saw them through previous changes in climatic shifts (Carey 2009).

Nonetheless, it is important to assess how shorebirds may be affected by near-term changes in the Arctic and elsewhere that result from changes in climate. If for no reason other than this one, identifying species or groups of species likely to be differentially effected will perhaps allow for some targeted conservation action to potentially offset the effects of threats later. Below, we outline how the potential and emerging effects of the changing arctic climate are affecting shorebirds or will likely affect them (see table 10.1 for life-history traits and habitats of key species).

Rapidly changing arctic seasonality has resulted in a shift in the phenology of biological processes and the altered timing of life-history activities within species. Spring snowmelt and subsequent green-up and flowering of plants are now occurring sooner in the Arctic, while fall lingers longer (Hinzman et al. 2005, Hoye et al. 2007, Post et al. 2009). Arctic lakes and rivers are breaking up earlier and freeze-up is occurring later; snowmelt is occurring earlier, and the growing season is beginning sooner and lasting longer. In some arctic regions there is evidence of earlier nesting by migratory birds, as snowmelt dates also average earlier in the spring (Liebezeit and Zack unpubl. data, Hoye et al. 2007, Meltofte et al. 2007). There is already significant evidence of earlier arrival and breeding of songbird species in temperate regions (Both et al. 2004). For the highly migratory shorebirds, seasonality shifts are less acute in their

Table 10.1. Life history, ecology, behavior, and migratory and conservation attributes of shorebirds typically encountered in arctic Alaska

Name	Mass[1]	Breeding habitat[2]	Mating system[3]	Parental care	Migration[4]	Winter habitat[5]	Conservation concern[6]
Black-bellied plover (*Pluvialis squatarola*)	180g	Dry tundra	Socially monogamous	Biparental	Throughout North and South America	Coastal beaches and estuaries	Moderate
American golden plover (*Pluvialis dominica*)	100–200+ g	Dry tundra	Socially monogamous	Biparental	Throughout North and South America	Variable, fields and coastlines	High
Semipalmated plover (*Charadrius semipalmatus*)	47 g	Gravel river banks	Socially monogamous	Biparental	Throughout North and South America	Coasts, lakes, and agricultural areas	Moderate
Upland sandpiper (*Bartramia longicauda*)	97–226g	Dry tundra	Socially monogamous	Biparental	Central South America	Grasslands and agricultural areas	High
Whimbrel (*Numenius phaeopus*)	310–493g	Dry tundra	Socially monogamous	Biparental	Throughout North and South America	Tidal flats, grasslands, meadows	High
Bar-tailed godwit (*Limosa lapponica*)	200–720 g	Dry tundra	Socially monogamous	Biparental	Australasia	Intertidal habitats	High
Ruddy turnstone (*Arenaria interpres*)	84–190 g	Well-drained, sparse tundra	Socially monogamous	Biparental	Americas, southern world	Coastal habitats	High
Red knot (*Calidris canutus*)	135 g	Foothills with low vegetation (western Alaska)	Socially monogamous	Biparental	Central and South America	Intertidal habitats	High

Species	Mass	Breeding habitat	Mating system	Parental care	Range	Wintering habitat	
Semipalmated sandpiper (*Calidris pusilla*)	21–32 g	Wet tundra	Socially monogamous	Biparental	Central and northern South America	Tidal mudflats	Moderate
Western sandpiper (*Calidris mauri*)	22–35 g	Wet tundra	Socially monogamous	Biparental	UnitedStates to northern South America	Coastal areas with sand	High
White-rumped sandpiper (*Calidris fuscicollis*)	40–60 g	Wet tundra	Socially monogamous	Biparental	Southern South America	Beaches, marshes, fields	Moderate
Baird's sandpiper (*Calidris bairdii*)	27–63 g	Sparse dry tundra	Socially monogamous	Biparental	South America	Montane meadows	Moderate
Pectoral sandpiper (*Calidris melanotos*)	65–98 g; males ca. 30% larger	Wet tundra	Polygynous, "promiscuous"	Uniparental; female care	South America	Interior grasslands and marshes	Moderate
Dunlin (*Calidris alpina*)	48–64 g	Wet tundra	Socially monogamous	Biparental	Arctic breeding *C.a.articola* to Asia	Coastal and interior wetlands	High
Stilt sandpiper (*Calidris himanotpus*)	50–70 g	Wet tundra	Socially monogamous	Biparental	Interior South America	Interior ponds, marshes	Moderate
Buff-breasted sandpiper (*Tryngites subruficollis*)	Male 57–78 g, Female 46–66 g	Wet tundra	Polygynous: lekking species	Uniparental; female care	Southern South America	Pampas grasslands	High
Long-billed dowitcher (*Limnodromus scolopaceus*)	90– 131 g	Wet tundra	Socially monogamous	Biparental; male care only after hatch	Southern United States, Central America	Coastal tidal flats and interior wetlands	High

(Continued)

Table 10.1. (Continued)

Name	Mass[1]	Breeding habitat[2]	Mating system[3]	Parental care	Migration[4]	Winter habitat[5]	Conservation concern[6]
Wilson's snipe (Gallinago delicata)	ca. 100 g	Wet habitats	Socially monogamous	Biparental; female incubates, both care for young	North America to northern South America	Wet habitats	Moderate
Red Phalarope (Phalaropus fulicaria)	ca. 45 g; f > m	Wet habitats near ponds	Polyandrous	Uniparental; male tends nest, young	Oceanic; subtropical and tropical	Pelagic	Moderate
Red-necked phalarope (Phalaropus lobatus)	30–40g; f > m	Wet habitats near ponds	Polyandrous	Uniparental; male tends nest and young	Oceans in tropical and subtropical Americas	Pelagic	Moderate

[1]From Poole (2005), or more recent literature. Mass variation reflects species more closely studied; single mass estimates represent a middle range.

[2]From Poole (2005) and personal observation. Broad categories are meant to convey the difference between species tending toward wet tundra habitats and species more closely associated with dry or upland arctic habitats.

[3]From Poole (2005). "Socially monogamous" are those species with a seasonal pair bond, but for which paternity is uncertain.

[4]Broad categories from Poole (2005).

[5]Broad categories from Poole (2005).

[6]After Brown et al. (2004); conservation concern category updated 2007: www.whsrn.org/about-shorebirds/shorebird-status.

more temperate and tropical wintering and stopover areas than in the Arctic. Thus, shorebirds are faced with differing rates of seasonal change across their range, and so may be particularly susceptible to "trophic mismatch" risks. Shorebirds, with their complicated migration calendar which spans multiple geographies, are being pushed to earlier breeding by climate change (see figure 10.3). It may become increasingly difficult for them to coordinate changes in physiology necessary for migration with arrival on the breeding grounds and securing of the nutrients they need to breed and produce eggs (Buehler and Piersma 2008). The core concern in the Arctic regarding the trophic mismatch is the risk of decoupling peak insect emergence from the hatching of young (Tulp and Schekkerman 2008). A lag between hatching of birds and the spike in insect productivity could dramatically affect hatchling survivorship, given the brief arctic breeding schedule. Shorebird species that have the longest migration pathways with the fewest stopover breaks, like bar-tailed godwits, would seem particularly susceptible to phenological mismatch since they require substantial replenishment upon arriving on their arctic breeding grounds. Uniparental breeders (table 10.1), like phalaropes, pectoral sandpipers, and buff-breasted sandpipers may feel the brunt of shifting phenology more than do biparental species, as one individual incubates and must gain resources while leaving its nests unattended. Some studies have found lower nest survivorship among uniparental species than among biparental breeders (Smith 2009, Liebezeit et al. 2011), and it is unknown whether climate change will play a larger role in this effect. A trophic mismatch could change migratory calendars of some shorebird species, increasing overlap with the migration of their predators, like peregrine falcons (*Falco peregrinus*), and thus increasing predation risk during migration (Butler et al. 2003, Lank et al. 2003).

Coastal shorelines are eroding dramatically in certain places on the Alaskan arctic coastline (Jones et al., 2008). The accelerated rate of arctic shoreline erosion results from the interaction of multiple climate-mediated changes. Changes in global climate patterns result in more intensive fall storms in arctic Alaska. Wave action in the Arctic Ocean is growing in intensity because of the increased exposure of free water as the extent of the ice cap continues to shrink. Erosion and rising sea levels are leading to breaching and subsequent inundation of coastal freshwater lakes, thereby replacing fresh with saline water and drastically changing near-shore lake ecology (Flint et al. 2008). North of Teshekpuk Lake on the Beaufort Sea coastline, wave action has eroded the coastline nearly a kilometer inland over the past 50 years, an estimated increase of more than 50% (Mars and Houseknecht 2007). Broadly along the Beaufort

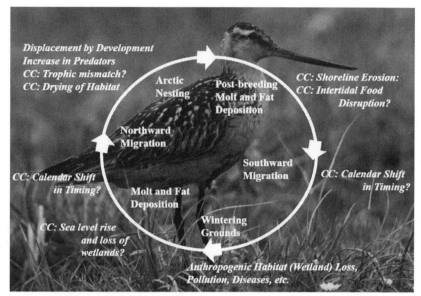

Figure 10.3. Annual life cycle of shorebirds and emerging conservation threats.

Sea, shorelines have eroded more than six meters per year, a rate that increased nearly 25% over the previous decade (Jones et al., 2008). Associated with this erosion is inundation of sea water into fresh-water bodies and tundra habitat by ocean storms.

After breeding, most shorebird species flock northward to the Arctic Ocean coastline, where they feed on invertebrates in productive coastal lagoons, river deltas, and wetlands before starting their long migration to points southward (Connors et al. 1984, Taylor et al. 2010). The eroding shorelines could affect the capacity of post-breeding shorebirds to gain the fat reserves they need for their southward migratory flights. Little is known about arctic intertidal ecology, and certainly nothing is known about how the changing climate and the resulting dramatic erosion affects primary productivity and shorebirds. Projected increases in temperature and precipitation in arctic Alaska could lead to increased sedimentation. An increased supply of sediment to river deltas may compensate for erosion or sea-level rise elsewhere. This is potentially significant for shorebirds because of the importance of river deltas during the post-breeding season (Taylor et al. 2010).

The warming of the Arctic is believed to be causing a drying trend in some regions driven by increased evaporation / precipitation ratios. In the Canadian High Arctic, previously permanent shallow lakes are drying up (Smith et al. 2005, Smol and Douglas 2007). Similarly, in the low tundra habitats of arctic Alaska, current general circulation models predict that the overall water balance (P-PET) is shifting to a negative trend (evapotranspiration exceeding precipitation; Martin et al. 2009). For regions like the immense wetland complex of western arctic Alaska, which are underlain by deep permafrost, the prospect of drying is complicated by several factors (Martin et al. 2009). The interaction of permafrost degradation, changes in spring melt, and uncertainties in the contribution of permafrost to surficial water amounts could lead to sudden lake draining and potential drying of wet tundra environs as water is channelized and redistributed. Higher nesting densities for some shorebird species can occur in habitat of higher heterogeneity (as compared to nearby areas) produced by localized thermokarst (Troy 2000), yet it is unknown whether region-wide thermokarst would benefit nesting birds.

Although a warming Arctic and a delayed freeze-up could improve reproductive success for some species (Soloviev et al. 2006), tundra drying in the productive low arctic tundra wetland complexes of Alaska, particularly in the vast western coastal plain where breeding shorebird diversity and density are highest (Johnson et al. 2007, Liebezeit et al. 2011), could result in negative impacts. Nesting habitat quality could be compromised or lost completely, particularly for species like phalaropes and long-billed dowitchers, which nest in the wettest tundra habitats (table 10.1). Species that nest in drier tundra, such as plovers and whimbrel, may remain unaffected by such changes or actually benefit by the creation of new, drier breeding habitat. Terrestrial land predators like arctic fox would potentially have more opportunities for nest predation, since birds nesting on islands could lose the natural protection of surrounding water. In addition to nesting habitat effects, many species forage on shoreline edges or on pond surfaces (as in the case of phalaropes) and so drying up of ponds could reduce foraging availability. Although western arctic Alaska wetlands will not likely dry up, given the immense reservoir of permafrost underneath (see Martin et al. 2009), the risk of wet tundra drying at the surface would change nesting habitat, prey availability, and susceptibility to nest predation. Widespread melting of permafrost could also result in paludification (formation of bogs and peat-dominated systems; Auerbach et al. 1997), potentially reducing habitat quality (and thus insect prey abundance) for wildlife, including nesting shorebirds.

The sedge-dominated tundra of the Arctic may change with dramatic increases in woody shrubs as plant succession is driven by the changing climate (Sturm et al. 2005, Hudson and Henry 2009). Over the past 50 years there has been an increasing amount of woody vegetation, including willows and alders, north of the Brooks Range and into the foothills of arctic Alaska as well as at other sites across the circumpolar regions, including the Seward Peninsula (Tape et al. 2006). There is also some indication that tundra fires are increasing in frequency and spatial scale, and such fires could quicken the succession to woody vegetation from sedge tundra (Racine and Jandt 2008).

Increased shrubiness in the Arctic would replace tundra dominated by low sedges with woody vegetation with a taller structure that would fundamentally change nesting habitat for all shorebirds and other migratory birds breeding in the Arctic. More vertical relief in vegetation could impair incubating shorebirds' ability to escape predators by limiting visibility. Further, the invertebrate community would likely change (Martin et al. 2009) in step with the vegetative community (i.e., movement of boreal fauna northward with more woody vegetation) and it is unknown whether shorebirds could readily switch to a new prey base. Although shorebirds may be negatively impacted by such changes, some arctic breeding songbird species may actually benefit from increased shrubiness, since they typically rely on shrub habitat in riparian areas for nesting.

Because of the warming climate, some boreal wildlife species are encroaching into arctic habitats and may directly or indirectly influence shorebirds. Red fox invasion is well documented in Scandinavia (Selås and Vik 2007) and Canada (Hersteinsson and Macdonald 1992) but not yet well documented in arctic Alaska, although anecdotal evidence is raising concern about the possibility of it replacing arctic fox (Pamperin et al. 2006). As arctic fox appear now to be the most important predator on shorebird nests (Liebezeit and Zack 2008), their potential replacement by red fox would leave unclear a net effect on predation pressure. Larger red foxes indeed are nest predators, but are perhaps less likely to bother with small shorebirds amid a landscape of waterfowl nests. A separate trophic issue occurring in the Arctic can have strong indirect effects on shorebirds. Lemming cycles are being disrupted, and in some cases are disappearing, in arctic environs (Ims and Fuglei 2006). When lemming numbers peak, predators, including would-be nest predators, redirect their predatory attention to the abundance of lemmings (Summers 1986, Bêty et al. 2001). For nesting birds, this effect can lead to higher nest productivity during such peaks (Liebezeit et al. 2011). Thus, declines in the cycling and abundance of lemmings would likely mean more consistent predation pressure on nesting shorebirds from year to year. Anecdotal evidence suggests that grizzly bears (*Ursus arctos*) are apparently more common now in arctic environs and that moose (*Alces alces*) are moving north into arctic settings, drawn by the higher levels of woody vegetation for forage. However, it is unclear whether increases in these larger species will affect shorebirds directly, or whether indirect interactions of predator and prey will affect them.

It should be noted that the observations and possibilities of climate-mediated effects described above would not occur in isolation. The real challenge is to imagine and confront the combined and likely synergistic effects arising from the changing climate. Further, it must be emphasized that our existing and "normal" challenges to wildlife conservation worldwide (e.g., habitat loss and degradation, habitat fragmentation, pollution, and hunting pressures) will remain and likely magnify with the changing climate. These challenges will likely interact and often act synergistically with emergent problems associated with the changing climate (see Rehfisch and Crick 2003 for examples).

A Conservation Future for Shorebirds with Respect to Climate Change in Arctic Alaska?

The Challenges

At present, our knowledge of how climate change is impacting and will continue to impact arctic nesting birds, and all wildlife, is limited by a paucity of basic information that can be gained only from targeted research and new modeling studies. At the same time, it has been only over the past few years that federal and state agencies have been mandated to prioritize climate change and incorporate it into strategic planning with respect to wildlife management in the Alaskan Arctic (e.g., Martin et al. 2009, CWCS 2006, WCS 2010). At the same time, nongovernmental organizations (NGOs) with a focus on arctic Alaskan conservation are also shifting focus to prioritize climate change into their conservation activities. Although there has been some coordinated effort primarily through data sharing, between government agencies and NGOs in examining how climate change is impacting wildlife, there is room for greater cooperation and synergy among these groups as well as with industry (namely the oil companies) in working together to find common ground on how to limit climate change effects on arctic Alaskan wildlife.

With shorebird conservation, there is the further complication that these birds only spend a few months of their life in arctic Alaska. The rest of their lives they remain in southerly wintering grounds and visit stop-over sites in migration. While there is some coordination in research and conservation efforts across flyways (e.g., the Western Hemisphere Shorebird Reserve Network, various flyway working groups), there is little activity focused on developing strategies to address climate change with respect to bird conservation across multiple geographies and flyways (see figure 10.3).

Unlike in temperate regions, there is little opportunity for habitat restoration in the Arctic. For example, forest management practices and even riparian restoration (e.g., Cooke and Zack 2009) allow a great deal of flexibility in the tools we can use to improve wildlife habitat relatively quickly. In the Arctic, disturbances to vegetation take decades and even centuries to recover (Emers et al. 1995, Cargill and Chapin 1987). Even with human intervention, habitat restoration efforts in the Arctic only allow slow successional response in vegetation (Harper and Kershaw 1996).

How to Move Forward and Work toward Solutions

It is clear that coordinated strategic planning efforts are necessary to develop the best land use practices and conservation measures for protecting shorebirds and other wildlife in the face of climate change in the Alaskan Arctic. Some initial efforts have been made (e.g., Martin et al. 2009, WCS 2010) to begin the groundwork of bringing experts together to identify the state of current knowledge, develop scenarios on how climate change could impact wildlife, and use rigorous analyses to identify the most vulnerable species. The next steps should include all stakeholders in more coordinated assessment and adaptation planning.

At the same time, more information is urgently needed to most effectively protect wildlife.Focused studies on particular effects of climate change on key species are needed (Martin et al. 2009, WCS 2010), and obviously the species to prioritize should include those that are recognized as being particularly vulnerable to climate change. For shorebirds this means, for example, that both phalarope species, which depend so highly on wet tundra for nesting and small ponds and arctic coastal areas for foraging prior to migration, require closer monitoring. (This is now being done by a new suite of studies across the North American Arctic: the Arctic Shorebird Demographic Network, in which we are participating with two sites.) Secondly, the development of an arctic-wide monitoring network in which biotic, climate, and hydrologic data are monitored over the long term will provide the best information to adequately understand the effects of climate change on species, with the emergent geographic variations. In the Arctic, candidate sites include places where a history of research can be built on. They include Barrow, Toolik Lake, Prudhoe Bay (with surrounding oilfields), and sites in the 1002 area of the Arctic National Wildlife Refuge. Within this framework, focused research on the phenotypic plasticity of shorebirds is needed to determine the range of habitat suitability for particular species and the environmental gradients (e.g., temperature and moisture) over which they can successfully nest and forage. Only in this way can we learn about the potential resiliency of shorebird populations challenged by a changing Arctic. In addition, more accurate predictive models for projections in precipitation and evapotranspiration are needed for the region, as well as finer-scale modeling of environmental systems (particularly soil and hydrologic regimes) in combination with biological systems (Martin et al. 2009).

The protection of large arctic wetlands as a buffer to the changing climate may be our best current option to protect breeding shorebirds and other wildlife in that climate. Protecting large regions helps by minimizing the loss of carbon that is already present in vegetation and soils while maintaining ecosystem integrity and reducing risks and impacts from extreme events. In arctic Alaska, this means affording permanent protection to key places in the biggest wetland complex in the NPR-A, a landscape that is still virtually undeveloped but which has large amounts of real estate, although significant portions are already leased for oil and gas exploration and development. A key region in the NPR-A, around Teshekpuk Lake, has long been designated as a "special area" for its wildlife importance (Derksen et al. 1979, King and Hodges 1979, Liebezeit et al. 2011), yet that designation is of uncertain standing with respect to potential development. Recent decisions to protect the lake from development and to delay leasing around it provide a short window of opportunity for making such protection permanent.

With better climate modeling and understanding of wildlife resiliency, we can begin to identify potential climate refugia for particular species. In a recent study, climate modeling enabled researchers to identify where arctic sea ice will likely remain in the decades to come, thus identifying likely refugia habitat for polar bears (Scott Bergen, personal communication). For shorebirds and other nesting birds it is still unknown whether and where potential climate refugia may lie. Anecdotal evidence suggests that the fog belt near the Arctic Ocean may act as a buffer against temperature increase and potentially offer respite to breeding shorebirds (WCS 2010). Also, the ice-rich loess (yedoma) abundant in the lower Brooks Range foothills just south of the coastal plain (Carter 1988) could be transformed into a new region of thaw lakes, as in the degraded yedoma landscape of the Seward Peninsula. This transformation, however, would take hundreds to thousands of years (Martin et al. 2009). Although not an issue for migratory birds, the identifying of corridors between arctic refugia is also needed to ensure movements of land-based megafauna, particularly caribou, for years to come.

Of course shorebird conservation cannot be relegated to the breeding grounds. Protection of remaining wetlands and restoration of wetlands important to wintering and stopover migrants should still be a key priority. As migratory shorebirds use a multitude of wetlands worldwide in their migration, there is a need to prioritize among them. Efforts to do so are underway (e.g., in association with the Western Hemisphere Shorebird Reserve Network initiatives; see www.whsrn.org). Yet it would seem that for coastal wetlands, identifying a capacity to withstand sea level rise by adjacent protected inland areas capable of becoming estuarine would be very important.

The expansion of human development in the region could exacerbate climate change impact if land-use planning is not considered carefully with appropriate empirical data. For example, if areas of arctic refugia are identified, development should not proceed in those areas if it is deemed that it could act synergistically with climate change to impact species further. However, there is room to work with the extractive industries to develop land use that will maximize protection of wildlife with development.

There is no set of magic bullets to insure the conservation future of shorebirds. There is a clear need for such protection in these changing times. Climate change and the "end of remote" are converging to disrupt arctic wildlife, including migratory shorebirds. Protection of key areas like the Teshekpuk Lake Special Area and the 1002 coastal plain region of the Arctic National Wildlife Refuge are, at minimum, buffers against changes to come. In the meantime, we need the opportunity to understand how best to conserve these master migrants in changing times.

ACKNOWLEDGMENTS
We thank our funders, including the Liz Claiborne/Art Ortenberg Foundation, the McCaw Foundation, the Kresge Foundation, the Disney Friends for Change Grant, the Neotropical Migratory Bird Conservation Act Grant (US Fish and Wildlife Service), the Wildlife Conservation Society, and private donors. We graciously thank Philip Martin, Rick Lanctot, and Steve Kendall of the US. Fish and Wildlife Service and Stephen Brown of the Manomet Center for Conservation Sciences for their input. We also thank anonymous reviewers and the editors of this book for their time and effort spent improving the manuscript.

LITERATURE CITED
Arctic Climate Impact Assessment (ACIA; 2004). *Impacts of Warming: Arctic Climate Impact Assessment*. Cambridge: Cambridge University Press.
Anisimov, O. A., D. G. Vaughan, T. V. Callaghan, C. Furgal, H. Marchant, T. D. Prowse, H. Vilhjálmsson, and J. E. Walsh. 2007. *Polar Regions (Arctic and Antarctic) Climate Change 2007: Impacts, Adaptation and Vulnerability. Contribution of Working Group II to the Fourth Assessment Report of the Intergovernmental Panel on Climate Change*, M. L. Parry, O. F. Canziani, J. P. Palutikof, P. J. van der Linden, and C. E. Hanson, eds., 653–85. Cambridge: Cambridge University Press.

Auerbach, N. A., M. D. Walker, and D. A. Walker. 1997. Effects of roadside disturbance on substrate and vegetation properties in arctic tundra. *Ecological Applications* 7:218–35.

Baker, A. J., and J. G. Strauch. 1988. Genetic variation and differentiation in shorebirds. *Acta XIX Congressus Internationalis Ornithilogici* 1639–45.

Ballard, W. B., M. A. Cronin, and H. A. Whitlaw. 2000. Caribou and oil fields. In *The Natural History of an Arctic Oil Field*, edited by J. C. Truett and S. R. Johnson, 85–104. San Diego: Academic Press.

Berteaux, D., D. Reale, A. G. McAdam, and S. Boutin. 2004. Keeping pace with fast climate change: Can arctic life count on evolution? *Integrative and Comparative Biology* 44:140–51.

Bêty, J., G. Gauthier, J.-F. Giroux, and E. Korpimäki. 2001. Are goose nesting success and lemming cycles linked? Interplay between nest density and predators. *Oikos* 93:388–400.

BLM/MMS. 1998. Northeast National Petroleum Reserve–Alaska. *Final Integrated Activity Plan /Environmental Impact Statement*. Vols. 1 and 2. Prepared by US Department of the Interior, Bureau of Land Management and Minerals Management Service, Anchorage. Washington: US Government Printing Office.

Both, C., A. V. Artemyev, B. Blaauw, R. J. Cowie, A. Dekhuijzen, T. Eeva, A. Enemar, L. Gustafsson, E. Ivankina, A. Järvinen, N. Metcalfe, N. E. Nyholm, J. Potti1, P.-A. Ravussin, J. J. Sanz, B. Silverin, F. Slater, L. Sokolov, J. Török, W. Winkel, J. Wright, H. Zang, and M. Visser. 2004. Large-scale geographical variation confirms that climate change causes birds to lay earlier. *Proceedings of the Royal Society of London B* 271:1657–62.

Buehler, D. M., and T. Piersma. 2008. Travelling on a budget: Predictions and ecological evidence of bottlenecks in the annual cycle of long-distance migrants. *Philosophical Transactions of the Royal Society B* 363:247–66.

Butler, R. W., R. C. Ydenberg, and D. B. Lank. 2003. Wader migration on the changing predator landscape. *Wader Study Group Bulletin* 100:130–33.

Cargill, S. M., and E. S. Chapin. 1987. Application of successional theory to tundra restoration: a review. *Arctic and Alpine Research* 19:366–72.

Carey, C. 2009. The impacts of climate change on the annual cycles of birds. *Proceedings of the Royal Society of London B* 364:3321–30.

Carter, L.D. 1988. *Loess and Deep Thermokarst Basins in Arctic Alaska. Proceedings of the Fifth International Conference on Permafrost, August 1988, Trondheim, Norway*, Vol. 1, 706–11. Trondheim: Tapir Publishers.

Connors, P.G. 1984. Ecology of shorebirds in the Alaskan Beaufort littoral zone. In *The Alaskan Beaufort Sea: Ecosystems and Environments*, edited by R. Barne, D. M. Shell, and E. Ramirez, 403–416. New York: Academic Press.

Cooke, H. A., and S. Zack. 2009. Use of standardized visual assessments of riparian and stream condition to manage riparian bird habitat in eastern Oregon. *Environmental Management* 44:173–84.

Coppack, T., and C. Both. 2002. Predicting life-cycle adaptation of migratory birds to global climate change. *Ardea* 90:369–78.

CWCS 2006 (Comprehensive Wildlife Conservation Strategy of Alaska). Alaska Department of Fish and Game. 2006. *Our Wealth Maintained: A Strategy for Conserving Alaska's Diverse Wildlife and Fish Resources*. Alaska Department of

Fish and Game, Juneau. Accessed at www.sf.adfg.state.ak.us/statewide/ngplan/NG_outline.cfm.

Day, R. H. 1998. *Predator Population and Predation Intensity on Tundra-Nesting Birds in Relation to Human Development*. Fairbanks: Alaska Biological Research Associates.

Derksen, D. V., M. W. Weller, and W. D. Eldridge. 1979. Distributional ecology of geese molting near Teshekpuk Lake, National Petroleum Reserve, Alaska. In *Management and Biology of Pacific Flyway Geese*, edited by R. L. Jarvis and J.C. Bartonek, 189–207. Corvallis: Oregon State University Book Stores.

Eberhardt, L. E., W. C. Hanson, J. L. Bengtson, R. A. Garrott, and E. E. Hanson. 1982. Arctic fox home range characteristics in an oil-development area. *Journal of Wildlife Management* 46:183–90.

Emers, M., J. C. Jorgenson, and M. K. Reynolds. 1995. Response of arctic tundra plant communities to winter vehicle disturbance. *Canadian Journal of Botany* 73:905–17.

Flint, P. L., E. J. Mallek, R. J. King, J. A. Schmutz, K. S. Bollinger, and D. V. Derksen. 2008. Changes in abundance and spatial distribution of geese molting near Teshekpuk Lake, Alaska: Interspecific competition or ecological change? *Polar Biology* 31:549–56.

Harper, K. A. and G. P. Kershaw. 1966. Natural revegetation on borrow pits and vehicle tracks in shrub tundra, 48 years following construction of the CANOL no. 1 pipeline, NWT, Canada. *Arctic and Alpine Research* 28:163–71.

Hersteinsson, P., and D. W. MacDonald. 1992. Interspecific competition and the geographic distribution of red and arctic foxes *Vulpes vulpes* and *Alopex lagopus*. *Oikos* 54:505–15.

Hinzman, L. D., N. D. Bettez, W. R. Bolton, F. S. Chapin III, M. B. Dyurgerov, C. L. Fastie, B. Griffith, R. D. Hollister, A. Hope, H. P. Huntington, A. M. Jensen, G.J. Jia, T. Jorgenson, D. L. Kane, D. R. Klein, G. Kofinas, A. H. Lynch, A. H. Lloyd, A. D. McGuire, F. E. Nelson, W. C. Oechel, T. E. Osterkamp, C. H. Racine, V. E. Romanovsky, R. S. Stone, D. A. Stow, M. Sturm, C. E. Tweedie, G. L. Vourlitis, M. D. Walker, D. A. Walker, P. J. Webber, J. M. Welker, K. S. Winker, and K.Yoshikawa. 2005. Evidence and implications of recent climate change in northern Alaska and other arctic regions. *Climatic Change* 72:251–98.

Høye, T. T., E. Post, H. Meltofte1, N. M. Schmidt, and M. C. Forchhammer. 2007. Rapid advancement of spring in the high Arctic. *Current Biology* 17:449–51.

Hudson, J. M .G., and G. H. R. Henry. 2009. Increased plant biomass in a high arctic heath community from 1981 to 2008. *Ecology* 90:2657–63.

Ims, R. A., and E. Fuglei. 2005. Trophic interaction cycles in tundra ecosystems and the impact of climate change. *BioScience* 55:311–22.

IPCC. 2007. Contribution of working group I to the fourth assessment report of the Intergovernmental Panel on Climate Change, 2007. S. Solomon, D. Qin, M. Manning, Z. Chen, M. Marquis, K. B. Averyt, M. Tignor, and H. L. Miller, eds. Cambridge and New York: Cambridge University Press.

Johnson, J. A., R. B. Lanctot, B. A. Andres, J. R. Bart, S. C. Brown, S. J. Kendall, and D. C. Payer. 2007. Distribution of breeding shorebirds on the arctic coastal plain of Alaska. *Arctic* 60:277–93.

Jones, B. M., Hinkel, K. M., C. D. Arp, and W. R. Eisner. 2008. Modern erosion rates and loss of coastal features and sites, Beaufort Sea coastline, Alaska. *Arctic* 61:361–72.

King, J. G., and J. I. Hodges. 1979. A preliminary analysis of goose banding on Alaska's arctic slope. In *Management and Biology of Pacific Flyway Geese*, edited by J. L. Jarvis and J. C. Bartonek, 176–88. Corvallis: Oregon State University Bookstores.

Lank , D. B., R. W. Butler, J. Ireland, and R. C. Ydenberg. 2003. Effect of predation danger on migration strategies of shorebirds. *Oikos* 103:303–19.

Lenton, T. M., H. Held, E. Kriegler, J. Hall, W. Lucht, S., Rahmstorf, and H. Schellnhuber. 2008. Tipping elements in the earth's climate system. *Proceedings of the National Academy of Sciences, USA* 105:1786–93.

Liebezeit, J. R., and S. Zack. 2008. Point counts underestimate the importance of arctic foxes as avian nest predators: Evidence from remote video cameras in arctic Alaskan oil fields. *Arctic* 61:153–61.

Liebezeit, J. R., S. J. Kendall, S. Brown, C. B. Johnson, P. Martin, T. L. McDonald, D. Payer, C. L. Rea, B. Streever, A. M.Wildman, and S. Zack. 2009. Influence of human development and predators on nest survival of tundra birds, arctic coastal plain, Alaska. *Ecological Applications* 19:628–44.

Liebezeit, J. R., G. C. White, and S. Zack. 2011. The importance of the Teshekpuk Lake region of the Alaskan arctic coastal plain for nesting birds. *Arctic* 64:32–44.

Mars, J. C., and D. W. Houseknecht. 2007. Quantitative remote sensing study indicates doubling of coastal erosion rate in past 50 yr along a segment of the arctic coast of Alaska. *Geology* 35:583–86.

Martin, P. D, J. L. Jenkins, F. J. Adams, M. T. Jorgenson, A. C. Matz, D. C. Payer, P. E. Reynolds, A. C. Tidwell, and J. R. Zelenak. 2009. Wildlife response to environmental arctic change: Predicting future habitats of arctic Alaska. Report of the Wildlife Response to Environmental Arctic Change (WildREACH): Predicting Future Habitats of Arctic Alaska Workshop, 17–18 November 2008. Fairbanks, US Fish and Wildlife Service. Accessed at www.arcus.org/alaskafws/.

McKinnon, L., P. A. Smith, E. Nol, J. A. Martin, F. I. Doyle, K. F. Abraham, H. G. Gilchrist, R. I. G. Morrison, and J. Bêt. 2010. Lower predation risk for migratory birds at high latitudes. *Science* 327:326.

Meltofte, H., T. Piersma, H. Boyd, B. McCaffery, B. Ganter, V. V. Golovnyuk, K. Graham, C. L. Gratto-Trevor, R. I. G. Morrison, E. Nol, H.-U. Rösner, D. Schamel, H. Schekkerman, M. Y. Soloviev, P. S. Tomkovich, D. M. Tracy, I. Tulp, and L. Wennerberg. 2007. Effects of climate variation on the breeding ecology of arctic shorebirds. *Meddelelser om Grønland-Bioscience* 59:1–48.

Mustin, K., W. J. Sutherland, and J. A. Gill. 2007. The complexity of predicting climate-induced ecological impacts. *Climate Research* 35:165–75.

NAS (National Academy of Sciences). 2003. Cumulative environmental effects of oil and gas activities on Alaska's North Slope. National Research Council of the National Academies. Washington: National Academies Press.

Pamperin, N. J., E. H. Follmann, and B. Petersen. 2006. Interspecific killing of an arctic fox by a red fox at Prudhoe Bay, Alaska. *Arctic* 59:361–64.

Piersma, T., and A. Lindstrom. 2004. Migrating shorebirds as integrative sentinels of global environmental change. *Ibis* 146:61–69.

Piersma, T., and P. Wiersma. 1996. Family Charadriidae (plovers). In *Handbook of the Birds of the World*, vol. 3, edited by J. del Hoyo, A. Elliott, and J. Sargatal, 384–442. Barcelona: Lynx Edicions.

Poole, A. F., P. Stettenheim, and F. B. Gill, eds. 2003. *The Birds of North America: Life Histories for the 21st Century*. Philadelphia: Academy of Natural Sciences. Washington: American Ornithologists' Union.

Post, E., M. C. Forchhammer, M. S. Bret-Harte, T. V. Callaghan, T. R. Christensen, B. Elberling, A. D. Fox, O. Gilg, D. S. Hik, T. T. Hoye, R. A. Ims, E. Jeppesen, D. R. Klein, J. Madsen, A. D. McGuire, S. Rysgaard, D. E. Schindler, I. Stirling, M. P. Tamstorf, N. J. C. Tyler, R. van der Wal, J. Welker, P. A. Wookey, N. M. Schmidt, and P. Aastrup. 2009. Ecological dynamics across the Arctic associated with recent climate change. *Science* 325:1355.

Racine, C., and R. Jandt. 2008. The 2007 "Anaktuvuk River" tundra fire on the arctic slope of Alaska: A new phenomenon? In *Ninth International Conference on Permafrost, Extended Abstracts*, edited by D. L. Kane and K. M. Hinkel, 247–48. Fairbanks: Institute of Northern Engineering, University of Alaska.

Rehfisch M., and H. Q. P. Crick. 2003. Predicting the impact of climatic change on arctic-breeding waders. *Wader Study Group Bulletin* 100:86–95.

Selas, V., and J. O. Vik. 2007. The arctic fox *Alopex lagopus* in Fennoscandia: A victim of human-induced changes in interspecific competition and predation? *Biodiversity and Conservation* 16:3575–83.

Smith, L. C., Y. Sheng, G. M. MacDonald, and L. D. Hinzman. 2005. Disappearing arctic lakes. *Science* 308:1429.

Smith, P. A. 2009. Variation in shorebird nest survival: Proximate pressures and ultimate constraints. PhD thesis, Carleton University, Ottawa, Canada.

Smol, J. P., A. P. Wolfe, H. J. B. Birks, M. S. V. Douglas, and V. J. Jones. 2005. Climate-driven regime shifts in the biological communities of arctic lakes. *Proceedings of the National Academy of Sciences, USA* 102:4397–4402.

Smol, J. P., and M. S. V. Douglas. 2007. Crossing the final ecological threshold in High Arctic ponds. *Proceedings of the National Academy of Sciences, USA* 104:12395–97.

Soloviev, M. Y., C. Minton, and P. S. Tomkovich. 2006. Breeding performance of tundra waders in response to rodent abundance and weather from Taimyr to Chukotka, Siberia. *Proceedings of the Waterbirds Around the World Conference, 2004*, Edinburgh.

Summers, R. W. 1986. Breeding production of dark-bellied brent geese *Branta bernicla bernicla* in relation to lemming cycles. *Bird Study* 33:105–8.

Sturm M., J. P. Schimel, G. J. Michaelson, J. M. Welker, S. F. Oberbauer, G. E. Liston, J. T. Fahnestock, and V. E. Romanovsky. 2005. Winter biological processes could help convert arctic tundra to shrubland. *BioScience* 55:17–26.

Tape, K., M. Sturm, and C. Racine. 2006. The evidence for shrub expansion in northern Alaska and the pan-Arctic. *Global Change Biology* 12:686–702.

Taylor, A. R., R. B. Lanctot, A. N. Powell, F. Huettmann, D. A. Nigro, and S. J. Kendall. 2010. Distribution and community characteristics of staging shorebirds on the northern coast of Alaska. *Arctic* 63:451–67.

Troy, D. M. 2000. Shorebirds, In *The Natural History of an Arctic Oil Field: Development and Biota*, edited by J. C. Truett and S. R. Johnson. San Diego: Academic Press.

Tulp, I. and H. Schekkerman. 2008. Has prey availability for arctic birds advanced with climate change? Hindcasting the abundance of tundra arthropods using weather and seasonal variation. *Arctic* 61:48–60.

WCS (Wildlife Conservation Society). 2010. New conservation priorities in a changing arctic Alaska. Workshop summary. Accessed at www.wcsnorthamerica .org/WildPlaces/ArcticAlaska/tabid/3640/Default.aspx.

Wennerberg, L., M. Klassenn, and A. Lindstrom. 2002. Geographic variation and population structure in the white-rumped sandiper *Calidris fuscicollis* as shown by morphology, mitochondrial DNA and carbon isotope ratios. *Oecologia* 131:380–90.

Island Species
with Nowhere to Go

Lisa Manne

Oceanic islands tend to be very special places—hotbeds of evolution, predator release, competitive release, and endemism. A recent map of global "endemism richness" shows that islands hold more endemic plant and vertebrate richness than similarly-sized mainlands, by nearly an order of magnitude (Kier et al. 2009). Islands have perhaps not been accorded the significance they deserve, because overall species richness on islands tends to be lower than that on mainlands, possibly as a result of past extinctions (Manne et al. 1999) and lower habitat diversity (Lack 1976). This oversight cannot be permitted to continue, as oceanic island species are already quite vulnerable to extinction processes that include climate, the widespread introduction of invasive species (Brooks et al. 2002), and massive reductions in habitat (Brooks et al. 1999).

What proportion of the world's vertebrates is endemic to islands? Only 14% of the world's mammal species are endemic to islands (Alcover et al. 1998; figure 10.1); this is partly due to the long history of human colonization and subsequent mammal extinctions of insular species. Of the approximately 8,225 reptile species worldwide, about 2,145 are endemic to oceanic islands (World Conservation Monitoring Centre of the United Nations Environment Programme [UNEP-WCMC] 2004; figure 11.1). Of a global total of 6,000 amphibian species, about 1,080 are endemic to islands; this total includes Australia, Taiwan, and Madagascar, though they are not true oceanic islands (IUCN et al. 2008; see figure 11.2). Finally, almost 30% of the approximately 11,000 bird species are endemic to islands, the highest proportion of any class of vertebreates (World Conservation Monitoring Centre of the United Nations Environment Programme [UNEP-WCMC] 2004).

Various characteristics of islands and island species render the climate change effects more acute, and thus render the island species more likely to be affected disproportionately by climate change (Ebi et al. 2006, Angeles et al. 2007, Cherian 2007). Such circumstances include the low-lying nature of some islands, which renders them vulnerable to rising sea level; the spread of disease vectors to islands, where rescue by

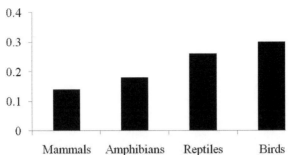

Figure 11.1. Proportion of total fauna that is endemic to oceanic islands. Proportions for birds and mammals from Alcover et al. (1998); proportions for reptiles and amphibians from the World Conservation Monitoring Centre of the United Nations Environment Programme (UNEP–WCMC), 2004. Note that these values are across all members of the taxon; proportions will be higher or lower for particular island chains.

noninfected individuals is less likely; the changing geographic range of resources, competitors or predators; developmental processes altered or disrupted by a warming climate (where a smaller gene pool of an insular population makes adaptation less likely); and interaction effects among one or more of these processes and other anthropogenic effects.

Below, I will assess the threat of rising sea levels to island vertebrate populations. Although all taxa are threatened by loss of habitat, different taxa are more or less vulnerable to the different processes. However, there are also indications that each taxon may display some degree of resilience to climate change impacts as well.

Rising Sea Levels

Sea level rise due to climate change has been documented by several different sources, though rises have been surprisingly spatially irregular. Sea levels are changing at a rate of between 0.3 and 1.8 mm per year, though these estimates are from different types of data (tide gauges versus remotely sensed data from satellite monitoring) and over different time scales (estimates from tidal gauges monitored between 1950 and 2000, throughout the 20th century, or from the past 70 years, etc.; see review by Bindoff et al. 2007). Rates of sea level rise are increasing due to accelerating glacier melt, which is occurring in many glacier systems (Pritchard et al. 2009, van den Broeke et al. 2009); thus, the impacts to low-lying islands are increasing in absolute effect and in speed of effect. Our current data do not allow us to predict how much island area will be lost by a particular date, but we can get an idea of which locations are particularly vulnerable, and which species might be hit particularly hard.

Figure 11.2. Locations of islands and coastal areas that are most vulnerable to rising sea levels lower than five meters in elevation (indicated by circles). White areas are typically covered by ice in August. Most coastal areas are lower than five meters elevation, unless they are rocky cliffs. For visual clarity, however, coastal areas were only marked with circles on this map if the area below five meters in elevation extended two or more pixels ($\geq$ 2 km) inland. Data from www.oceandots.com, the UNEP island directory: islands.unep.ch, and Hastings et al. 1999.

In figure 11.2 I have mapped low-lying islands entirely below 5 m of elevation, and coastal areas where the land below 5 m in elevation extends inland for more than 2 km. These are the areas most vulnerable to rising sea level.

The particular species for which this will be an issue are those that occupy and/or breed in only these areas. An interesting case study is provided by the endangered Bermuda petrel (*Pterodroma cahow*). While its decline is due to nonclimate factors, its continued survival is dependent on management policies to counter climate change effects. Thought to be extinct, this petrel (commonly called "cahow" due to the sound of its call) was rediscovered in the 1950s (Collar 1992). It is restricted to Bermuda, a group of about 140 islands and islets. Once widespread throughout Bermuda (Collar 1992 and references therein), this burrow-nesting species has suffered predation by introduced mammals, habitat destruction by feral pigs, deforestation (leading to soil thinning, and thus contributing to removal of nesting habitat), and nest competition with white-tailed tropic birds. All of these processes caused the Bermuda petrel's relegation to marginal islets and consequent near extinction (Collar 1992). A recovery program was initiated in 1961; activities included the establishment of the Castle Harbour Islands as a national park, and translocations of birds to reestablish a breeding colony on Nonsuch Island (one of the Castle Harbour Islands). Also, as there is not enough soil on many of the islands to support burrows, the Bermudan government has constructed artificial concrete burrows for the birds there (Wingate 1985). Toads (*Bufo marinus*) and rats (*Rattus rattus* and *Rattus norvegicus*) are being removed from cahow islands—the former because they can occupy nesting burrows, and the latter because they eat chicks (Dobson and Madeiros 2009). Two-thirds of the cahow population now uses these artificial burrows, and chicks are fledging successfully from them. The population is now growing slowly and steadily, having fledged 35 to 40 young each in 2003, 2005, 2008, and 2009 (IUCN 2010). Early in 2009, the first documented chick since 1620 hatched without human assistance from the new Nonsuch Island colony (IUCN 2010). However, higher sea levels and severe storms are flooding the burrows (artificial and natural), typically resulting in nest failure. Conservationists have been able to translocate some individuals to Nonsuch Island or to locations on existing breeding islets that are less prone to rising sea level and storms, and for now this seems to be helping.

Limited Dispersal for Insular Populations

An important risk factor for island-dwelling reptiles undergoing climate change seems to be temperature-dependent sex determination. This situation will represent a conservation problem for many reptiles in warming conditions, but in the island environment there may be limited scope for dispersing to a cooler location (Simmons et al. 2004); there can only be dispersal upslope, assuming that adequate elevational gradients exist on a given island. Moreover, many endangered reptiles on islands are already restricted to high elevation sites (Bahir and Surasinghe 2005).

The impacts of temperature on the two tuatara species (*Sphenodon* spp.) recognized for their unique evolutionary history has been studied in some detail. Tuataras were once widespread through all of New Zealand, but they are now restricted to predator-free islands near shore. For several egg-laying reptile species, the sex ratio of the offspring is determined by the sand temperature during a critical period in embryonic development. For the rarest tuatara species (*Sphenodon guntheri*), it is estimated that if climate warming occurs at the extreme warm end of current predictions, at *all* current nest sites clutches will be entirely male by the mid-2080s (Mitchell et al. 2008). It is unknown whether the tuatara can evolutionarily adapt to this situation (see Austin et al. this volume). Possible behavioral adaptations are to lay eggs later in the season (deemed unlikely by Mitchell et al. 2008), or to choose shaded nest sites (an option whose likelihood is unknown).

Sea turtles, while essentially marine species, need land to breed, and islands provide important nesting areas. These turtles face a similar risk of skewed offspring sex ratios with climate warming; research over the past 150 years shows that there has been a consistent warming trend at turtle nests (Hays et al. 2003), while change in social policy to address these endangering processes lags far behind (Bickford et al. 2010). Some estimates of sand temperature change show that nearly 100% of sea turtle hatchlings will be female by 2070, due to the warming of primarily open beaches (Fuentes et al. 2010). A recent empirical assessment of the temperatures of Guadeloupe hawksbill sea turtle (*Eretmochelys imbricata*) nests allowed researchers to specify the contribution of warmer and cooler beach areas to the sex ratio of turtle offspring (Kamel and Mrosovsky 2006). Interestingly, turtle nests in forested (cooler) areas contributed a significant number of males, and nests in more open (warmer) areas contributed a significant number of females to the offspring sex ratio. This research suggests some direction to how resource managers might formulate helpful policies for sea turtles (see below).

A negative effect of isolation is natural variation in resource levels. In a bad resource year, island populations may crash. This type of effect has been demonstrated on Barro Colorado Island, Panama. An El Niño followed by a mild and dry season resulted in lowered fruit production; mammalian frugivores suffered a mass famine (Wright et al. 1999). This effect could not be explained away by a lack of predators (which might lead to density dependence in an unregulated frugivore population). In such circumstances, island populations are less able to migrate short distances to better conditions, and so they suffer the generally lower habitat diversity found on islands. Mainland frugivore populations living in conditions of higher habitat diversity should have better capacity to withstand drought years and other vagaries of resource availability.

Much of the research on climate change effects on island populations of mammals relies on studies of arctic and antarctic mammals. There is growing consensus that ice-dependent marine mammals in the Arctic will fare worse than pelagic feeders, if expected warming scenarios are realized (Bluhm and Gradinger 2008). Ice-dependent feeders are more sensitive to climate warming (Laidre et al. 2008) because of the limited forage opportunities afforded by ice, and the dwindling area of melting ice (Joughin et al. 2003).

Synergistic Threats to Insular Populations

Island populations face many of the same threats from a changing climate as do mainland populations. In some cases, however, insularity complicates species responses to climate change. Recall that one in six global amphibian species is restricted to islands (figure 11.1); one in three amphibian species is already threatened with extinction, and the current extinction rate is 200 times the background rate (McCallum 2007). This group's life history—the fact that many amphibian species have both terrestrial and aquatic developmental stages—makes it particularly vulnerable to changes in climate and the likely changes in habitat that will accompany them.

To understand how climate will affect a species, we must understand how it affects it in both life stages. This is made more difficult when there is a density-dependent effect on the species' reaction to climate. A study of the Pacific tree frog (*Pseudacris regilla)* on Vancouver Island, British Columbia, showed that rising temperature correlated with increased tadpole survival when the tadpoles were at low density, and decreased survival when they were at high density (Govindarajulu and Anholt 2006).

While research on the effects of climate change on amphibian island endemics is somewhat lacking, one clear conservation issue stands out from the rest: the effect of the chytrid fungus *Batrachochytrium dendrobatidis* on island populations. This fungus is cosmopolitan and many species have been exposed to it. However, the warming climate is altering this interaction (Pounds et al. 2006), potentially making the fungus's damaging effects too great for many amphibian populations to sustain (Pounds et al. 2006). An assessment of amphibians in Trinidad and Tobago found that while populations of the critically endangered frog *Mannophryne olmonae* on the lowland of Tobago were unlikely to be made more susceptible to *B. dendrobatidis* due to warming climate, the presence of high montane environments on Trinidad brought increased risk of chytridiomycosis for *M. olmonae* on that island (Alemu et al. 2008). Amphibian declines on Puerto Rico are attributed to an interaction between chytridiomycosis and warming climate (Joglar et al. 2007); this situation will be made more severe if part of the effect of climate is an increased frequency of drought, causing frogs to aggregate around remaining water bodies. While this effect is not unique to islands (Pounds et al. 2006), the ability of any island population to recover will be hampered by its isolation and the relative low dispersal ability of frogs due to inhospitable habitat. Thus, a rescue effect for island populations is unlikely.

Interestingly, a field study of several frog species in Australia documented the effects of *B. dendrobatidis* in different frog species over time. Retallick et al. (2004) found that not all species exposed to the fungus showed incidence of infection. Of the species that were infected, the incidence did not change over time, remaining constant at 18% (*Taudactylus eungellensis*) and 28% (*Litoria wilcoxii/jungguy*), and the species did not decline during the period of the study. It appears that some species or even guilds of amphibians, for reasons that remain unknown, can sustain some level of chytrid infection. This may be because *B. dendrobatidis* grows and reproduces at temperatures in 4° C to 25° C; growth stops at 28° C, and 50% mortality occurs at 30° C. The climate tolerances of the chytrid fungus may be overwhelmed in Australia; so although Australia is more reasonably regarded as continental, the implication is that frogs on islands sharing its climate may be better able to withstand this cosmopolitan fungus.

Birds are arguably the most studied of all vertebrate taxonomic groups, so the literature on endemic island species is larger for birds than for the other taxonomic groups (figure 11.1). However, members of one group of endemic island birds will disproportionately suffer the effects of climate change: the seabirds. A crucial reason why there are so many endemic breeding seabirds on remote islands is that by breeding on those islands, seabirds escape the high predation pressure that exists on mainlands (Burger and Gochfeld 1994).

Seabirds require different habitats in which to nest (terrestrial areas) and forage (marine areas). Thus, they endure climate warming in two quite different habitats, and will presumably need to make adjustments to warming in each of them. In terrestrial areas they will endure the same issues as other mainland species, in that land conditions will change and many species may need to move their nesting areas to higher elevations. However, climate change in the marine realm will cause a cascade of effects that will likely ultimately lead to decreased food quality, changing distribution of food sources, compromised reproductive success, and interaction between multiple damaging effects.

The net effects of sea surface temperature (SST) change (a convenient index of "climate change") on seabirds have mainly been manifest as changes in prey distribution and quality, rather than as a direct physiological effect of changing temperature, though some research indicates that the strength of the effect depends on the foraging habitat and/or the trophic level of the main prey item of the particular seabird species studied (Kitaysky and Golubova 2000, Inchausti et al. 2003)

Many studies have shown a correlation between changing climate and decreased seabird abundance. Fewer snow petrels (Antarctica, *Pagodroma nivea*) bred when sea ice cover was extensive (Olivier et al. 2005). Abundance of Cassin's auklet (*Ptychoramphus aleuticus*) varied with winter Southern Oscillation Index (SOI), SST, and year-dependent annual variation (Lee et al. 2007). The Southern Oscillation index is a convenient index of climate change, calculated from seasonal or monthly variations in air pressure for the area extending from Tahiti to Darwin, Australia. A long period of negative SOI values is usually associated with a warming of the central and eastern tropical Pacific Ocean, lowered rainfall in eastern and northern Australia, and lowered intensity of the Pacific trade winds: an El Niño year. Sustained positive SOI values indicate a La Niña year, in which trade winds are strong and the waters north of Australia are warmer than average.

A relationship between SST and foraging success has been documented for sooty terns (*Onychoprion fuscatus*) of the Great Barrier Reef (Erwin and Congdon 2007). Parent foraging success and chick growth for wedge-tailed shearwater *Puffinus pacificus* correlates negatively with within-year and between-year variations in SST (Peck et al. 2004). Though a study of the effects of climate on fulmar population sizes shows that climate change effects can take time to manifest (Thompson and Ollason 2001), studies over longer time frames have shown a negative correlation between fulmar (*Fulmarus glacialis*) survival and the winter North Atlantic oscillation between 1962 and 1995 (Grosbois and Thompson 2005).

Rising sea surface temperature (SST) has been correlated with a 90% decline of sooty shearwaters; the authors implicate a change in the underlying food web (Veit et al. 1997). Changes in the underlying food web were again the likely mechanism of a correlation between changing SST and abundances of two murre species (Irons et al. 2008).

Decreased food availability or quality has critical effects on seabird breeding success. Warmer temperatures may induce prey switching, sometimes to prey at lower trophic levels. This phenomenon is known as the "junk food hypothesis" (Gremillet et al. 2008), as the lower trophic levels generally provide much less nutritional value per unit biomass.

Low-quality prey is typically characterized by its low lipid or energy content. There are a number of studies, both correlative and manipulated, that demonstrate the effects of a low- or high-lipid diet. Developing seabirds fed diets low in lipids suffered nutrition stress and lower cognitive abilities, which perhaps negatively affect future recruitment (Kitaysky et al. 2006). High-lipid diets are important to seabirds, resulting in increases in egg size (Sorensen et al. 2009), fledging weight (Dahdul and

Horn 2003), breeding success (Gremillet et al. 2008), chick survival or chick provisioning rate (Litzow et al. 2002), bone development (Sears and Hatch 2008), nestling growth (Romano et al. 2006), and colony success (Wanless et al. 2005). A switch from a high-energy food to a low-energy food does not necessarily lead to poorer chick condition; if prey are abundant enough, parents can increase the feeding frequency to make up the deficit in prey type (Eilertsen et al. 2008), assuming that the prey are within easy reach of the breeding colony. However, if the nutritional deficit is too large, the parents will not be able to bridge it.

The studies above spanned relatively short time scales. Effects of lower-quality diet have been demonstrated in longer-term studies as well. Stable isotope ratios for the last 50 years indicate that marbled murrelets (*Brachyramphus marmoratus*) have been feeding lower on the food chain (on krill) in more recent years (Becker and Beissinger 2006), and that population size can be closely approximated by the amount of fish in the murrelet diet (Norris et al. 2007). This switch from fish to krill has been implicated in the decline of this endangered seabird. A similar mechanism is operating in the following example, though the end effects are at the community level. In the late 1970s, following a change from a cool cycle to a warm cycle, forage fish (e.g., capelin) collapsed, groundfish and salmon became more abundant, and seabird and mammal populations declined (Anderson and Piatt 1999). The system has yet to recover from this reorganization; the authors speculate that this is because the groundfish resource requirements now outweigh the resource requirements of birds, mammals, and humans combined (Anderson and Piatt 1999).

The respective timing of prey and seabird life-history processes may also impact breeding success (Hipfner et al. 2008). Cassin's auklets have reduced breeding success in warm-water years; Hipfner et al. (2008) showed that the likely underlying mechanism is that copepod abundance peaked three weeks early in warmer years. The decreased availability of copepod in those years resulted in lowered auklet offspring survival, and also in lowered offspring biomass (Hipfner 2008), presumably because the auklets switched to a lower-quality prey source.

Climate warming is expected to result in poleward movement of fish stocks (Murawski 1993, Roessig et al. 2004, Lehodey et al. 2006). In some locations, dwindling fish resources will result in increased contact between seabirds and fishing operations. Seabird mortality has long been associated with long-line fisheries (Gilman et al. 2008), and in some cases it has been grossly underestimated (Brothers et al. 2010). However, reduction in seabird mortality is relatively easy to accomplish. After introduction of regulations encouraging fishing methods meant to reduce seabird bycatch rates, Gilman et al. (2008) found that the bycatch was reduced 67% by a variety of fishing methods, with one method (side netting with 60 g weights) resulting in no seabird bycatch at all. However, this was in one fishery only; disappointingly, measures of implementation worldwide show that many longline fishermen do not employ seabird avoidance measures, even though they are efficient and should result in reduced loss of bait (Gilman et al. 2005).

These results are important to the question of indirect climate effects on island seabird populations: change of location and intensity in fishing (resulting from fish stocks redistributing themselves as climate warms), and lower-quality food for nestlings and adults. Since many seabird species already live in "marginal" circumstances, breeding on far islands, these climate effects will be large. The link with climate is not an island effect per se, but since seabirds have evolved to take advantage of isolation, requiring long forage flights (and now, with lower diet quality, even longer forage flights), climate changes will affect them strongly.

Island species suffer the same problems as mainland species, and one such issue is that of coping with introduced species. Many seabird populations have been profoundly negatively impacted by introduced mammals (Blackburn et al. 2005, Russell and Le Corre 2009), usually via their predation on seabird eggs and chicks. However, this need not result in the extinction of nesting seabird populations. Quilfeldt et al. (2008) report that thin-billed prions (*Pachyptila belcheri*), largely restricted to New Island in the Falkland Islands, have managed to coexist with three introduced mammal species (cats, mice, and rats). The prion does comprise a portion of the rat and mouse diet, a larger proportion during the chick-rearing period, but rat and mouse populations are relatively small, and areas that provide cover for rodent populations also remain small (Quillfeldt et al. 2008). Direct management of rodent populations might be difficult, but management of areas that provide them with cover should be well within human abilities.

Changing temperatures can also affect changes in reproductive success for island populations of seal species that are not ice-dependent but instead breed on land. The Southern Oscillation Index (SOI) explains "some of" the variation in population size and reproductive rate for the southern elephant seal, *Mirounga leonina* (McMahon et al. 2009); cooler years yield higher pup body mass in northern elephant seals, reflecting higher forage quality, as well as higher foraging success of pregnant females (Le Boeuf 2005).

The Importance of Long-Term Monitoring
of Insular Populations

Like the tuatara, sea turtles and other large reptiles live long and are slow to reproduce. These characteristics make it difficult for researchers to detect trends in their physiological condition, demographics, or abundance dynamics. In a long-term monitoring study of *Sphenodon punctatus*, Moore et al. (2007) revealed a 50-year decline in body condition. However, because of seasonal and annual variation in body condition, the declining trend was not obvious until more than 22 years of data had been accumulated and analyzed. This result underscores the need for long-term monitoring to detect effects that take more time to materialize, particularly for populations that may be more vulnerable to changing climate.

Similar results emphasizing the importance of long-term monitoring have been found for sea turtle breeding cycles and their relationship to climate factors. Green sea turtles (*Chelonia mydas*) do not breed every year; the females spend more than a year accumulating stores of body fat in preparation for yolk production, migration to breeding grounds, and egg laying (Limpus and Nicholls 2000). Since the numbers of female turtles at known breeding beaches vary between years, sometimes by more than an order of magnitude, attempts to link those numbers with climate have had mixed results. However, researchers have now realized the importance of the females' preparations for breeding, and have noted that the numbers of females on a beach represents the numbers of females that have made preparations to breed about two years earlier. Thus, the numbers of breeding green turtles on Australian beaches can be correlated with the Southern Oscillation Index (SOI) after a two-year time lag (Limpus and Nicholls 2000).

Likewise, conservation actions for endangered seabirds are limited by a lack of knowledge of life history, particularly in the non-breeding season. The Bermuda petrel has only recently been documented in "new" parts of its non-breeding range: North Carolina (Wingate et al. 1998) and the Azores (Bried and Magalhaes 2004). Without this knowledge of species movement, it is difficult to conceive of adequate comprehensive conservation plans.

The research described above argues for monitoring to influence management policy to prevent or ameliorate the forseeable negative impact of climate warming. However, other research shows that some populations can adapt very quickly to a changing environment. Monitoring would reveal this happy circumstance as well, thus allowing conservation monies to be spent on more urgent situations. Fossil data indicates that morphological evolution, which reflects adaptation to a selective force, can occur very rapidly in mammals (Hill et al. 2008), and can in fact be faster on islands than in mainland areas (Millien 2006). Millien (2006) interprets this fast evolution as evidence that mammals do have the capacity to adapt quite quickly to rapidly changing conditions.

Management Options to Increase Resilience to Climate Change

For all taxa there are conservation actions that can help mitigate the effects of climate change. Low-lying island nations are uniting to share expertise and discuss strategies. In 2008 the Commonwealth of the Bahamas hosted a Global Island Partnership (GLISPA) event, where representatives from islands across the globe met to discuss general conservation issues, including problems of climate change and sea level rise. As part of this initiative, the Caribbean islands announced their plan to protect 10% of their marine resources by 2012. Similarly, the IUCN hosted a conference in Reunion in July 2008 to discuss biodiversity and climate change in the European Union's overseas entities. (More information can be found at http://www.iucn.org/about/union/secretariat/offices/europe/places/overseas/.) Thus, even in these most vulnerable locations actions are being taken by local conservationists, research scientists, and government entities to ameliorate the effects of climate change.

Coastal management will be important for beach-nesting sea turtles, and possibly for other beach-nesting reptiles such as tuataras. Coastal sites are more prone to erosion—and nests laid in beach sand may be in danger from tidal overwash—from the increased storm frequency expected to accompany climate warming. Sea turtle nests are thus in danger of tidal overwash, complete washout, or warming to the point where only female turtles can be produced.

Tidal overwash results in some level of arrested sea turtle embryo development (Caut et al. 2010), depending on the frequency and level of tidal overwash events. To counter this possibility, researchers have been examining the effects of relocating so-called "doomed" sea turtle nests, either further away from the high tide line or to hatcheries. Interestingly, while more experienced female turtles tend to choose better nest sites which are dryer or will not be eroded over the nesting season, over long time scales 97% of females at least once choose a site that will be eroded or inundated (Pfaller et al. 2009). The effects of moving nests can be measured in different ways (e.g., difference in hatch success or clutch size), and these differences are not consistent across studies. In some cases, short-distance nest relocation does not affect incubation duration or nest temperature (Tuttle and Rostal 2010), hatchling size (Pintus et al. 2009, Tuttle and Rostal 2010), or hatch success (Garcia et al. 2003, Tuttle and Rostal 2010). However, some researchers find that hatch success is greatest in nests that remain in situ (Almeida and Mendes 2007, Pintus et al. 2009, Medicci et al. 2010). Other indices of nest success (emergence percentage and recruitment percentage) tend to be higher in nests that have not been moved (Medicci et al. 2010).

Few studies have examined the impact of sea turtle nest relocation on nest sex ratio (but see Pintus et al. 2009, who found that nest relocation did not alter sex ratio). As above, the temperature of the nest is predictive of nest sex ratio, and sea turtle hatchling sex ratios are even now skewed strongly toward females. Exciting new research shows that if nests are moved to hatcheries, incubator technology can very precisely control the ratio of males to females (Lopez-Correa et al. 2010). Nests incubated by the method introduced in Lopez-Correa et al. (2010) resulted in hatch success between 74% and 96% with a mean success of 89%. While intensive management of sea turtle nests is probably not a viable long-term option for most conservation groups (Garcia et al. [2003] show that results are poor if funds are low), incubating nests and releasing hatchlings may be a feasible short-term alternative. This is particularly true if the endangering circumstance can be amelorated with time (see below).

Vegetated beaches provide cool microhabitats and protection against erosion. Habitat protection and restoration can maintain or increase the resilience of certain beach-nesting reptiles to climate change impact, if the impact includes skewed sex ratios or loss of nests to erosion. It is known that hawksbill turtle (*Eretmochelys imbricata*) nests are more numerous and successful in vegetated areas than in less vegetated areas (Ficetola 2007). Where natural disturbance is high or coastal plants have been lost to anthropogenic process, restoration of native beach plants can have positive impacts on the sex ratios of beach-nesting reptiles. However, natural regeneration of beach plants may be slow, as coastal seed banks are quite variable and have lower species diversity than areas above ground (Leicht-Young et al. 2009). Thus, a long-term plan for reptile nesting areas might be to restore vegetation communities in the long term while employing incubators in hatcheries for reptile nests prone either to loss to erosion or to heavily skewed sex ratios.

Finally, if island habitats should go so far as to become unsuitable for reptiles, there is a small body of research showing that ecosystem management can work quite well to establish or reestablish viable populations. In one assisted colonization experiment, transplants of island species to neighboring islands reversed this unsuitability over time. Transplantation of iguana (*Cyclura cychlura*) populations from Leaf Cay to Alligator Cay, Bahamas, had improved population growth over the founder population (Knapp 2001) and retained genetic diversity over the short term (10 years; Knapp and Malone 2003). Yet such assisted colonization risks impact on the recipient community (see Popescu and Hunter, this volume), which may be difficult to quantify. Elsewhere in the Bahamas, a reforestation project was initiated to recover lost populations of reptiles (Joglar et al. 2007). Native forest seedlings were planted in a deforested area and monitored for two years. Study sites were established there, and species richness and community structure were studied for 12 months. The change in microclimate conditions favored early-colonizing reptile species. The existence of these early colonizers then facilitated the colonization of larger predatory reptiles (Joglar et al. 2007). Joglar et al. (2007) state that the process was assisted by the close proximity of intact forest; even so, the success of this "restoration via community succession" is remarkable and encouraging.

Summary

Island species are subject to the same risks from climate change as their mainland counterparts, but their insularity also brings added risks and complications. While the importance of policy to ameliorate or lessen the impact of climate change cannot be overstated, many species and/or systems have shown a remarkable capacity to adapt or recover from very large disturbances (e.g., iguana populations in the Bahamas, mammal adaptation rates from the fossil record). Further, recent research shows that innovations in reproductive technology can provide short-term aid to lessen skew of turtle sex ratios, and even a one-time planting of native plants resulted in the rebuilding of natural community structure in Puerto Rican reptile communities. The key message from these case studies is that even a small input of help from humans can result in large benefits for island biodiversity.

LITERATURE CITED

Alcover, J. A., A. Sans, and M. Palmer. 1998. The extent of extinctions of mammals on islands. *Journal of Biogeography* 25:913–18.

Alemu, J. I., M. N. E. Cazabon, L. Dempewolf, A. Hailey, R. M. Lehtinen, R. P. Mannette, K. T. Naranjit, and A. C. J. Roach. 2008. Presence of the chytrid fungus *Batrachochytrium dendrobatidis* in populations of the critically endangered frog *Mannophryne olmonae* in Tobago, West Indies. *Ecohealth* 5:34–39.

Almeida, A. d. P., and S. L. Mendes. 2007. An analysis of the role of local fishermen in the conservation of the loggerhead turtle (*Caretta caretta*) in Pontal do Ipiranga, Linhares, ES, Brazil. *Biological Conservation* 134:106–12.

Anderson, P. J. and J. F. Piatt. 1999. Community reorganization in the Gulf of Alaska following ocean climate regime shift. *Marine Ecology Progress Series* 189:117–23.

Angeles, M. E., J. E. Gonzalez, D. J. Erickson, and J. L. Hernandez. 2007. Predictions of future climate change in the Caribbean region using global general circulation models. *International Journal of Climatology* 27:555–69.

Bahir, M. M., and T. D. Surasinghe. 2005. A conservation assessment of the Sri Lankan Agamidae (Reptilia: Sauria). *Raffles Bulletin of Zoology* 2:407–12.

Becker, B. H., and S. R. Beissinger. 2006. Centennial decline in the trophic level of an endangered seabird after fisheries decline. *Conservation Biology* 20:470–79.

Bickford, D., S. D. Howard, D. J. J. Ng, and J. A. Sheridan. 2010. Impacts of climate change on the amphibians and reptiles of Southeast Asia. *Biodiversity and Conservation* 19:1043–62.

Bindoff, N. L., J. Willebrand, V. Artale, A. Cazenave, J. Gregory, S. Gulev, K. Hanawa, C. L. Quéré, S. Levitus, Y. Nojiri, C. K. Shum, L. D. Talley, and A. Unnikrishnan. 2007. Observations: Oceanic climate change and sea level. In S. Solomon, D. Qin, M. Manning, Z. Chen, M. Marquis, K. B. Averyt, M. Tignor, and H. L. Miller, eds., *Climate Change 2007: The Physical Science Basis: Contribution of Working Group I to the Fourth Assessment Report of the Intergovernmental Panel on Climate Change.* Cambridge: Cambridge University Press.

Blackburn, T. M., O. L. Petchey, P. Cassey, and K. J. Gaston. 2005. Functional diversity of mammalian predators and extinction in island birds. *Ecology* 86:2916–23.

Bluhm, B. A., and R. Gradinger. 2008. Regional variability in food availability for arctic marine mammals. *Ecological Applications* 18:S77–96.

Bried, J., and M. C. Magalhaes. 2004. First Palearctic record of the endangered Bermuda petrel *Pterodroma cahow*. *Bulletin of the British Ornithologists' Club* 124:202–6.

Brooks, T., S. Pimm, V. Kapos, and C. Ravilious. 1999. Threat from deforestation to montane and lowland birds and mammals in insular South-east Asia. *Journal of Animal Ecology* 68:1061–78.

Brooks, T. M., R. A. Mittermeier, C. G. Mittermeier, G. A. B. da Fonseca, A. B. Rylands, W. R. Konstant, P. Flick, J. Pilgrim, S. Oldfield, G. Magin, and C. Hilton-Taylor. 2002. Habitat loss and extinction in the hotspots of biodiversity. *Conservation Biology* 16:909–23.

Brothers, N., A. R. Duckworth, C. Safina, and E. L. Gilman. 2010. Seabird Bycatch in Pelagic Longline Fisheries Is Grossly Underestimated when Using Only Haul Data. *PLoS ONE* 5:e12491.

Burger, J. and M. Gochfeld. 1994. Predation and effects of humans on island-nesting seabirds. In *Seabirds on Islands: Threats, Case Studies and Action Plans*, edited by D. N. Nettleship, J. Burger, and M. Gochfeld. Cambridge: Birdlife International.

Caut, S., E. Guirlet, and M. Girondot. 2010. Effect of tidal overwash on the embryonic development of leatherback turtles in French Guiana. *Marine Environmental Research* 69:254–61.

Cherian, A. 2007. Linkages between biodiversity conservation and global climate change in small island developing States (SIDS). *Natural Resources Forum* 31: 128–31.

Collar, N. J. 1992. *Threatened Birds of the Americas: The ICBP/IUCN Red Data Book.* Cambridge, MA: Smithsonian Institution Press.

Dahdul, W. M., and M. H. Horn. 2003. Energy allocation and postnatal growth in captive Elegant Tern (*Sterna elegans*) chicks: Responses to high- versus low-energy diets. *The Auk* 120:1069–81.

Dobson, A. F., and J. Madeiros. 2009. Threats facing Bermuda's breeding seabirds: Measures to assist future breeding success. In Fourth International Partners in Flight Conference, February 13–16, 2008: Tundra to Tropics, McAllen, Texas, 223–26.

Ebi, K. L., N. D. Lewis, and C. Corvalan. 2006. Climate variability and change and their potential health effects in small island states: Information for adaptation planning in the health sector. *Environmental Health Perspectives* 114:1957–63.

Eilertsen, K., R. T. Barrett, and T. Pedersen. 2008. Diet, growth and early survival of Atlantic Puffin (*Fratercula arctica*) chicks in North Norway. *Waterbirds* 31:107–14.

Erwin, C. A., and B. C. Congdon. 2007. Day-to-day variation in sea-surface temperature reduces sooty tern *Sterna fuscata* foraging success on the Great Barrier Reef, Australia. *Marine Ecology-Progress Series* 331:255–66.

Ficetola, G. F. 2007. The influence of beach features on nesting of the hawksbill turtle *Eretmochelys imbricata* in the Arabian Gulf. *Oryx* 41:402–5.

Fuentes, M. M. P. B., M. Hamann, and C. J. Limpus. 2010. Past, current and future thermal profiles of green turtle nesting grounds: implications from climate change. *Journal of Experimental Marine Biology and Ecology* 383:56–64.

Garcia, A., G. Ceballos, and R. Adaya. 2003. Intensive beach management as an improved sea turtle conservation strategy in Mexico. *Biological Conservation* 111:253–61.

Gilman, E., N. Brothers, and D. R. Kobayashi. 2005. Principles and approaches to abate seabird by-catch in longline fisheries. *Fish and Fisheries* 6:35–49.

Gilman, E., D. Kobayashi, and M. Chaloupka. 2008. Reducing seabird bycatch in the Hawaii longline tuna fishery. *Endangered Species Research* 5:309–23.

Govindarajulu, P. P. and B. R. Anholt. 2006. Interaction between biotic and abiotic factors determines tadpole survival rate under natural conditions. *Ecoscience* 13:413–21.

Gremillet, D., L. Pichegru, G. Kuntz, A. G. Woakes, S. Wilkinson, R. J. M. Crawford, and P. G. Ryan. 2008. A junk-food hypothesis for gannets feeding on fishery waste. *Proceedings of the Royal Society B* 275:1146–56.

Grosbois, V. and P. M. Thompson. 2005. North Atlantic climate variation influences survival in adult fulmars. *Oikos* 109:273–90.

Hastings, D. A., P. K. Dunbar, G. M. Elphingstone, M. Bootz, H. Murakami, H. Maruyama, H. Masaharu, P. Holland, J. Payne, N. A. Bryant, T. L. Logan, J.-P. Muller, G. Schreier, and J. S. MacDonald. 1999. *The Global Land One-Kilometer Base Elevation (GLOBE) Digital Elevation Model, Version 1.0*. National Oceanic and Atmospheric Administration.

Hays, G. C., A. C. Broderick, F. Glen, and B. J. Godley. 2003. Climate change and sea turtles: A 150-year reconstruction of incubation temperatures at a major marine turtle rookery. *Global Change Biology* 9:642–46.

Hill, M. E., M. G. Hill, and C. C. Widga. 2008. Late Quaternary bison diminution on the Great Plains of North America: Evaluating the role of human hunting versus climate change. *Quaternary Science Reviews* 27:1752–71.

Hipfner, J. M. 2008. Matches and mismatches: Ocean climate, prey phenology and breeding success in a zooplanktivorous seabird. *Marine Ecology-Progress Series* 368:295–304.

Inchausti, P., C. Guinet, M. Koudil, J. P. Durbec, C. Barbraud, H. Weimerskirch, Y. Cherel, and P. Jouventin. 2003. Inter-annual variability in the breeding performance of seabirds in relation to oceanographic anomalies that affect the Crozet and the Kerguelen sectors of the Southern Ocean. *Journal of Avian Biology* 34:170–76.

Irons, D. B., T. Anker-Nilssen, A. J. Gaston, G. V. Byrd, K. Falk, G. Gilchrist, M. Hario, M. Hjernquist, Y. V. Krasnov, A. Mosbech, B. Olsen, A. Petersen, J. B. Reid, G. J. Robertson, H. Strom, and K. D. Wohl. 2008. Fluctuations in circumpolar seabird populations linked to climate oscillations. *Global Change Biology* 14:1455–63.

IUCN. 2010. *Pterodroma cahow. IUCN 2010. IUCN Red List of Threatened Species.* Version 2010.3. <www.iucnredlist.org>. IUCN.

IUCN, C. International, and NatureServe. 2008. An analysis of amphibians on the 2008 IUCN Red List <www.iucnredlist.org/amphibians>.

Joglar, R. L., A. O. Alvarez, T. M. Aide, D. Barber, P. A. Burrowes, M. A. Garcia, A. Leon-Cardona, A. V. Longo, N. Perez-Buitrago, A. Puente, N. Rios-Lopez, and P. J. Tolson. 2007. Conserving the Puerto Rican herpetofauna. *Applied Herpetology* 4:327–45.

Joughin, I., E. Rignot, C. E. Rosanova, B. K. Lucchitta, and J. Bohlander. 2003. Timing of recent accelerations of Pine Island glacier, Antarctica. *Geophysical Research Letters* 30:1706–1170.

Kamel, S. J., and N. Mrosovsky. 2006. Deforestation: Risk of sex ratio distortion in hawksbill sea turtles. *Ecological Applications* 16:923–31.

Kier, G., H. Kreft, T. M. Lee, W. Jetz, P. L. Ibisch, C. Nowicki, J. Mutke, and W. Barthlott. 2009. A global assessment of endemism and species richness across island and mainland regions. *Proceedings of the National Academy of Sciences, USA* 106:9322–27.

Kitaysky, A. S. and E. G. Golubova. 2000. Climate change causes contrasting trends in reproductive performance of planktivorous and piscivorous alcids. *Journal of Animal Ecology* 69:248–62.

Kitaysky, A. S., E. V. Kitaiskaia, J. F. Piatt, and J. C. Wingfield. 2006. A mechanistic link between chick diet and decline in seabirds? *Proceedings of the Royal Society B* 273:445–50.

Knapp, C. R. 2001. Status of a translocated *Cyclura iguana* colony in the Bahamas. *Journal of Herpetology* 35:239–48.

Knapp, C. R. and C. L. Malone. 2003. Patterns of reproductive success and genetic variability in a translocated iguana population. *Herpetologica* 59:195–202.

Lack, D. 1976. *Island Biology Illustrated by the Land Birds of Jamaica*. Berkeley: University of California Press.

Laidre, K. L., I. Stirling, L. F. Lowry, O. Wiig, M. P. Heide-Jorgensen, and S. H. Ferguson. 2008. Quantifying the sensitivity of arctic marine mammals to climate-induced habitat change. *Ecological Applications* 18:S97–125.

Le Boeuf, B. J. a. D. E. C. 2005. Ocean climate and seal condition. *BMC Biology* 3:9–18.

Lee, D. E., N. Nur, and W. J. Sydeman. 2007. Climate and demography of the planktivorous Cassin's auklet *Ptychoramphus aleuticus* off northern California: implications for population change. *Journal of Animal Ecology* 76:337–47.

Lehodey, P., J. Alheit, M. Barange, T. Baumgartner, G. Beaugrand, K. Drinkwater, J.-M. Fromentin, S. R. Hare, G. Ottersen, R. I. Perry, C. Roy, C. D. V. D. Lingern, and F. Werner. 2006. Climate variability, fish, and fisheries. *Journal of Climate* 19:5009–30.

Leicht-Young, S., N. B. Pavlovic, R. Grundel, and K. J. Frohnapple. 2009. A comparison of seed banks across a sand dune successional gradient at Lake Michigan dunes (Indiana, USA). *Plant Ecology* 202:299–308.

Limpus, C. and N. Nicholls. 2000. ENSO regulation of indo-Pacific green turtle populations. *Applications of Seasonal Climate Forecasting in Agricultural and Natural Ecosystems* 21:399–408.

Litzow, M. A., J. F. Piatt, A. K. Prichard, and D. D. Roby. 2002. Response of pigeon guillemots to variable abundance of high-lipid and low-lipid prey. *Oecologia* 132:286–95.

Lopez-Correa, J., M. A. Porta-Gandara, J. Gutierrez, and V. M. Gomez-Munoz. 2010. A novel incubator to simulate the natural thermal environment of sea turtle eggs. *Journal of Thermal Biology* 35:138–42.

Manne, L. L., T. M. Brooks, and S. L. Pimm. 1999. Relative risk of extinction of passerine birds on continents and islands. *Nature* 399:258–61.

McCallum, M. L. 2007. Amphibian decline or extinction? Current declines dwarf background extinction rate. *Journal of Herpetology* 41:483–91.

McMahon, C. R., M. N. Bester, M. A. Hindell, B. W. Brook, and C. J. A. Bradshaw. 2009. Shifting trends: detecting environmentally mediated regulation in long-lived marine vertebrates using time-series data. *Oecologia* 159:69–82.

Medicci, M. R., J. Buitrago, and H. J. Guada. 2010. Reproductive biology of the "Cardon" turtle (*Dermochelys coriacea*) in beaches of the peninsula of Paria, Venezuela, during nesting seasons 2000–2006. *Interciencia* 35:263–70.

Millien, V. 2006. Morphological evolution is accelerated among island mammals. *PLOS Biology* 4:1863–68.

Mitchell, N. J., M. R. Kearney, N. J. Nelson, and W. P. Porter. 2008. Predicting the fate of a living fossil: How will global warming affect sex determination and hatching phenology in tuatara? *Proceedings of the Royal Society B* 275:2185–93.

Moore, J. A., J. M. Hoare, C. H. Daugherty, and N. J. Nelson. 2007. Waiting reveals waning weight: Monitoring over 54 years shows a decline in body condition of a long-lived reptile (tuatara, *Sphenodon punctatus*). *Biological Conservation* 135:181–88.

Murawski, S. A. 1993. Climate change and marine fish distributions: forecasting from historical analogy. *Transactions of the American Fisheries Society* 122:647–58.

Norris, D. R., P. Arcese, D. Preikshot, D. F. Bertram, and T. K. Kyser. 2007. Diet reconstruction and historic population dynamics in a threatened seabird. *Journal of Applied Ecology* 44:875–84.

Olivier, F., J. A. v. Franeker, J. C. S. Creuwels, and E. J. Woehler. 2005. Variations of snow petrel breeding success in relation to sea-ice extent: Detecting local response to large-scale processes? *Polar Biology* 28:687–99.

Peck, D. R., B. V. Smithers, A. K. Krockenberger, and B. C. Congdon. 2004. Sea surface temperature constrains wedge-tailed shearwater foraging success within breeding seasons. *Marine Ecology Progress Series* 281:259–66.

Pfaller, J. B., C. J. Limpus, and K. A. Bjorndal. 2009. Nest-site selection in individual loggerhead turtles and consequences for doomed-egg relocation. *Conservation Biology* 23:72–80.

Pintus, K. J., B. J. Godley, A. McGowan, and A. C. Broderick. 2009. Impact of clutch relocation on green turtle offspring. *Journal of Wildlife Management* 73:1151–57.

Pounds, J. A., M. R. Bustamante, L. A. Coloma, J. A. Consuegra, M. P. L. Fogden, P. N.

Foster, E. L. Marca, K. L. Masters, A. Merino-Viteri, R. Puschendorf, S. R. Ron, G. A. Sánchez-Azofeifa, C. J. Still, and B. E. Young. 2006. Widespread amphibian extinctions from epidemic disease driven by global warming. *Nature* 439:161–67.

Pritchard, H. D., R. J. Arthern, D. G. Vaughan, and L. A. Edwards. 2009. Extensive dynamic thinning on the margins of the Greenland and Antarctic ice sheets. *Nature* 461:971–75.

Quillfeldt, P., I. Schenk, R. A. R. McGill, I. J. Strange, J. F. Masello, A. Gladbach, V. Roesch, and R. W. Furness. 2008. Introduced mammals coexist with seabirds at New Island, Falkland Islands: Abundance, habitat preferences, and stable isotope analysis of diet. *Polar Biology* 31:333–49.

Retallick, R. W. R., H. McCallum, and R. Speare. 2004. Endemic infection of the amphibian chytrid fungus in a frog community post-decline. *PLOS Biology* 2:e351.

Roessig, J. M., C. M. Woodley, J. J. Cech, and L. J. Hansen. 2004. Effects of global climate change on marine and estuarine fishes and fisheries. *Reviews in Fish Biology and Fisheries* 14:251–75.

Romano, M. D., J. F. Piatt, and D. D. Roby. 2006. Testing the junk-food hypothesis on marine birds: effects of prey type on growth and development. *Waterbirds* 29:407–14.

Russell, J. C. and M. Le Corre. 2009. Introduced mammal impacts on seabirds in the Îles Éparses, Western Indian Ocean. *Marine Ornithology* 37:121–29.

Sears, J. and S. A. Hatch. 2008. Rhinoceros auklet developmental responses to food limitation: an experimental study. *Condor* 110:709–17.

Simmons, R. E., P. Barnard, W. R. J. Dean, G. F. Midgley, W. Thuiller, and G. Hughes. 2004. Climate change and birds: Perspectives and prospects from southern Africa. *Ostrich* 75:295–308.

Thompson, P. M., and J. C. Ollason. 2001. Lagged effects of ocean climate change on fulmar population dynamics. *Nature* 413:417–20.

Tuttle, J., and D. Rostal. 2010. Effects of nest relocation on nest temperature and embryonic development of loggerhead sea turtles (*Caretta caretta*). *Chelonian Conservation and Biology* 9:1–7.

Van den Broeke, M., J. Bamber, J. Ettema, E. Rignot, E. Schrama, W. J. van de Berg, E. van Meijgaard, I. Velicogna, and B. Wouters. 2009. Partitioning recent Greenland mass loss. *Science* 326:984–86.

Veit, R. R., J. A. McGowan, D. G. Ainley, T. R. Wahl, and P. Pyle. 1997. Apex marine predator declines ninety percent in association with changing oceanic climate. *Global Change Biology* 3:23–28.

Wingate, D. B. 1985. The restoration of Nonsuch Island as a living museum of Bermuda's pre–colonial terrestrial biome. In *International Council for the Preservation of Birds Technical Publication No. 3*, edited by P. J. Moor, 225–38.

Wingate, D. B., T. Hass, E. S. Brinkley, and J. B. Patteson. 1998. Identification of Bermuda petrel. *Birding* 30:18–36.

World Conservation Monitoring Centre of the United Nations Environment Programme (UNEP-WCMC). 2004. Species data. World Conservation Monitoring Centre.

Wright, S. J., C. Carrasco, O. Calderon, and S. Paton. 1999. The El Niño Southern Oscillation variable fruit production, and famine in a tropical forest. *Ecology* 80:1632–47.

12

Retreat of the American Pika: Up the Mountain or into the Void?

Chris Ray, Erik Beever, and Scott Loarie

As the climate warms, the ranges of many species may track more favorable climates by shifting poleward or toward higher elevations. For good dispersers in mountainous terrain, an upslope range shift may be a quick and common response to a warming climate, and the ranges of alpine mammals seem especially likely to retreat upslope (Guralnick 2007, Loarie 2009). Unfortunately, an upslope range shift is likely to involve tradeoffs: the climate may improve with elevation, but the area and connectivity of available habitats may not (Sekercioglu et al. 2008, Guralnick et al. in press). As ranges shrink and connectivity is lost, remnant populations become less resilient to any threat, climatic or otherwise (Gilpin and Soulé 1986). To paraphrase the popular press, a warming climate may force cold-adapted alpine species to walk a narrowing plank toward extinction.

The American pika (*Ochotona princeps*) has become an icon of such climate-mediated endangerment (Krajick 2004, Lee 2005, Nijhuis 2005, Tolmé 2006, Blakemore 2007, Lagorio 2007, Siegler 2007, Frey 2007, Webb 2007, Unrau 2008, Bertrand 2008). A relic of the last ice age (Mead 1987, Grayson 2005), this "rock rabbit" is currently restricted to some of the most remote habitats in western North America—places where rock piles dominate the landscape (Smith and Weston 1990). In this chapter we review evidence that a recent shift in this species' range is related to climate and, more importantly, we summarize the relative support for alternative mechanistic hypotheses that may explain this range shift. Because the consequences of climate change appear so dire for alpine species (Pounds et al. 1999, Root et al. 2003, Hijmans and Graham 2006, Parmesan 2006, Pimm et al. 2006, Sekercioglu et al. 2008, Trivedi et al. 2008), it is especially important that ecologists identify the mechanisms that drive any impact on these species. Only a mechanistic understanding can fully support the development of effective strategies for adaptation, mitigation, conservation, monitoring, and management. Furthermore, because climate-mediated threats can be addressed only through unprecedented cooperation at the global scale, demonstrating links between climatic threats to humans and to other species may be necessary for prompting appropriate action.

There has long been evidence that climate limits the distribution of the American pika (Merriam 1894; Grinnell 1917; Smith 1974a; MacArthur and Wang 1973, 1974), and reason to hypothesize that projected trends in climate could cause a dramatic contraction of its range (Murphy and Weiss 1992, McDonald and Brown 1992). Because pikas do not hibernate, they maintain a high metabolic rate to compensate for heat loss from their small (approximately 150-gram) bodies during winter. As a consequence, the resting body temperature of *O. princeps* is near its lethal maximum (MacArthur and Wang 1973). A narrow thermal tolerance limits surface activity in this species, especially during summer days (MacArthur and Wang 1974, Smith 1974a), which is precisely when these animals must actively gather and cache food for the winter (Huntly et al. 1986, Smith and Weston 1990, Dearing 1997). Individuals mitigate thermal stress by inhabiting the voids between rocks in taluses or lava beds—voids that provide refugia from summer heat as well as relatively dry storage for cached food and protection from predators (Smith and Weston 1990).

Evidence for recent effects of climate on the distribution of the pika (*O. princeps* unless otherwise noted) comes from several sources. Moritz et al. (2008) documented a pattern of upslope range retraction for pikas and other species in the Sierra Nevada mountains of California by resampling the "Yosemite transect" first sampled by Joseph Grinnell in the early 1900s. Along that transect, the lower limit of the pika's range appears to have risen more than 150 meters during the past century. Upslope range retraction suggests that pikas and several other Sierran species have been lost where summer temperatures have risen (Moritz et al. 2008). A different perspective on the mechanism(s) driving this range retraction comes from demographic studies of both *O. princeps* and its northern relative, *O. collaris*. In both species, survival over the winter can be lower where the snowpack is of insufficient depth or quality to provide adequate insulation from subzero temperatures (Tapper 1973, Morrison and Hik 2007). Studies have also documented effects of the timing and length of the frost-free season on survival and recruitment in both species (Tapper 1973, Smith 1978, Kreuzer and Huntly 2003, Morrison and Hik 2007).

The majority of evidence for effects of climate on *O. princeps* derives from studies within the Great Basin, a vast region of basin-and-range topography between the Sierra Nevada and the Rocky Mountains. In this region, the species has been losing ground for at least 10,000 years: the minimum elevation of the pika's distribution has retracted upslope by nearly 800 meters since the last glacial maximum, eliminating some populations and isolating others (Grayson 2005). Genetic analyses support this pattern of recent demographic contraction, showing an unprecedented decline in effective population size throughout the range of *O. princeps* following the last glacial maximum (Galbreath et al. 2009).

The rate of pika losses within the Great Basin has dramatically accelerated since prehistoric times. Historical losses were documented by surveying 25 sites in the Great Basin that were initially sampled primarily by E. Raymond Hall in the early 1900s (Beever et al. 2003). As along the Yosemite transect, the lowest records of pikas at these sites has risen more than 150 meters during the past century (Grayson 2005), and additional losses within the past decade put the current rate of uphill retreat at more than 13 meters per year (Beever et al. 2011). In this dataset, the lower elevational limit of the pika's distribution is racing uphill much faster than any other climate-related range retraction reported in the literature to date (cf. Kullman 2002, Parmesan and Yohe 2003, Wilson et al. 2005, Parmesan 2006, Moritz et al. 2008, Lenoir et al. 2008, Raxworthy et al. 2008). Below, we recount our attempt to infer the driver(s) of this changing distribution.

Quantifying Patterns of Range Retraction

Pikas were first recorded at these 25 sites between 1898 and 1990 (Beever et al. 2003). Each site was surveyed during two recent periods, first during 1994–99 (Beever et al. 2003) and second during 2003–8 (Beever et al. 2010). During each period, sites of apparent extirpation were surveyed multiple times. There are many ways to define species loss (Rickart 2001, Parmesan and Yohe 2003, Walther et al. 2005, Wilson et al. 2005, Root and Schneider 2006), and for this analysis, we used local extinction—extirpation—as our primary metric of species response. Extirpations can be difficult to document (Diamond 1987, Kéry 2002, Rowe 2005, Shoo et al. 2006), but the pika offers more conspicuous evidence of occupancy than most other mammals. Detection of species presence is facilitated by the pika's open habitat (which increases sightings) as well as its characteristic food caches, territorial vocalizations, and distinctive scat (Smith 1974b, Smith and Weston 1990, Beever et al. 2003). Detection probability for pikas has been estimated repeatedly at more than 0.90 (Ray and Beever 2007, Beever et al. 2008, Beever et al. 2010, Rodhouse et al. 2010). Each site was surveyed within a radius of three kilometers, which exceeds the dispersal distance of most individuals (Hafner and Sullivan 1995; Peacock 1997a, b; Peacock and Smith 1997). This generous search area resulted in a conservative estimate of the number of extirpa-

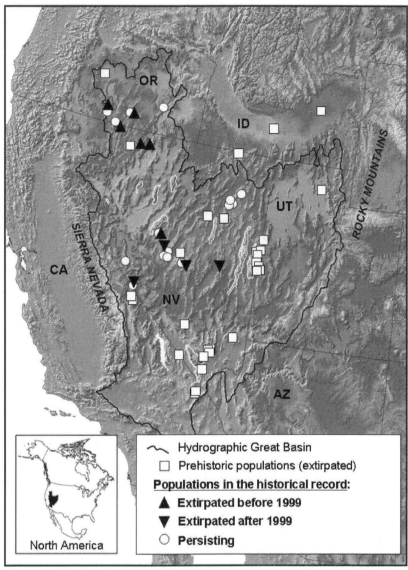

Figure 12.1. Records of the American pika (*Ochotona princeps*) within the hydrographic Great Basin, including sites with evidence of prehistoric occupancy and 25 populations in the historical record. Locations in the historical record were surveyed repeatedly for pika signs during the 1990s and 2000s. Surveys encompassed all pika habitat (taluses) within 3 km of the historical record. Ten extirpations were documented, including four since 1999. No recolonizations have been detected during any survey of these 10 locations.

tions: four of the populations we classify as "persisting" have experienced upslope range retraction of at least 200 meters (Beever et al. 2010).

The pattern of extirpation observed (figure 12.1) demonstrates continuing loss of connectivity among persisting populations (sensu Grayson 2005) and accelerating loss of populations. The mean and median dates of historical records for these 25 sites coincide at 1933. Using 1933, 1999, and 2008 as proxy dates for defining the periods between surveys, the rate of extirpation increased from about one population per decade before 1999 to about four per decade after 1999. Similarly, the rate of upslope range retraction also increased from about 13 meters per decade before 1999 to about 145 meters per decade after 1999 (Beever et al. 2011). The minimum elevation of occupied taluses rose nearly as much between 1999 and 2008 as during the much longer period prior to 1999. For the 15 populations persisting through 2008, the minimum elevation of occupied taluses in historical records was 2,375 ± 94 m (mean ± SE), rising to 2,475 ± 93 m by 1999 and 2,560 ± 97 m by 2008. Including populations that eventually were extirpated, the minimum elevation of occupied taluses was 2,286±74 m in historical records ($n = 25$) and 2,465 ± 75 m by 1999 ($n = 19$). Only six of the 25 sites surveyed exhibited evidence of long-term stability in the lower elevational limit of occupancy (figure 12.2). It is unlikely that the extirpations we report are the temporary result of metapopulation dynamics, because site recolonization has

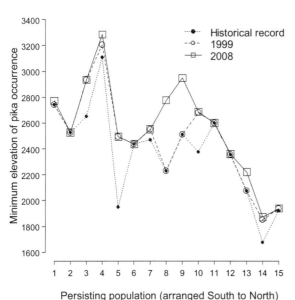

Figure 12.2. Variation in the upslope retraction of Great Basin pika populations studied by Beever et al. (in review), after excluding sites of extirpation ($n = 10$). Populations at these 15 sites were originally recorded between 1898 and 1956. Historical records likely represent locations at which pikas were encountered opportunistically, and may be biased upward relative to the minimum elevations actually occupied by historical populations.

Legend:
- Historical record
- 1999
- 2008

Y-axis: Minimum elevation of pika occurrence
X-axis: Persisting population (arranged South to North)

never been detected despite multiple surveys at each site between 1994 and 2008 (Beever et al. 2010; Beever et al. 2011).

Identifying Correlates and Inferring Drivers of Extirpation

In an exploratory analysis, Beever et al. (2003) explained the pattern of historical extirpations through 1999 in relation to numerous potential predictor variables, including amount of habitat, habitat connectivity, direct anthropogenic effects, and climate. Each of these variables was defined in terms of proxy data that were readily available or easily collected. As a proxy for potential heat stress experienced by pikas at each site, Beever et al. (2003) used PRISM estimates of average daily maximum temperatures in August during 1961–90 (AugMaxT), modeled to a resolution of approximately four kilometers (Daly et al. 1994). As a proxy for connectivity with a climatic refugium or potential source of population rescue, Beever et al. (2003) determined the maximum elevation of talus within three kilometers of each site (MaxElev). A related predictor (MaxElevR) was later defined (Beever et al. 2010) as the residual from a regression of MaxElev on latitude. The relationship between MaxElev and latitude clearly explains much of the variance in persistence across the basin (figure 12.3), suggesting that MaxElevR may approximate the integrated effects of climate and habitat connectivity: more southerly populations require higher local refugia in order to withstand a warmer, drier climate. The extent of available habitat was approximated in two ways. First, the amount and distribution of talus were classified at the scale of the surrounding mountain range (RangeHab = 1 if amount and connectivity were relatively high, 0 otherwise; sensu Beever 1999). Second, the amount of habitat within 800 m of the historical site (LocalHab) was quantified through ground surveys (Beever 1999). Proxies for anthropogenic impacts included a classification of long-term grazing by livestock (Grazing = 1 if present, 0 if absent; sensu Beever 1999) and a quantification of accessibility approximated as the straight-line distance to the nearest road passable by vehicles without four-wheel drive (DistRoad).

In the original analysis, five of the above proxies (AugMaxT, MaxElev, RangeHab, Grazing, and DistRoad) were combined into models designed to represent alternative explanations for the pattern of pika persistence/extirpation observed through 1999 (Beever et al. 2003). These five predictor variables were selected from among more than a dozen measured variables chosen for their potential to affect pika dynamics. In order to reduce the number of potential models constructed, the number of potential predictor variables was pruned using correlation analyses (to identify redundant covariates) and regression analyses (to identify poor univariate

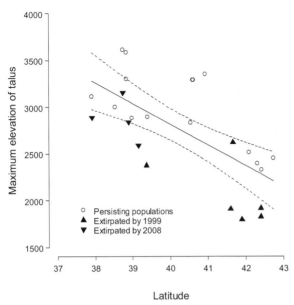

Figure 12.3. Relationship between latitude (decimal degrees) and a metric of elevational refuge, the elevation (m) of the highest talus within a pika's dispersal range (3 km), for 25 pika populations in the historical record. The strong linear relationship between elevational refuge and latitude ($y = 11652 - 221x$; adjusted $R^2 = 0.425$; $F = 18.75$; $df = 1,23$; $P < 0.001$) reflects a similar relationship between latitude and the maximum elevation of mountain ranges in the Great Basin. Pikas persist within the basin primarily where the residuals of this regression are positive. Residuals of this regression were used in models as a predictive index of connectivity with an upslope refuge or source of immigration (MaxElevR).

predictors). The use of statistical considerations to cull candidate models can weaken an information-theoretic analysis (Burnham and Anderson 2002). We emphasize this point because the conundrum encountered in this analysis—too many potential predictor variables—is likely to arise in many studies, especially given the many variables that can be derived from electronic databases, geographic information systems, remote sensing, and climate models. Until key mechanisms are better understood, exploratory analyses are often necessary, but a priori selection of alternative models and their constituent predictors will strengthen inference in any analysis.

Despite the exploratory nature of the first analysis, the primary result was corroborated by subsequent analyses. The most important predictor of pika persistence in the Great Basin through 1999 was the maximum elevation of local habitat (Beever et al. 2003). In subsequent analyses

Table 12.1. Confronting models developed in the 1990s with data on persistence from three periods (historical–1999, historical–2008, and 1999–2008) reveals consistent support for a metric of elevational refuge (MaxElevR) as a predictor of both long-term and recent population dynamics. For each period, the response variable was modeled using the same set of $m = 21$ candidate models. All models with strong support ($\Delta AIC_c < 2$) are summarized here by predictor(s), log likelihood (LL), number of fitted parameters (K), and relative support statistics including AIC_c (Akaike's information criterion, corrected for small sample size), ΔAIC_c (the difference between AIC_c in the focal model and in the model with lowest/best AIC_c), and Akaike weight (the relative weight of evidence in favor of each model). Predictors are as described in figure 12.4. Of 25 historical populations, 19 persisted through 1999, reducing sample size (n) for the third analysis below. The signs of all fitted coefficients were as expected: for example, persistence was positively related to MaxElevR, RangeHab, and DistRoad, and negatively related to Grazing and AugMaxT.

Best models (predictors)	LL	K	AIC_c	ΔAIC_c	Weight
Persistence of historical populations through 1999 (n = 25, m = 21)					
MaxElevR, RangeHab	−6.57	3	20.292	0.000	0.203
MaxElevR	−8.33	2	21.204	0.912	0.129
MaxElevR, RangeHab, DistRoad	−5.60	4	21.207	0.916	0.129
RangeHab, DistRoad	−7.53	3	22.207	1.915	0.078
MaxElevR, RangeHab, Grazing	−6.14	4	22.275	1.983	0.075
Persistence of historical populations through 2008 (n = 25, m = 21)					
MaxElevR, RangeHab, Grazing	−5.22	4	20.430	0.000	0.222
MaxElevR, Grazing	−6.73	3	20.594	0.164	0.204
MaxElevR, RangeHab	−7.02	3	21.179	0.749	0.152
MaxElevR	−8.64	2	21.829	1.399	0.110
Persistence of 1999 populations through 2008 (n = 19, m = 21)					
MaxElevR	−5.48	2	15.700	0.000	0.254
MaxElevR, RangeHab	−4.61	3	16.811	1.111	0.146
MaxElevR, Grazing	−4.72	3	17.036	1.336	0.130
MaxElevR, AugMaxT	−4.84	3	17.274	1.574	0.116

(Beever et al. 2011), MaxElev was replaced by MaxElevR, and the otherwise original set of candidate models was used to predict pika persistence across three periods of time (table 12.1). Across all periods, all but one of the models with strong support included MaxElevR, and models based solely on MaxElevR were strongly supported in each period.

Because the presence of an elevational refuge was so strongly supported as a predictor of persistence in the Great Basin, other predictors with apparent support may have merely "hitchhiked" (Anderson 2008) on the support garnered by MaxElevR in multivariate models. However, we would expect the relative support for hitchhiking predictors to vary unpredictably among datasets, which is not what we observed. Support for models based on RangeHab declined steadily over time, being strongest for dynamics through 1999, when one of the best models was based on RangeHab and DistRoad—the only strongly supported model that did not include MaxElevR. By the period 1999–2008, RangeHab was included in only one of the best models of persistence. The pattern of support for models based on Grazing suggested a lagged effect. For models of persistence during 1999–2008, the value of Grazing was based on data from the same period (i.e., data that were not lagged). For models of persistence from historical records through either 1999 or 2008, the value of Grazing was based on data through 1999, and Grazing was best supported as a predictor for models of persistence through 2008 (when persistence data were somewhat lagged). Models based on DistRoad were only supported for patterns of persistence through 1999. AugMaxT did not appear among supported models until after 1999. This recent support for a putative metric of heat stress, combined with recently increasing support for a metric of elevational refuge (MaxElevR), suggests a rise in the importance of climate as a factor in recent pika dynamics. The evolution of support for each individual predictor is shown in figure 12.4.

Improving Models of Climate-Mediated Extirpation

Given the recent support for effects of climate on pika dynamics in this study and others (Kreuzer and Huntly 2003, Smith et al. 2004, Morrison and Hik 2007), we wanted to test for these effects more directly and mechanistically. To improve the climate data, Beever et al. (2010) used thermal sensors to record subsurface temperatures in the talus within several currently or formerly occupied pika territories at each study site. Because permission to install sensors was not obtained for one of the original 25 sites, sensor data from the site of a recently documented population at a comparable latitude (Beever et al. 2008) were substituted in our analysis. Sensors at each of these 25 sites recorded temperatures every two to four hours during 2005 and 2006. These data were used not only to characterize recent temperatures, but also to model temperatures within each site over the past six decades (1945–2006). These hindcasts were developed by regressing data from the in-situ sensors on data from weather stations within the Historical Climate Network. Although distances between HCN stations and study sites were often more than 40 km, a strong correlation was found between data recorded at each study site and at least one HCN station (e.g., $0.87 \leq r \leq 0.98$ for daily mean temperatures). For each study site, the best relationship between in-situ and HCN data from 2005–6 was

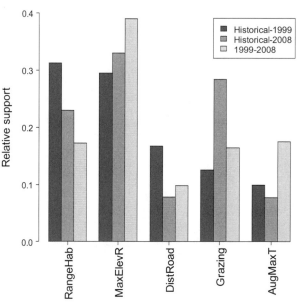

Figure 12.4. Relative support for potential predictors of pika persistence when confronted with data from three different time periods. Observed persistence within each time period was predicted using the same set of candidate models, in which each logistic regression model was based on up to four of five predictors. Within each period, the relative support for each predictor p was calculated as the sum of Akaike weights (table 12.1) over all models containing p divided by the number of models containing p, and rescaled for presentation such that the relative support for all predictors sums to unity. Predictors (defined further in text): RangeHab = presence (1) or absence (0) of extensive talus within a mountain range surrounding the historical site; MaxElevR = residual maximum elevation of accessible talus versus latitude, a metric of upslope refuge (see figure 12.3); DistRoad = distance (km) to the nearest road passable by vehicles without four-wheel drive; Grazing = presence (1) or absence (0) of long-term livestock grazing at the historical site prior to 1999 (or between1999 and 2008 for models of 1999–2008 persistence); AugMaxT = average daily maximum temperature for August 1–31, 1961–90 (or 1971–2000 for models of 1999–2008 persistence).

used to hindcast metrics of subsurface temperature for each year, stretching back to 1945, based on HCN data. Each hindcast was divided into two periods, roughly 0–30 ybp (years before present) and 30–60 ybp, in order to characterize both the prevailing climate (e.g., summer temperature averaged over both periods) and climate change (e.g., the difference between the two 30-year periods in average summer temperature).

Taking advantage of more detailed data on the relevant microclimate, Beever et al. (2010) posited alternative ways in which pikas might experience thermal stress in the basin, including chronic heat stress (high average summer temperature), acute heat stress (the number of daily maxima above 28° C), and acute cold stress (the number of daily minima below 0° or –5° C). For acute heat stress, there were prior data suggesting specific thresholds relevant to pikas (MacArthur and Wang 1973, Smith 1974a). This was not true for acute cold stress, which is one reason why multiple thresholds were considered. The choice of 0° C as a threshold for stress may appear odd for this cold-adapted species, especially because pikas tend to occur only where ambient temperatures remain below 0° C for more than 180 days per year (Hafner 1993, 1994). However, our thresholds apply to conditions beneath the talus, where we hypothesize that pikas in suitable sites are protected from subzero ambient temperatures by an insulating snowpack.

These three types of thermal stress were predicted to affect populations in three different ways. First, thermal stress may have a cumulative effect on individuals over time, or the long-term average of a stress metric may contain less noise. In either case, extirpation should occur at sites exhibiting higher metrics of thermal stress in the prevailing climate (0–60 ybp). These may be sites at the edge of the bioclimatic envelope of the species' range. Second, populations may be stressed by climate change. In this case, extirpations should occur at sites exhibiting larger changes in stress metrics between periods (0–30 versus 30–60 ybp). Sites exhibiting more dramatic changes in climate may or may not lie within the species' bioclimatic envelope today. Third, populations may be stressed by changes in climate that are more recent than can be addressed using a model with a 30-year horizon. In this case, extirpations might be explained best by climate observed during 2005–6. Predictors based on recent data may also be favored if populations are stressed by aspects of the microclimate that are not well modeled over the long term, or using data from distant weather stations.

In all, we selected 12 candidate predictors of climate-mediated extirpation: four metrics of stress (chronic and acute heat stress, and two thresholds of acute cold stress), each calculated from data on recent climate (2005–6), prevailing climate (1945–2006), and climate change (1945–75 versus 1976–2006). Of these 12 variables, the best predictor was a metric of chronic heat stress, AveSummerT, the average temperature during summer months. The best supported version of AveSummerT was calculated from observed data on recent microclimates. Assuming that our long-term models of microclimate were adequate, this result suggested that recent summer temperature was the most influential factor affecting pika extirpations. However, Beever et al. (2010) also asked whether there were increasingly detrimental effects of extreme cold, a trend suggested by the initial analysis: the number of daily minima below some threshold (ColdDays) was better supported as a predictor when the threshold was –5° C than when it was 0° C. In a post-hoc analysis to investigate this trend, the number of days with minimum temperatures below –10° C was by far the best predictor of local extinction—better even than recent summer temperature.

The results of this post-hoc analysis suggested that, ironically, this cold-adapted species may be stressed most by cold temperatures, at least within the Great Basin. Temperature profiles recorded by in-situ sensors show that extirpations occurred primarily where seasonal snow cover was brief (≤ 2 weeks) relative to the period during which pikas might be exposed to ambient temperatures well below freezing (Beever et al. 2010). Winter snowpack has been in decline throughout the western United States since 1950 (Mote et al. 2005, Lundquist et al. 2007), a period which largely overlaps our best predictor of extirpation: ColdDays in the prevailing climate (0–60 ybp). We have not been able to test cold-stress hypotheses directly, due to a lack of sufficiently high-resolution information on the spatiotemporal distribution and depth of snow cover in these topographically complex habitats. Even more problematic is the fact that the insulating properties of the snowpack eliminate most of the daily variation in subsurface temperatures. Our hindcasts were derived from models based on data collected during periods when in-situ temperature sensors were not covered by snow (Beever et al. 2010).

Revising Models of Extirpation for Pikas in the Great Basin

Taken together, the analyses of Beever et al. (2003, 2010) provide a rich set of predictors for modeling pika persistence in the Great Basin. In both analyses, the best models were far superior to a "null" model defined as no effect of predictor variables. However, it is not possible to compare the relative support for each predictor between these analyses, due to a change in the population of sites analyzed: one of the 25 historical sites used in the first analysis was replaced with a new site for the second analysis. By dropping the new site from our analyses, we were able to model the remaining 24 historical sites using all previously described predictor variables. Fifteen of these 24 sites harbored pikas at least through 1999, allowing one analysis of recent dynamics (1999–2008) and another analysis of long-term dynamics (historical–2008). For each analysis, a separate set of candidate models was proposed.

Candidate models of recent dynamics emphasized potential effects of climate (MaxElevR, ColdDays and AveSummerT), habitat (RangeHab, LocalHab and, again, MaxElevR), and anthropogenic effects (Grazing), as well as effects of the size of the persisting population in the 1990s

(Pop1990s) and latitude (Latitude). By including both Latitude and the residual variable MaxElevR among our predictors, we were able to separate effects of latitude and refuge that may have been confounded in the first analysis (figure 12.3). Candidate models of long-term dynamics emphasized a similar set of potential effects, omitting only Latitude and Pop1990s but adding AvePrecip, the average annual precipitation at each site based on PRISM values scaled to approximately 800 m resolution for the period 1971–2000. Because much of the precipitation in the basin falls as snow (Mote et al. 2005), we expected AvePrecip to represent at least a portion of any effect of snowpack on persistence.

Candidate predictors overlapped substantially between these new analyses, which helped us examine the evolution of support for certain predictors as pikas have responded to their environments during the past several decades. Regardless of the period modeled, every top model was based on a combination of predictors drawn from climate-mediated hypotheses and habitat-mediated hypotheses of extirpation (table 12.2).

Table 12.2. Confronting new models with new data on persistence from two periods (historical-2008 and 1999-2008), reveals continuing support for a metric of elevational refuge (MaxElevR) and new support for a metric of acute cold-stress (ColdDays) as predictors of both long- and short-term dynamics. We tested m = 22 new candidate models of persistence through 2008, and m = 21 new candidate models of persistence during 1999-2008. Relative AIC_c values suggest that these new models are much better supported than previous models ($\Delta AIC_c > 2$; *cf.* AIC_c values between top models in tables 12.1 and 12.2), confirming the benefit of incorporating microclimatic data in models of pika persistence. Persistence during 1999-2008 was predicted perfectly by the top (three-parameter) model as well as by several other models with four fitted parameters. Four-parameter models were not supported in our analysis because an increase in fitted parameters from three to four raises AIC_c by 3.363 ($\Delta AIC_c > 2$) for models based on a sample this small (n = 18). Predictors are as described in figure 12.5. The signs of all fitted coefficients are as expected: e.g., persistence was positively related to MaxElevR, RangeHab and LocalHab, and negatively related to ColdDays and AveSummerT.

Best models (predictors)	LL	K	AIC_c	ΔAIC_c	Weight
Persistence of historical populations through 2008 (n = 24, m = 22)					
MaxElevR, LocalHab, ColdDays	−3.21	4	16.531	0.000	0.288
MaxElevR, ColdDays	−5.35	3	17.900	1.369	0.145
AveSummerT, RangeHab, ColdDays	−4.12	4	18.343	1.812	0.117
MaxElevR, RangeHab, ColdDays	−4.19	4	18.483	1.952	0.109
Persistence of 1999 populations through 2008 (n = 18, m = 21)					
MaxElevR, ColdDays	−1.80E−09	3	7.714	0.000	0.439

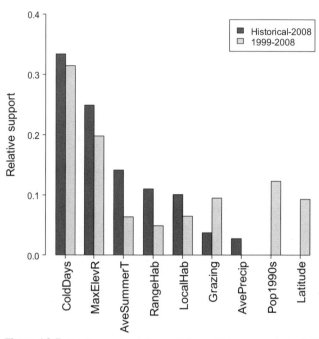

Figure 12.5. Relative support for predictors within new models of pika persistence in the Great Basin. This analysis was similar to that described in figure 12.4 except that each updated dataset on observed persistence (historical–2008 and 1999–2008) was predicted using a unique set of a priori models (see table 12.2 for top models). Overlap in several predictors between model sets allows visual evaluation of the temporal evolution of support for those predictors. Potential predictors were chosen on the basis of performance in previous information-theoretic analyses, or on the basis of logical arguments. For example, AveSummerT (mean daily temperature observed during June 1–September 30, 2005–6) and ColdDays (number of daily minima below −10° C estimated for 1945–2006) garnered best support among a set of purely thermal metrics used to predict persistence of these populations through 2008 (Beever et al. 2010). In contrast, logic supports the consideration of Pop1990s (local population size estimated during 1990s surveys) and LocalHab (local amount of inhabitable talus) as density-related predictors of population persistence during 1999–2008. Other predictors are as defined in figure 12.4.

The best models of recent persistence explained the data perfectly, perhaps due to low sample size. However, the predictors in the top model, ColdDays and MaxElevR, garnered highest support across all models in both periods (figure 12.5). Other predictors, including AveSummerT, received relatively minor support. Although latitude appeared closely related with the pattern of extirpation after 1999 (figure 12.1), it garnered little support as a predictor despite inclusion in both univariate and multivariate models. The poor support for AvePrecip may appear to undermine the hypothesis that reduced snow cover is contributing to the effect of ColdDays. However, AvePrecip may be a poor metric of snow cover during colder periods, especially given the potential for extreme spatial variation in snow accumulation and wind-driven redistribution in complex mountain terrain.

Conclusions and Management Implications

In general, we found robust support for hypotheses that pika persistence in the Great Basin is affected by the microclimates that individuals experience and by the local availability of a climatic refugium in the form of talus habitat at relatively high elevation. These results were robust across different datasets on persistence, representing different periods of time and different subsets of historically occupied sites analyzed. Results were even robust to changing the status of one site in 2008—a site on the cusp of extirpation, where only one pika has been observed over the past two years (Beever et al. 2011). Given such a small dataset, testing the effects of such changes seems advisable.

But what do the results really mean? Our models were hypothesis-driven, but it is important to note that other hypotheses could have generated a similar set of candidate models. For example, postulating effects of climate on the spread of disease among pikas could have led us to consider a model of population persistence based on positive effects of MaxElevR and negative effects of ColdDays. However, we have proposed climatic mechanisms that may govern the dynamics of this species, and we have found support for those mechanisms using data that are relatively easy to collect. Further, we have shown that the most recent patterns of extirpation offer somewhat higher support for climatic predictors of persistence. Without the ability to conduct in-situ experiments, this process of evaluating and reevaluating the relative support for predictors derived from observational data must be considered as a relatively powerful way to fill—or to suggest—gaps in our understanding.

Our results also suggest that extirpations of this "cold-adapted" species may be more likely where temperatures get very cold than where average summer temperatures are warmer. This result may explain the persistence of pikas at relatively low elevations and latitudes outside the Great Basin (Beever 2002, Ray and Beever 2007, Simpson 2009, Millar and Westfall 2010, Rodhouse et al. 2010). Low-latitude and low-elevation sites may experience relatively mild winter conditions that do not expose pikas to acute cold stress. If pikas at these sites are also able to use talus and lava-bed microclimates to escape summer heat, then persistence at lower elevations and latitudes might be expected. For these reasons, spatially and bioclimatically peripheral populations of the American pika may be relatively persistent.

This exercise has brought us a step closer to a mechanistic model of range-shift in response to climate change. In fact, Loarie et al. (in review) have projected the pika's response to several climate scenarios using a hypothesis-driven model similar to models outlined here. Modeling persistence as a function of climate at 97 sites scattered widely across the pika's current range, average annual temperature was identified as the best predictor of annual persistence. Specifically, sites with warmer annual temperatures were less likely to support pika persistence. Loarie et al. project a substantial probability (> 0.5) that climate change resulting from even low-emission scenarios will extirpate pikas from 55% (95% CI = 35%–66%) of currently suitable habitat by 2100. Although these results do not account for potential recolonization, they may be conservative for several reasons, including conservatism inherent in the response variable. Generally, "extirpation" is not reported until there are no pikas left at a site, so declining populations were not accounted for in this model. The model was certainly conservative in predicting the pattern of persistence in the Great Basin from 1999 to 2007—data that were reserved for model evaluation. All three sites where extirpation was documented from 1999 to 2007 were projected to maintain persisting populations with high probability ($p > 0.9$). The model's qualitative projections were relatively good, however, as these three sites were projected to be among those least likely to support persistence during this period. In any case, the projected rates of extirpation seem unlikely to be offset substantially by recolonization, given the lack of observed recolonization noted above.

Models of species persistence based on mechanistic hypotheses should be preferable to ecological niche models when it comes to projecting effects of climate change. Niche models based on current presence/absence data are well suited to predicting the current distribution of suitable habitats for a taxon (Guisan and Zimmermann 2000), but distributional patterns based on current conditions cannot necessarily be projected onto future conditions, for a variety of reasons including adaptive responses and uncoordinated changes in predictor variables (Ree et al. 2005, Araújo et al. 2005, Araújo and Rahbek 2006). The Bayesian approach adopted by Loarie et al. (in review) not only allows the fit and projection of a potentially complex mechanistic model, but also allows uncertainty in parameter estimates to be projected through the model in order to appropriately expand prediction intervals. Mechanistic models also allow us to identify particular sources of uncertainty that can be reduced through further research. Furthermore, the general concurrence of several independent projections for a shrinking distribution of *O. princeps* (Galbreath et al. 2009, Trook 2009, Calkins et al. in press, Loarie et al. in review) is striking.

The pika's prospects may appear bleak, especially in the Great Basin. In our analyses to date, climate appears to be the strongest driver of extirpation and upslope range retraction in this species. Livestock grazing is the only stressor we have identified that is clearly amenable to management actions. Given these results, the potential for human intervention to alter the pika's prospects will depend on several unknowns, discussed below.

Grazing garnered relatively weak support as a predictor of extirpation in our studies, but its effects may be larger where talus patches are smaller and the talus edge-to-area ratio is larger. For example, during Great Basin surveys in both the 1990s and 2000s, analysis of the variance in pika density revealed a significant interaction between grazing and the amount of talus available; pika density was lower at grazed sites with smaller talus patches ($p < 0.05$; E. A. Beever, unpublished data). However, any apparent effects of grazing may be spurious: in this dataset, long-term grazing is more common at lower elevations where pikas are also more exposed to potential thermal stress.

Several other variables amenable to management have yet to be analyzed. Among these, disease is an obvious candidate for driving extirpations in a warming world (e.g., Pounds et al. 2006), and strategies for managing diseases in wildlife do exist (e.g., Delahay et al. 2009). Little is known about disease in North American pikas, but diseases such as sylvatic plague often affect Asian pika populations (Biggins and Kosoy 2001), and plague is suspected as a factor in the recent decline of the Ili pika in western China (Li and Smith 2005). Plague has altered North American mammal communities dramatically since its introduction to the continent, circa 1900 (Biggins and Kosoy 2001). Outbreaks of plague in North America have been tied to climate metrics (Snäll et al. 2008, Ben Ari et al. 2008) and should be considered among the factors speculated to affect pika extirpations (Wilkening et al. 2011).

Without information on potential stressors other than those analyzed here, it is not entirely possible to classify *O. princeps* according to its fate, per Brodie et al. (this volume). By some arguments, the pika may be nearly doomed. Although assisted dispersal or colonization could be implemented to expand or regain the species' range, its cost may be relatively high for this species, which is quite susceptible to stress-related mortality during handling (MacArthur and Wang 1973, Wunder 1992). Even if successful protocols for assisted movement were developed, rescue or recolonization of extinction-prone populations could require ongoing translocation efforts. Furthermore, because *O. princeps* occurs mainly in association with glacial, periglacial, and other rock-ice features (Hafner 1994, Millar and Westfall 2010), the number of previously occupied habitats that could support long-term persistence is undoubtedly dwindling (IPCC 2007). Pika population growth might be improved by restricting livestock from some locations, but arguments above suggest that the impacts of livestock are low relative to other factors, and may be limited to sites with higher potential for edge effects. If disease were shown to be an important factor in pika decline, the implementation of disease management strategies might improve pika population growth, possibly over broad regions. However, disease management would be difficult at best for such remote populations with poorly characterized disease dynamics (Hudson et al. 2002, Collinge and Ray 2006, Ostfeld et al. 2008, Delahay et al. 2009). Finally, the potential for northward expansion of the species' range is complicated by the distribution of a similar species, *O. collaris*, which already occupies taluses throughout central and southeastern Alaska, the Yukon, and northwestern British Columbia (MacDonald and Jones 1987). Moreover, it is unclear whether northern habitats offer suitable future refugia for pikas; recent trends suggest that *O. collaris* may also be suffering effects of climate change (Smith et al. 2004, Morrison and Hik 2008).

On the other hand, the pika could potentially be saved by innovative intervention to improve persistence. If it were possible to develop successful protocols for assisted movement, it should be relatively straightforward to identify suitable current and future refugia for recolonization. The rock-ice features that pikas prefer are relatively insulated and should persist longer in a warming climate than more exposed glacial and periglacial features (Millar and Westfall 2008). Genetic management of translocation efforts appears particularly feasible in this species, for at least two reasons. First, patterns of diversity and evolutionary dynamics are relatively well characterized for *O. princeps* (Galbreath et al. 2009). Additional inference regarding patterns of gene flow at any scale is quite feasible, given recent success in extracting genetic data from pika scat (Mary Peacock, personal communication) and snared hair (Henry and Russello 2011), both easily collected through noninvasive means. Second, genetic diversity in *O. princeps* appears robust to effects of population bottlenecks such as colonization events. Even where local extinctions are frequent, there is evidence that genetic diversity can be maintained in pika metapopulations (Peacock and Smith 1997, Peacock and Ray 2001, Ray 2001). Finally, the conservation of relatively intact ecosystems for the maintenance of managed populations should be facilitated by the fact that pika habitats are generally remote and many occur within national parks and other protected lands.

Before implementing aggressive management strategies, we should further evaluate evidence for the mechanisms of climate-mediated endangerment supported by the studies described here. Because the pika is still broadly distributed, there is ample opportunity to study how its physiology and behavior affect its demographic response to microclimate and microhabitat. If we knew more about how pikas use different habitats to mitigate physiological limitations and improve demographic success, we might be in a better position to predict whether they're headed up the mountain or down into voids in the talus to escape climate-related stress. Research priorities in this system include further characterization of (1) microhabitats and microclimates preferred by pikas, including structure of the rock-void matrix, available vegetation, and period of snow cover; (2) pika activity budgets and foraging strategies in relation to microclimate and microhabitat; (3) variation in pika physiology, including body temperature and metabolic rate, as well as metrics of stress, immune function, and exposure to disease; and (4) demographic rates in relation to factors outlined in items 1 through 3.

Implications for Other Species

If declines in the winter snowpack are driving extirpations, then the plight of the pika is a particularly relevant omen for other species in western North America—such as ourselves—who depend on the montane snowpack as a primary source of water. The pika may also be an appropriate indicator species for the many plants and animals that are strongly affected by the timing and quality of the montane snowpack.

The pika is a suitable model for projecting the potential nature of climate-mediated range shifts, especially in montane systems, for both practical and logical reasons. Pikas and their habitats are relatively easy to survey. Patterns of patch occupancy are not complicated by individual movement patterns, because pikas do not migrate seasonally or relocate after establishing a territory (Smith and Weston 1990). The temperature data generated for these pika studies will likely be useful in distributional studies of other species, given the high correlation between temperatures above and below the surface of the talus, as measured by paired sensors (Beever et al. 2010). This high correlation also suggests that the variation in microclimates available to pikas provides a suitable model for the variation available to other species inhabiting these montane landscapes.

Logical arguments for using the pika as a sentinel species center on its ecology. Most importantly, the recent range shift described here appears to be controlled by both temperature and precipitation, the major determinants of geographic range in most species. Effects of precipitation are also implicated in evidence that historical pika populations in the southern Rocky Mountains have persisted in all but the driest sites (Erb et al. 2011), and that pikas in the southern Sierra Nevada are distributed primarily within taluses supporting subsurface ice features (Millar and Westfall 2010). Indirect evidence for the importance of precipitation derives from studies suggesting that pikas are more common where the vegetation community is dominated by forbs rather than xeric-adapted grasses (Ray and Beever 2007, Rodhouse et al. 2010, Wilkening et al. 2011). A similar suite of responses is expected or observed for other montane species (e.g., Murphy and Weiss 1992, Root et al. 2003, Epps et al. 2004, Parmesan 2006, Raxworthy et al. 2008, Sekercioglu et al. 2008). Finally, because the pika uses subsurface habitats to persist where surface conditions would preclude its survival (MacArthur and Wang 1973, 1974; Smith 1974a), it may serve as a model for the complexity of factors that interact to determine a species' resilience to climate change, including behavioral adaptations involving the use of microclimates. Upslope range shifts may be less pronounced in species that normally make use of favorable microclimates, such as those found in burrows. More generally, species with sufficient plasticity might avoid an upslope retreat by altering their behaviors to make better use of available microclimates. Still, spending more time in a comfortable burrow may not compensate for energy lost to climate-mediated degradation of food resources on the surface. The potential for alternative responses to climate change depends in part on the mechanism(s) underlying a species' sensitivity to climate.

ACKNOWLEDGMENTS
Several of our analyses would not have been possible without the exceptional data on pika microclimates gathered by Jennifer Wilkening, whose continued dedication to discovering the mechanics of pika persistence is much appreciated. Our interpretations of the data were shaped through conversations with Jennifer and many other colleagues.

LITERATURE CITED
Anderson, D. R. 2008. *Model Based Inference in the Life Sciences: A Primer on Evidence*. New York: Springer.
Araújo, M. B., R. G. Pearson, W. Thuiller, and M. Erhard. 2005. Validation of species-climate impact models under climate change. *Global Change Biology* 11:1504–13.
Araujo, M. B., and C. Rahbek. 2006. How does climate change affect biodiversity? *Science* 313:1396–97.

Beever, E. A. 1999. Species- and community-level responses to disturbance imposed by feral horse grazing and other management practices. PhD thesis, University of Nevada.

Beever, E. A. 2002. Persistence of pikas in two low-elevation national monuments in the western United States. *Park Science* 21:23–29.

Beever, E. A., P. F. Brussard, and J. Berger. 2003. Patterns of apparent extirpation among isolated populations of pikas (*Ochotona princeps*) in the Great Basin. *Journal of Mammalogy* 84:37–54.

Beever, E. A., C. Ray, P. W. Mote, and J. L. Wilkening. 2010. Testing alternative models of climate-mediated extirpation. *Ecological Applications* 20:164–78.

Beever, E. A., C. Ray, J. L. Wilkening, P. W. Mote, and P. F. Brussard. 2011. Contemporary climate change alters the pace and drivers of extinction. *Global Change Biology* 17:1–17.

Beever, E. A., J. L. Wilkening, D. E. McIvor, S. S. Weber, and P. F. Brussard. 2008. American pikas (*Ochotona princeps*) in northwestern Nevada: A newly discovered population at a low-elevation site. *Western North American Naturalist* 68:8–14.

Ben Ari, T., A. Gershunov, K. L. Gage, T. Snäll, P. Ettestad, and K. L. Kausrud. 2008. Human plague in the USA: The importance of regional local climate. *Biology Letters* 4:737–40.

Bertrand, P. 2008. Professor suggests method to move threatened species. *Daily Texan*, July 22.

Biggins, D., and M. Kosoy. 2001. Influences of introduced plague on North American mammals; implications from ecology of plague in Asia. *Journal of Mammalogy* 82:906–16.

Blakemore, Bill. 2007. Route to extinction goes up mountains, scientists say. Accessed at abcnews.go.com/WN/GlobalWarming/story?id=3155909&page=1.

Burnham, K. P., and D. R. Anderson. 2002. *Model Selection and Multimodel Inference: A Practical Information-Theoretic Approach*. New York: Springer.

Calkins, M. T., E. A. Beever, K. G. Boykin, J. K. Frey, and M. C. Andersen. In press. Not-so-splendid isolation: Modeling climate-mediated range collapse of a montane mammal (*Ochotona princeps*) across numerous ecoregions. *Ecography*. Accessed at onlinelibrary.wiley.com/journal/10.1111/(ISSN)1600-0587/earlyview.

Collinge, S. K., and C. Ray, eds. 2006. *Disease Ecology: Community Structure and Pathogen Dynamics*. Oxford: Oxford University Press.

Daly, C., R. P. Neilson, and D. L. Phillips. 1994. A statistical-topographic model for mapping climatological precipitation over mountainous terrain. *Journal of Applied Meteorology* 33:140–58.

Delahay, R. J., G. C. Smith, and M. R. Hutchings, editors. 2009. *Management of Disease in Wild Mammals*. Tokyo: Springer. Dearing, M. D. 1997. The function of hay piles of pikas (*Ochotona princeps*). *Journal of Mammalogy* 78:1156–63.

Diamond, J. M. 1987. Extant unless proven extinct or extinct unless proven extant? *Conservation Biology* 1:77–79.

Erb, L. P., C. Ray, and R. Guralnick. 2011. On the generality of a climate-mediated shift in the distribution of the American pika (*Ochotona princeps*). *Ecology* 92:1730–35.

Epps, C. W., D. R. McCullough, J. D. Wehausen, V. C. Bleich, and J. L. Rechel. 2004.

Effects of climate change on population persistence of desert-dwelling mountain sheep in California. *Conservation Biology* 18:102–13.

Frey, D. 2007. Climate change pushing pika off mountains. *Aspen Daily News*, Carbondale, CO, February 15.

Galbreath K., D. Hafner, and K. Zamudio. 2009. When cold is better: Climate-driven elevation shifts yield complex patterns of diversification and demography in an alpine specialist (American pika, *Ochotona princeps*). *Evolution* 63:2848–63.

Grayson, D. K. 2005. A brief history of Great Basin pikas. *Journal of Biogeography* 32:2103–11.

Gilpin, M. E., and M. E. Soulé. 1986. Minimum viable populations: The processes of species extinctions. In *Conservation Biology: The Science of Scarcity and Diversity*, edited by M. Soulé, 13–24. Sunderland, MA: Sinauer Associates.

Grinnell, J. 1917. Field tests of theories concerning distributional control. *American Naturalist* 51:115–28.

Guisan, A., and N. E. Zimmermann. 2000. Predictive habitat distribution models in ecology. *Ecological Modeling* 135:147–86.

Guralnick, R. P. 2007. Differential effects of past climate warming on mountain and flatland species' distributions: A multispecies North American mammal assessment. *Global Ecology and Biogeography* 16:14–23.

Guralnick, R., L. P. Erb, and C. Ray. In press. Mammalian distributional response to climatic change: A review and research prospectus. In *Ecological Consequences of Climate Change: Mechanisms, Conservation, and Management*, edited by E. A. Beever and J. Belant. London: Taylor and Francis.

Hafner, D. J. 1993. North-American pika (*Ochotona princeps*) as a Late Quaternary biogeographic indicator species. *Quaternary Research* 39:373–80.

———. 1994. Pikas and permafrost: Post-Wisconsin zoogeography of *Ochotona* in the southern Rocky Mountains, USA. *Arctic and Alpine Research* 26:375–82.

Hafner, D. J., and R. M. Sullivan. 1995. Historical and ecological biogeography of nearctic pikas (Lagomorpha: Ochotonidae). *Journal of Mammalogy* 76:302–21.

Henry, P., and M. A. Russello. 2011. Obtaining high-quality DNA from elusive small mammals using low-tech hair snares. *European Journal of Wildlife Research* 57:429–35.

Hijmans, R. J., and C. H. Graham. 2006. The ability of climate envelope models to predict the effect of climate change on species distributions. *Global Change Biology* 23:1–10.

Hudson, P. J., A. Rizzoli, B. T. Grenfell, H. Heesterbeek, and A. P. Dobson, eds. 2002. *The Ecology of Wildlife Diseases*. New York: Oxford University Press.

Huntly, N. J., A. T. Smith, and B. L. Ivins. 1986. Foraging behavior of the pika (*Ochotona princeps*), with comparisons of grazing versus haying. *Journal of Mammalogy* 67:139–48.

Intergovernmental Panel on Climate Change (IPCC). 2007. *Climate Change 2007: The Physical Science Basis*. New York: Cambridge University Press.

Kéry, M. 2002. Inferring the absence of a species: A case study of snakes. *Journal of Wildlife Management* 66:330–38.

Krajick, K. 2004. All downhill from here? *Science* 303:1600–1602.

Kreuzer, M. P., and N. J. Huntly. 2003. Habitat-specific demography: Evidence for source-sink population structure in a mammal, the pika. *Oecologia* 134:343–49.

Kullman, L. 2002. Rapid recent range-margin rise of tree and shrub species in the Swedish Scandes. *Journal of Ecology* 90:68–77.

Lagorio, C. 2007. Climate and the animal kingdom: What happens to polar bears and pikas when seasons are mild. *CBS Evening News*, January 5.

Lee, S. 2005. Experts explore weather-wildlife link as behavior changes. *Great Falls Tribune*, January 3.

Lenoir, J., J. C. Gégout, P. A. Marquet, P. De Ruffray, and H. Brisse. 2008. A significant upward shift in plant species optimum elevation during the 20th century. *Science* 320:1768–71.

Li, W-D., and A. T. Smith. 2005. Dramatic decline of the threatened Ili pika *Ochotona iliensis* (Lagomorpha: Ochotonidae) in Xinjiang, China. *Oryx* 39:30–34.

Loarie, S. R., P. B. Duffy, H. Hamilton, G. P. Asner, C. B. Field, and D. D. Ackerly. 2009. The velocity of climate change. *Nature* 462:1052–55.

Loarie, S. R., C. B. Field, C. Ray, E. A. Beever, P. B. Duffy, K. Hayhoe, J. L. Wilkening, and J. S. Clark. In review. Climate threats to the American pika: Modeling historical persistence for 21st-century projections.

Lundquist J. D., D. R. Cayan, and M. D. Dettinger. 2007. Surface temperature patterns in complex terrain: Daily variations and long-term change in the central Sierra Nevada, California. *Journal of Geophysical Research* 112:D11124, doi10.1029/2006JD007561.

MacArthur, R.A., and L.C.H. Wang. 1973. Physiology of thermoregulation in the pika, *Ochotona princeps*. *Canadian Journal of Zoology* 51:11–16.

MacArthur, R. A., and L.C.H. Wang. 1974. Behavioral thermoregulation in the pika, *Ochotona princeps*: a field study using radio-telemetry. *Canadian Journal of Zoology* 52:353–58.

Mead, J. I. 1987. Quaternary records of pika, *Ochotona*, in North America. *Boreas* 16:165–71.

MacDonald, S. O., and C. Jones. 1987. *Ochotona collaris*. *Mammalian Species* 281:1–4.

McDonald , K. A., and J. H. Brown. 1992. Using montane mammals to model extinctions due to global change. *Conservation Biology* 6:409–15.

Merriam, C. H. 1894. Laws of temperature control of geographic distribution of terrestrial mammals and plants. *National Geographic* 6:229–38.

Millar, C. I., and R. D. Westfall. 2008. Rock glaciers and periglacial rock-ice features in the Sierra Nevada; classification, distribution, and climate relationships. *Quaternary International* 188:90–104.

Millar, C. I., and R. D. Westfall. 2010. Distribution and climatic relationships of the American pika (*Ochotona princeps*) in the Sierra Nevada and western Great Basin, U.S.A.; Periglacial landforms as refugia in warming climates. *Arctic, Antarctic, and Alpine Research* 42:76–88.

Morrison, S. F., and D. S. Hik. 2007. Demographic analysis of a declining pika *Ochotona collaris* population: Linking survival to broad-scale climate patterns via spring snowmelt patterns. *Journal of Animal Ecology* 76:899–907.

Moritz, C., J. L. Patton, C. J. Conroy, J. L. Parra, G. C. White, and S. R. Beissinger. 2008. Impact of a century of climate change on small-mammal communities in Yosemite National Park, USA. *Science* 322:261–64.

Mote, P. W., A. F. Hamlet, M. P. Clark, and D. P. Lettenmaier. 2005. Declining mountain snowpack in western North America. *Bulletin of the American Meteorological Society* 86:39–49.

Murphy, D. D., and S. B. Weiss. 1992. The effects of climate change on biological diversity in western North America: Species losses and mechanisms. In *Global Warming and Biological Diversity*, edited by R. L. Peters and T. J. Lovejoy, 355–68. New Haven: Yale University Press.

Nijhuis, M. 2005. In the Great Basin, scientists track global warming. *High Country News*, Paeonia, CO, October 17.

Ostfeld, R. S., F. Keesing, and V. T. Eviner, eds. 2008. *Infectious Disease Ecology: Effects of Ecosystems on Disease and of Disease on Ecosystems*. Princeton, NJ: Princeton University PressJ.

Parmesan, C. 2006. Ecological and evolutionary responses to recent climate change. *Annual Review of Ecology, Evolution and Systematics* 37:637–69.

Parmesan, C., and G. Yohe. 2003. A globally coherent fingerprint of climate change impacts across natural systems. *Nature* 421:37–42.

Peacock, M. M. 1997a. Determining natal dispersal patterns in a population of North American pikas (*Ochotona princeps*) using direct mark-resight and indirect genetic methods. *Behavioral Ecology* 8:340–50.

Peacock, M. M. 1997b. The effect of habitat fragmentation on dispersal patterns, mating behavior, and genetic variation in a pika (*Ochotona princeps*) metapopulation. *Oecologia* 112:524–33.

Peacock, M., and C. Ray. 2001. Dispersal in pikas (*Ochotona princeps*): Combining genetic and demographic approaches to reveal spatial and temporal patterns. In *Dispersal: Causes, Consequences and Mechanisms of Dispersal at the Individual, Population and Community Level*, edited by J. Clobert, E. Danchin, J. D. Nichols, and A. A. Dhondt, Oxford: Oxford University Press.

Peacock, M. M., and A. T. Smith. 1997. The effects of habitat fragmentation on dispersal patterns, mating behavior, and genetic variation in a pika (*Ochotona princeps*) metapopulation. *Oecologia* 112:524–33.

Pimm, S., P. Raven, A. Peterson, C. H. Sekercioglu, and P. R. Ehrlich. 2006. Human impacts on the rates of recent, present, and future bird extinctions. *Proceedings of the National Academy of Sciences* 103:10941–46.

Pounds, J. A., M. R. Bustamante, L. A. Coloma, J. A. Consuegra, M. P. L. Fogden, P. N. Foster, E. La Marca, K. L. Masters, A. Merino-Viteri, R. Puschendorf, S. R. Ron, G. A. Sanchez-Azofeifa, C. J. Still, and B. E. Young. 2006. Widespread amphibian extinctions from epidemic disease driven by global warming. *Nature* 439:161–67.

Pounds, J. A., M. P. L. Fogden, and J. H. Campbell. 1999. Biological response to climate change on a tropical mountain. *Nature* 398:611–15.

Ray, C. 2001. Maintaining genetic diversity despite local extinctions: Effects of population scale. *Biological Conservation* 100:3–14.

Ray, C., and E. A. Beever. 2007. Distribution and abundance of the American pika (*Ochotona princeps*) in Lava Beds National Monument. Report to United States National Park Service, Lava Beds National Monument, Tulelake, CA.

Raxworthy, C. J., R. G. Pearson, N. Rabibisoa, A. M. Rakotondrazafy, J. Ramanamanjato, A. P. Raselimanana, S. Wu, R. A. Nussbaum, and D. Stone. 2008. Extinction vulnerability of tropical montane endemism from warming and upslope displacement: A preliminary appraisal for the highest massif in Madagascar. *Global Change Biology* 14:1703–20.

Ree, R. H., B. R. Moore, C. O. Webb, and M. J. Donoghue. 2005. A likelihood framework for inferring the evolution of geographic range on phylogenetic trees. *Evolution* 59:2299–2311.

Rickart, E.A. 2001. Elevational diversity gradients, biogeography, and the structure of montane mammal communities in the intermountain region of North America. *Global Ecology and Biogeography* 10:77–100.

Rodhouse, T. J., E. A. Beever, L. K. Garrett, K. M. Irvine, M. Munts, C. Ray and M.R. Shardlow. 2010. Distribution of American pikas in a low-elevation lava landscape: Conservation implications from the range periphery. *Journal of Mammalogy* 91: 1287–99.

Root, T. L., J. T. Price, K. R. Hall, S. H. Schneider, C. Rosenzweig, and J. A. Pounds. 2003. Fingerprints of global warming on wild animals and plants. *Nature* 421: 57–60.

Root, T. L., and S. H. Schneider. 2006. Conservation and climate change: The challenges ahead. *Conservation Biology* 20:706–8.

Rowe, R. J. 2005. Elevational gradient analyses and the use of historical museum specimens: A cautionary tale. *Journal of Biogeography* 32:1883–97.

Sekercioglu, C. H., S. H. Schneider, J. P. Fay, and S. R. Loarie. 2008. Climate change, elevational range shifts, and bird extinctions. *Conservation Biology* 22:140–50.

Shoo, L.P., S.E. Williams, and J. Hero. 2006. Detecting climate change induced range shifts: Where and how should we be looking? *Austral Ecology* 31:22–29.

Siegler, K. 2007. Pikas: In Colorado's high country, some early warning signs of the dire consequences of climate change are emerging. Aspen Public Radio, August 20.

Simpson W. G. 2009. American pikas inhabit low-elevation sites outside the species' previously described bioclimatic envelope. *Western North American Naturalist* 69:243–50.

Smith, A. T. 1974a. The distribution and dispersal of pikas: Influences of behavior and climate. *Ecology* 55:1368–76.

———. 1974b. The distribution and dispersal of pikas: Consequences of insular population structure. *Ecology* 55:1112–19.

———. 1978. Comparative demography of pikas (*Ochotona*): Effect of spatial and temporal age-specific mortality. *Ecology* 59:133–39.

Smith A. T., W. Li, and D. Hik. 2004. Pikas as harbingers of global warming. *Species* 41:4–5.

Smith, A. T., and M. L. Weston. 1990. *Ochotona princeps*. *Mammalian Species* 352:1–8.

Snäll, T., R. O'Hara, C. Ray, and S. Collinge. 2008. Climate-driven spatial dynamics of plague among prairie dog colonies. *American Naturalist* 171:238–48.

Tapper, S. C. 1973. The spatial organisation of pikas (*Ochotona*), and its effect on population recruitment. PhD thesis. University of Alberta.

Tolmé, P. 2006. No room at the top. *National Wildlife Magazine* 44:22–30.

Trivedi, M. R., P. M. Berry, M. D. Morecroft, and T. P. Dawson. 2008. Spatial scale affects bioclimate model projections of climate change impacts on mountain plants. *Global Change Biology* 14:1089–1103.

Trook, J. 2009. Effects of climate change on the distribution of American pika (*Ochotona princeps*) in the western United States. MS thesis. University of Idaho.

Unrau, J. 2008. Pika populations decimated in Ruby Range. *Whitehorse Star*, June 11.

Walther, G. R., S. Beissner, and C. A. Burga. 2005. Trends in the upward shift of alpine plants. *Journal of Vegetation Science* 16:541–48.

Webb, D. 2007. Pika's decline may serve as a warning about global warming. *Aspen Times*, Glenwood Springs, CO, February 15.

Wilkening, J. L., C. Ray, E. A. Beever, and P. F. Brussard. 2011. Modeling contemporary range retraction in Great Basin pikas (*Ochotona princeps*) using data on microclimate and microhabitat. *Quaternary International* 235:77–88.

Wilson, R. J., D. Gutiérrez, J. Gutiérrez, D. Martínez, R. Agudo, and V. J. Monserrat. 2005. Changes to the elevational limits and extent of species ranges associated with climate change. *Ecology Letters* 8:1138–46.

Wunder, B. A. 1992. Morphophysiological indicators of the energy state of small mammals. In *Mammalian Energetics: Interdisciplinary Views of Metabolism and Reproduction*, edited by T. Tomasi and T. Horton, 83–104. Ithaca, NY: Cornell University Press.

Sensitivity of High Arctic Caribou Population Dynamics to Changes in the Frequency of Extreme Weather Events

Joerg Tews, Rebecca Jeppesen, and
Carolyn Callaghan

Temperature and precipitation, alone and in conjunction with one another, are considered to be the most important environmental variables for terrestrial ecosystems (Lovejoy 2008). Changes in temperature and precipitation can be quantified through longer-term averages (e.g., mean annual rainfall) as well as through changes in the frequency and intensity of outliers or extreme events (e.g., recurrence and intensity of floods). An increase in climatic variability, such as an increase in the frequency and intensity of extreme weather events, is often predicted with climatic change (Marshall et al. 2008, Beniston et al. 2007).

Definitions of what is considered an extreme weather event may vary, but they generally include severe temperatures and/or precipitation, and their consequences. Severely low temperatures (Nicholls 2008, Jentsch et al. 2007, Altwegg et al. 2006) are frequently considered as extreme events, especially when they are examined in the context of seasonal averages. Abnormally high temperatures (Welbergen et al. 2008, Beniston et al. 2007, Schneider and Lane 2006) can lead to heat waves or droughts (Marshall et al. 2008, Trentberth 2008) when coupled with very little to no precipitation for an extended period of time. Intense precipitation (Frei and Schar 2001, Haylock and Nicholls 2000) can lead to flooding, and it is often caused by tropical storms or hurricanes, often considered some of the most severe extreme weather events (Tejada-Cruz and Sutherland 2005, Webster et al. 2005). Other, less well known extreme weather events include "rain on snow" events leading to runoff damage and flooding (McCabe et al. 2007) and the formation of ground ice in arctic ecosystems (Putkonen and Roe 2003).

On a regional scale, extreme weather events may pose one of the greatest threats of climate change (Beniston et al. 2007). However, there are a number of difficulties associated with such a prediction. The variability inherent in natural systems makes it difficult to predict extreme climatic events with confidence (Hulme 2005), given their relative rarity. In addition, the severity and potential impacts of extreme weather are disproportionate to the relatively small spatial and temporal scales that they affect (Jentsch et al. 2007). Statistically, infrequent climatic events will affect mean values, resulting in wider confidence intervals with an increase in the number of extreme events that occur in a given location (Frei and Schar 2001). Rare and extreme events were therefore often excluded from historic climatic analysis, on the basis of the assumption that they were noninformative outliers (Jentsch et al. 2007).

Extreme events are often identified as such on the basis of their positions as outliers in the tails of a statistical probability distribution (Haylock and Nicholls 2000) calculated from mean temperature and precipitation data (Nicholls 2008) compiled for a given reference time frame, be it years, decades, or longer (Jentsch et al. 2007). The primary defining characteristics of extreme events are rarity and intensity (IPCC 2001). Rarity is often quantified in the context of "recurrence intervals," while intensity is associated with values that deviate greatly from the mean (Beniston et al. 2007), relative to the norm observed at a particular location. Frei and Schar (2001) differentiate between "intense" events, which exceed a given threshold roughly once every 30 days, and "extreme" events, which have a return interval of approximately one year.

The concepts of rarity and intensity are open to interpretation, and while it is commonly accepted that extremes are at the high and low ends of a range of values for a given variable (Nicholls 2008) there is less agreement as to how high or low these values must be relative to the mean before they are considered extreme. The IPCC (2001) has suggested the 10th or 90th percentiles, while Haylock and Nicholls (2000) use the 5th and 95th percentiles, and Jentsch et al. (2007) identify a potential quality of extreme events as surpassing previously established minimum or maximum values.

Some measure of duration must also be included when defining certain types of extreme weather events. Short-term changes in minimum or maximum temperatures or precipitation amounts can be useful when classifying extreme cold or the probability of flash flooding, but longer time scales must be considered when dealing with events such as droughts, which would not be evident with analyses on daily or even weekly time scales (Nicholls 2008).

Changes in the frequency and/or intensity of extreme weather events may be indicative of changes in mean values of climatic parameters, and vice versa (Nicholls 2008, Jentsch et al. 2007). For example, an increase in the number of extremely hot days could be reflective of an increase in mean temperature, while the same logic suggests that an increase in average temperature could result in more days with a temperature above some given threshold, leading to heat waves or droughts (Trentberth 2008, Schneider and Lane 2006). In these types of cases, it may be practical to use the change in mean values as a basis for quantifying changes in extremes (Nicholls 2008). On the other hand, an increase in mean temperature could result in a correlated rise in atmospheric water vapor, leading to changes in the frequency and intensity of extreme precipitation events (Trentbert 2008, Schneider and Lane 2006) or tropical storms (Webster et al. 2005), which depend on warm sea surface temperatures for their formation (Nicholls 2008, Webster et al. 2005). In these situations, the overall shape of the statistical distribution may transform abruptly, meaning that changes in average values are poor indicators of changes in the frequency of extreme events.

Ecological disturbance is often considered important in promoting spatially heterogeneous habitat and biodiversity when it occurs intermittently and with intermediate intensity (Tejada-Cruz and Sutherland 2005). Disturbance has been defined as a comparatively discrete event, resulting in a disruption of the resource availability or physical structure of an ecosystem (Smith and Smith 2001). Jentsch et al. (2007) characterize discrete events as abrupt, defined in terms of magnitude or extremity of an event over its duration. They suggest that magnitude and extremity must be interpreted relative to the life spans of the organisms affected; a concept also put forth by Picket and White (1985) who propose that temporal scale also be interpreted in such a context. Such definitions, however, do not clearly differentiate between extreme events and disturbance.

Definitions of disturbance in the literature are often closely associated with a particular species or community and its resource levels and resilience, whereas the characterization of a weather event as extreme refers to the spatial and/or temporal pattern and magnitude of an abiotic variable. For example, heat waves and lightning could be considered extreme weather events that could lead to fire as a disturbance agent affecting individuals, populations, and communities. An excessive amount of rainfall over a short period of time may not itself affect food accessibility in riparian areas, but it could lead to flooding, which can have such negative consequences. Thus, community impacts of the floods—including temporary loss of riparian habitat and more permanent impacts such as localized erosion, as well as any mediated implications for higher trophic levels—can all be attributed to the extreme weather. In other words, cascading effects of ecosystem disturbance to wildlife habitat and other critical resources are often attributed to extreme weather.

The structure and composition of ecological communities have often been found to be affected by extreme weather events and their associated disturbances (e.g. Jentsch et al. 2007), although how these events affect ecological processes in comparison to usual climatic variability is not yet fully understood (Altwegg et al. 2006). This information gap may have important implications for wildlife in the face of increasingly frequent extreme weather events. The frequency and intensity of such events may play a larger role than changes in mean values in determining resistance and resilience of many species (Hulme 2005). Given that extreme weather may accelerate changes in the system already occurring because of changes in mean values (Jentsch et al. 2007), the amount of time a species has to adapt to a new and changing environment will be further decreased. The scale of responses will vary between systems with the spatial and temporal extent of the extreme events (Marshall et al. 2008), and it will depend on the systems' constituent species. The resistance and resilience of wildlife to the effects of extreme weather may be mediated by their associated trophic interactions, and dictated to a large extent by their specific physiological tolerances and life histories (Parmesan 2006).

The concept of differing physiological responses to extreme weather is illustrated by a study by Welbergen et al. (2008) on a community of flying fox fruit bats (*Pteropus* spp.) in Australia. During a heat wave on a single day when temperatures reached an all-time high, 5% to 6% of a *P. alecto* colony (more than 1,450 individuals) died as a direct result of the extreme heat. Concerning differential impacts within a species, Altwegg et al. (2006) found evidence in their study of barn owls (*Tyto alba*) in Switzerland that age-specific demographic parameters can also be affected by extreme events. Based on more than 60 years of data, they found that although individuals of any age and condition were affected by harsh winters, variability in the adult survival rate increased with these extreme events while that of juveniles was unaffected. Following from these results, Altwegg et al. (2006) suggested that many population studies of short duration might not incorporate extreme weather events, thus leading to miscalculations in survival variability: a suggestion that could have dire implications in determining optimal management practices.

Bauer et al. (2008) studied the potential effects of climate change on the pink-footed goose (*Anser brachyrhynchus*), a European migratory species. They found that changing values of climatic means affected the conditions at commonly used stopover sites, which could have direct implications for the birds' energy budgets and indirect implications by changing the availability of food resources. In addition, they suggest that more frequent extreme weather events may have interacting effects on energy expenditure, which could result in further modifications of typical behaviour.

An example of direct effects of extreme weather patterns on population demography can be found in High Arctic caribou. In some winters in the High Arctic, snow conditions may deviate greatly from average conditions through extremely low or high snowfall, unusual amounts of ice formation, or unusual winds during snow cover accumulation. In winters with severe snow cover conditions, increased forage inaccessibility may lead to increased mortality, increased emigration, or reductions in recruitment and productivity (COSEWIC 2004 and references therein). Even if enough resources are available, temporal or spatial inaccessibility of forage can result in lower survival and fecundity rates. In extreme years this can result in population die-offs of as much as 85% within one winter season, as observed in Peary caribou (*Rangifer tarandus pearyi*), a caribou subspecies living in the High Arctic (Miller and Gunn 2003a, 2003b, Tews et al. 2007a, b). Such similar population-level consequences have also been analyzed for Svalbard reindeer (*Rangifer tarandus platyrhynchus*) in a study by Kohler and Aanes (2004). The authors found that 80% of the variance in observed reindeer population growth rate was related to mean winter ground ice. Although less pronounced, similar density-independent population declines have been hypothesized for other arctic ungulate species, such as musk oxen (*Ovibos moschatus*; e.g., Gunn et al. 1991)

The aforementioned studies show that the consequences of extreme weather events on wildlife can often be lethal. However, those events do not always push the system beyond its resilience threshold (Jentsch et al. 2007). One of the most extreme types of weather, tropical cyclones, has the potential to destroy the physical structure of entire forest ecosystems in a very short period of time. Tejada-Cruz and Sutherland (2005) compared avian species richness and density in cloud forests in southern Mexico before and after a major hurricane passed over the area. They found that although overall community structure changed across sample plots, the majority of this change was observed during the first year following the storm and could be attributed to forested patches that became gaps. Interestingly, no difference in bird species richness, density estimates, or detection probability was observed between pre- and post-hurricane periods.

The inconsistent results of these studies serve only to illustrate the variety of responses and tolerances of different species to extreme weather events. Although it is possible that extreme weather may cause severe short-term effects on relatively small spatial scales, there is currently little information available regarding the potential longer-term impact and the response of various species to such abrupt changes in the environment (Jentsch et al. 2007). As climate continues to change and extreme events increase in frequency and intensity, the potential lasting implications will become more apparent and thus need to be evaluated. However, we are very often faced with large uncertainties, especially when dealing with the effects on the demography of a species.

This uncertainty has two primary facets. First, there is general uncertainty about anticipated extreme weather patterns and their spatial and temporal scales. Second, even if knowledge about the direct consequences of extreme weather events (e.g., percent of population die-off after an event) is available, we are often confronted with minimal knowledge about vital rates of wildlife populations—that is, rates of fecundity and survival—and how these may vary in space and time. If we want to make sound assessments of potential population trends under future climate change, sufficient knowledge of demographic rates of a species is often required. However, such knowledge is often very limited.

Here we present results from a population modeling analysis on climate change-related die-offs in Peary caribou with demographic uncertainty. In this study we address demographic and ecological uncertainty by analyzing a large set of randomized demographic models that are within a realistic range for High Arctic caribou populations. With the results of this study we show that even with this uncertainty, it is possible to assess relative difference of population viability for scenarios of extreme weather events under future climate change.

Demographic Simulations

This study is based on extensive empirical work (see, e.g., COSEWIC 2004) and previous modeling analysis (Tews et al. 2006; Tews et al. 2007a, b) of population dynamics in Peary caribou, an endemic *Rangifer tarandus* subspecies inhabiting the Canadian High Arctic. The range of Peary caribou is large and relatively unaffected by human activity (COSEWIC 2004). Peary caribou occur in six populations across the Canadian High Arctic, connected by occasional emigration-immigration events across the sea ice (Miller et al., 1977; Miller, 1998; Gunn and Dragon, 2002). Total population size was recently estimated at about 7,000 individuals (COSEWIC 2004). However, these overall estimates are based on aerial surveys from different years and regions using different survey techniques, and are therefore of limited quality. Less than 5% of the total caribou range is vegetated (Gould et al. 2003), and only a fraction of vegetated areas are accessible from mid- to late winter (usually November throughMay), as Peary caribou search for forage on ridges or other topographical exposures where snow is absent, shallow, or relatively soft (Larter and Nagy 2001). In extreme years with severe winter weather, this limited forage may become inaccessible for several consecutive weeks until snow conditions improve.

Peary caribou population die-offs are well documented. The Bathurst Island complex (BIC) population declined from about 3,600 caribou in 1961 to 270 in 1974, increased to more than 3,000 by 1994, and then crashed to less than 250 until 2001 after three severe winters during 1995–97. Between 1961 and 2001 the four reported population die-offs in the BIC were 70%, 27%, 75%, and 86% respectively, with a mean of 64% (Miller and Gunn 2003a, b; see table 1 in Tews et al. 2006a).

Healthy Peary caribou cows may breed as yearlings and produce their first calves as two-year-olds, yet first calf production at three years is more common. Males and females are reproductively capable up to 13 years, and live up to 15 years. Pregnancy rates and calf survival are strongly affected by nutritional status, which in turn depends on weather conditions, as well as forage quality and quantity (Thomas et al. 1976, Thomas and Broughton 1978, Thomas 1982, Larter and Nagy 2000b). In severe winters, yearling recruitment can drop to zero and annual pregnancy rates can vary extremely (COSWIC 2004). Up to 80% of adult females may produce calves in a given year (Gunn et al. 1998), although calf production reached 97% calves during the 1990s on Banks Island, another population in the Canadian High Arctic (COSEWIC 2004).

Knowledge of annual variation in and rates of fecundity and survival for different life stages is relatively poor. We therefore used a generic, female-only, two-stage-based Leslie matrix model to simulate population dynamics under different scenarios of extreme weather events. The model is based on a pre-calving census; i.e., each year prior to calving, total recruits, deaths and numbers of caribou that move to the adult stage are calculated. Fecundity of the first stage class (i.e., individuals that are almost one year of age at census) is calculated by multiplying parturition rate (parturition at one year) by first-year survival (survival zero to one year) and again by the sex ratio (see table 13.1). The same applies to the fecundity rate for the second stage class (i.e., individuals that are almost two years of age and older at census) except that here a higher parturition rate is used (parturition at two or more years). All vital rates were applied

Table 13.1. Model parameters

Parameter	Range / value
Simulation time	50 years
Initial population size (adults)	900
Initial population size (1- to 2-year-olds)	100
Carrying capacity	1,000
Coefficient of variation of carrying capacity	10%
Maximum growth rate R_{max}	1.3
Density dependence model	Scramble competition (Ricker equation)
Sex ratio	50% female
Parturition (1 year)	0.01–0.1
Parturition (2+ years)	0.75–1.0
Survival (0–1 year)	0.1–0.3
Survival (1–2 years)	0.75–0.9
Survival (2+ years)	0.8–0.95
Extreme event mortality	Normal distribution, $\mu = 0.6$; see methods
Annual probability of extreme event	0.025 / 0.05 / 0.1 / 0.2 / 0.4

as probabilistic values. For example, if for an individual in the first-stage class a random number in the interval zero to one fell below the survival rate, the individual survived and moved to the adult stage. This way, demographic stochasticity was also accounted for. To incorporate environmental stochasticity, we assumed a coefficient of variation (CV) of 10% for all vital rates as well as the carrying capacity (K).

For a given demographic model, parameter values for the vital rates (see table 13.1) were randomly selected from a predefined parameter range. For any extreme event scenario (see below) that we generated, we selected 1,000 randomized demographic models from these parameter envelopes. Each demographic model was then replicated 1,000 times for a total of 50 years. Thus, we ran a total of one million simulation runs (1,000 demographic models × 1,000 replicate runs) for each extreme event scenario. For our analysis we used BWCSim1.1, a free simulation tool available through the Caribou Viability Assessment Portal (CVAP) (http://www.norecaconsulting.com/simu_tools.php5) for studying nonspatial population viability and population parameter sensitivity of caribou.

We incorporated density dependence as scramble competition (Ricker equation) with a maximum growth rate of $R_{max} = 1.3$ (at $N_{t-1} = 0$) at which populations may increase in the absence of density dependence. Thus, in the model, annual growth of a population is a function of its density in the previous year under the assumption of

$$R_t = R_{\text{max}} * e^{\dfrac{-\ln(R_{\text{max}})* N_{t-1}}{K}} \qquad \text{[eq. 1]}$$

with t as the annual time step, R_t as growth rate, N_t as population size, and K as carrying capacity. All simulation runs started with a total female population of 100 almost one-year-olds (stage 1) and 900 adults (almost two years and older). In the computing algorithm, each year's total population growth (based on the calculations of the stage matrix; R_s) was compared to population growth predicted by the density dependence function (R_{DD}). If R_s was above or below R_{DD}, individuals in both stage classes were probabilistically either removed or added to fit the total population growth predicted by the density dependence model (based on the annual ratio between R_s and R_{DD}).

We ran five extreme event scenarios in which we changed the annual probability of a mortality event striking a population. Simulated annual probabilities and resulting average return intervals (i.e., the number of years for die-off to occur again) were 0.4 (2.5 years), 0.2 (5 years), 0.1 (10 years), 0.05 (20 years), and 0.025 (40 years), respectively. Current average return intervals for the BIC populations are 10 years, with a mean population die-off of 64%. We assumed a near-normal distributed probability density function for die-off severity. If in a given year a population is affected by an extreme event, the probability distribution from 10% to 90% die-off (affecting mortality in the entire population) is 0.005, 0.005, 0.01, 0.05, 0.19, 0.49, 0.19, 0.05, and 0.01, respectively. Thus, on average in 86% of the "extreme" years, die-off will be 50%, 60%, or 70%.

Local traditional ecological knowledge (TEK) suggests that die-offs are generally lower for smaller population sizes: spatial heterogeneity of resources and snow conditions allow higher survival if only a small number of caribou are present on the range and caribou may still be able to utilize the sparsely distributed resources during extreme winters (M. A. D. Ferguson, personal communication). This would lead to density-dependent effects on (primarily) density-independent die-offs and lower average die-off rates. However, the analysis of this relationship is not the focus of this study; it will be dealt with in a separate modeling analysis.

Simulation Results and Discussion

In a first step we analyzed the dataset of 1,000 demographic models without any extreme event to detect the most important demographic model parameter among five vital rates. The survival rate of adults (more than two years of age) showed the highest correlation when each vital rate (as a dependent variable) was plotted against extinction risk ($N = 0$; figure 13.1). The second most important parameter was first-year survival (zero to one year). Surprisingly, parturition was of lower importance. However, since fecundity is the product of first-year survival, parturition and sex ratio, using the product of first-year survival and parturition as a dependent variable, resulted in a stronger correlation that was almost identical to adult survival (not shown). Parturition of almost two-year olds (i.e., the youngest stage) was of low importance because of significantly lower rates and smaller population numbers (about 10% to 15% of the total population for stable age distributions). Survival rates for one- to two-year-olds was also of very minor importance, again as a result of the relatively small cohort size.

We then introduced extreme events and looked at how extinction probability changes with variation in adult survival (the most important vital rate; figure 13.2). When the probability for an extreme event was very low (with an average 40-year return interval), resulting extinction probabilities were only slightly higher than in model runs without die-offs (compare figure 13.2, $P = 0.025$, with bottom panel in figure 13.1). However, this changed dramatically for higher average return intervals (figure 13.2). For example, when a die-off strikes a population on average 10 times over the course of a 50-year simulation run ($P = 0.2$, 5-year return interval), all demographic models showed extinction risks of more than 65%, even when adult survival rates were very high. For an annual probability of $P = 0.4$, all demographic models resulted in extinction after 50 years (not shown).

The distribution of extinction risks for all 1,000 demographic models with an average five-year extreme event return was highly biased to high risk, with almost all models (99.5%) resulting in 90% extinction risk or higher (see lower panel in figure 13.3). However, extinction risk was more evenly distributed for models with higher return times (figure 13.3). Interestingly, the most noticeable shift occurred between annual probabilities of 0.05 and 0.0025, corresponding to average return intervals of 20 to 40 years. This apparently sensitive parameter range is significantly lower than the currently estimated average return interval of 10 years.

Absolute risks of extinction are of very limited value for management purposes, so we calculated changes in extinction probability for all simulated extreme event scenarios relative to no die-off (figure 13.4). To do so, we averaged extinction probabilities across all 1,000 demographic models (a total of one million simulation runs). This resulted in one extinction probability value for each extreme event scenario. We found that any

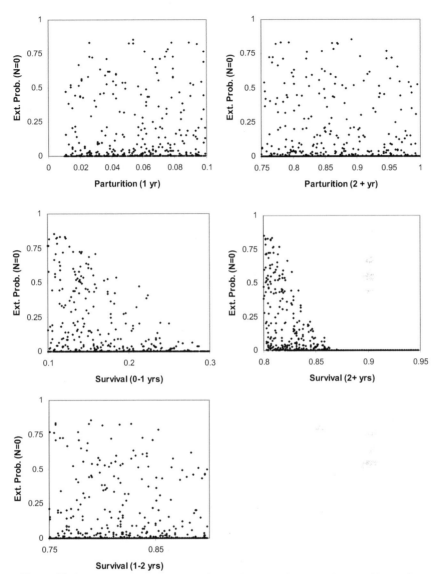

Figure 13.1. Extinction probability (N = 0) as a function of changes in parturition and survival. Each point represents the average extinction risk of 1,000 replicate simulation runs for one demographic model scenario. A total of 1,000 demographic scenarios were selected by randomly choosing parameter values of five different vital rates within a "realistic" range (see table 13.1) resulting in a total of one million simulations. Each simulation was run for 50 years with a starting population of 1,000 female caribou.

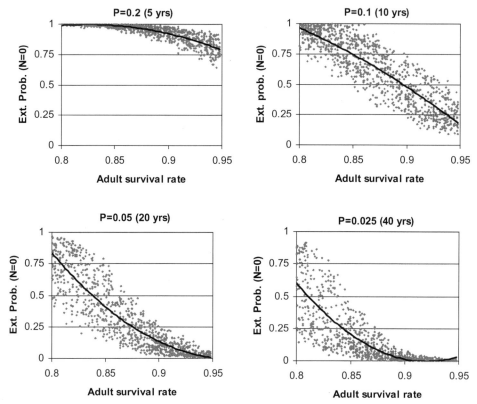

Figure 13.2. Extinction probability (N = 0) as a function of adult survival for four different scenarios of extreme events. For the upper left panel, for example, annual probability of a die-off was 0.2, corresponding to an average return interval of five years. Each point represents the average extinction probability of 1,000 replicate simulation runs. Each simulation was run for 50 years with a starting population of 1,000 female caribou.

increase in current frequencies beyond the current estimates (P = 0.1, or every 10 years on average) will further increase extinction risk. Most importantly, however, this relative change is far lower than any potential changes in the lower range of extreme event probabilities (figure 13.4). For example, average risk of extinction over 50 years (all demographic models included) increased from 5.6% to 58.7% when annual probabilities were increased from 0.0 to 0.05 (i.e., an extreme year every 20 years, on average). This corresponds to more than a tenfold increase. However, when currently estimated probabilities of 0.1 (with an average simulated extinction risk of 85.6%) were doubled to 0.4, the risk of extinction increased to 99.9%.

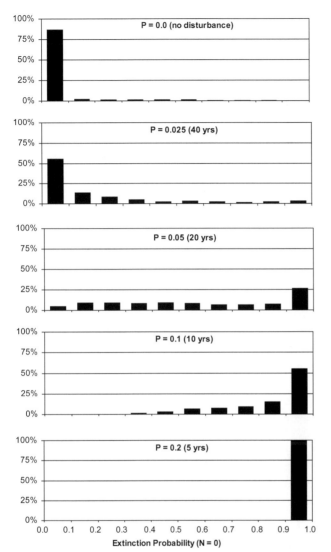

Figure 13.3.
Simulated frequency distributions of extinction probabilities ($N = 0$) for 1,000 demographic models (1,000 replicates) for $P = 0.0$ (no die-off), $P = 0.025$ (40-year average return interval), $P = 0.05$, $P = 0.1$, and $P = 0.2$.

Assuming that our simulated demographic models are "realistic" and caribou population die-offs in the High Arctic occur with a severity like the one assumed in our analysis, any further deterioration with ongoing climate change may have only minor effects on population persistence, as the current status may be already highly critical. We found that most significant effects on population dynamics occur in ranges lower than the currently observed frequencies. However, these findings are based on the

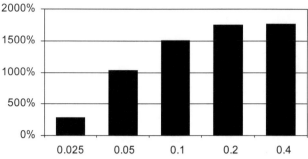

Figure 13.4. Relative changes in extinction probability ($N = 0$) for five extreme event scenarios (annual probabilities of 0.025, 0.05, 0.1, 0.2, 0.4) relative to a model with no die-off. For example, for an annual probability of 0.05 (average return interval of 20 years) the probability of extinction increased from 0.056 to 0.587. That is, it increased more than tenfold (1,039%) compared to no die-off. Average extinction probabilities for $P = 0.0, 0.025, 0.05, 0.1, 0.2,$ and 0.4 were 0.056, 0.159, 0.587, 0.856, 0.991, and 0.999, respectively.

assumption that die-off severity (the distribution of die-off classes in our model) will not change with regional warming in the Arctic.

As shown in a previous study (Tews et al. 2007a), a potential increase in the severity of extreme events may pose an even greater threat to population persistence than a comparable increase in frequency. These results were based on a two-factorial experiment in which different sets of parameter combinations for severity (percentage of die-off) and frequency (probability of occurrence) were simulated (instead of our using probability values for different severity classes for a single extreme event, as in the study presented here). It is most likely that the lower importance of changes in frequency is due to a relatively high maximum growth rate and recovery potential of caribou. This particularly applies if frequencies of extreme events stay below a certain threshold and allow sufficient time for population recovery. Our study suggests that frequencies observed in the most recent decades are most likely above such a threshold, and that High Arctic Peary caribou populations may be in a highly critical state.

Any potential reduction in the frequency of extreme winters may have significant positive effects on Peary caribou populations. This is particularly true for any reductions in the temporal autocorrelation of events. For example, three out of four extreme winters that were experienced by Peary caribou in the Bathurst Island complex over the course of four decades occurred during three years in the 1990s, with a devastating effect on population numbers. It might therefore be worth investigating additional effects of temporal autocorrelation that are believed to have overall negative effects on population persistence.

Synthesis

High Arctic caribou populations are subject to extreme winter weather events that make forage inaccessible, leading to their starvation (Miller and Gunn, 2003). The population-level severity of these extreme events is governed by a combination of temperature, precipitation, wind conditions, and topography. Our modeling analysis showed how population dynamics may be affected by simulated changes in the frequency of extreme winter weather. Due to the low quality and quantity of demographic data, we simulated population-level consequences by implementing scenarios of extreme events for a wide range of realistic, randomly assigned demographic models. Our results showed that significant improvements for longer-term population persistence would depend on these events occurring much less often.

Anticipated trends of climate change in the Arctic indicate that extreme winter weather events may become more frequent. This can have far-reaching ecological consequences for humans and arctic ecosystems alike: caribou, an integral part of arctic food webs, are traditionally harvested by local Inuit, who depend on them as part of their diet, integrity, and culture. The Inuit of Resolute Bay and Grise Fjord have already carefully monitored and restricted their own harvest of Peary caribou. The current harvest policies will leave not much room for further climate change mitigation, or management options for maintaining resilience.

It is very likely that change in extreme events may be ecologically more important than change in longer-term averages of climate variables. But how is it possible to define at which level the impact of precipitation, temperature, or any other relevant climate variable should be considered "extreme" rather than "normal?" From a general ecological point of view, we suggest considering an event as "extreme" if it causes total population loss of more than half the total population growth in a given time period (where no density-independent mortality events have been observed).

For example, in a 10-year period with an average annual growth rate of 5% and no extreme events, an initial population of 1,000 individuals would increase to 1,628, thus representing a total population growth of 62.8%. If an event strikes this population after that time and kills 32% of the population (resulting in 1,107 individuals) we suggest "extreme." Thus, the defined threshold depends on the total population growth within a certain time frame, independent of population size. It also considers the frequency of an extreme event. For example, if an exogenous mortality event strikes a population on average every five years, taking the above example would lower the threshold to 13.8%.

Another important question that needs to be addressed is: How can we generically characterize and quantify extreme weather events? Based on our literature review and case study, we deem six factors as important:

1. Thresholds: At which level is an event "extreme?"
2. Frequency: How often do extreme weather events occur?
3. Severity: How severe are the population-level effects of a single event?
4. Temporal pattern of severity: Does the event result in immediate consequences, or is there a time lag?
5. Temporal autocorrelation: Do the events occur in a regular pattern, or are they correlated over time?
6. Spatial pattern and autocorrelation: What is the event's spatial scale? Does it affect all local or regional populations simultaneously, or only a few?

In this chapter we have mostly discussed weather events in the sense of direct abiotic effects, in which extreme environmental conditions have an immediate and lethal effect on a local population. However, weather events or trends that result from climate change may also appear as indirect, longer-term effects from interspecific interactions such as predation, competition, and resource availability (Barbraud and Weimerskirch 2001, Durant et al. 2003, Pounds et al. 2006), especially in the Arctic (Post et al. 2009). For example, an extreme event may directly increase the mortality rate of juveniles and decrease the overall population abundance of one species. If that species is consumed by another, and the predator then must face reduced resource availability (assuming that it is not capable of switching to another prey species), this would result in indirect (biotic) effects. Hence, indirect consequences have the potential to vastly increase or decrease the proportion of species affected by climate change beyond the numbers predicted by direct effects.

ACKNOWLEDGMENTS
J. T. would like to thank the editors for their kind invitation to contribute this chapter, and Environment Canada's Landscape Science and Technology Division for providing financial support to develop the software tool BWCSim1.1.

LITERATURE CITED
Altwegg, R. A. Roulin, M. Kestenholz, and L. Jenni. 2006. Demographic effects of extreme winter weather in the barn owl. *Oecologia* 149:44–51.

Barbraud, C. and H. Weimerskirch. 2001. Emperor penguins and climate change. *Nature* 411:183–86.

Bauer, S., M. Van Dinther, K.A. Hogda, M. Klaassen, and J. Madsen. 2008. The consequences of climate-driven stop-over sites changes on migration schedules and fitness of arctic geese. *Journal of Animal Ecology* 22:654–60.

Beniston, M., D. Stephenson, O. Christensen, C. Ferro, C. Frei, S. Goyette, K. Halsnaes, T. Holt, K. Jylhä, B. Koffi, J. Palutikof, R. Schöll, T. Semmler, and K. Woth 2007. Future extreme events in European climate: an exploration of regional climate model projections. *Climatic Change* 81:71–95.

Berteaux, D., M. Humphries, C. Krebs, M. Lima, A. McAdam, N. Pettorelli, D. Réale, T. Saitoh, E. Tkadlec, R. Weladji, and N. Stenseth. 2006. Constraints to projecting the effects of climate change on mammals. *Climate Research* 32:151–58.

Borrvall, C., B. Ebenman, and T. Jonsson. 2000. Biodiversity lessens the risk of cascading extinction in model food webs. *Ecology Letters* 3:131–36.

Charmantier,A. R. McCleery, L. Cole, C. Perrins, L. Kruuk, and B. Sheldon. 2008. Adaptive phenotypic plasticity in response to climate change in a wild bird population. *Science* 320:800–803.

COSEWIC, 2004. Assessment and update status report on the Peary caribou (*Rangifer tarandus pearyi*) and barren-ground caribou (*Rangifer tarandus groenlandicus*) dolphin and union population in Canada. Committee on the status of endangered wildlife in Canada, Ottawa. Accessed at www.sararegistry.gc.ca/status/status e.cfm.

Durant, J. M., T. Anker-Nilssen, and N. C. Stenseth. 2003. Trophic interactions under climate fluctuations: the Atlantic puffin as an example. *Proceedings of the Royal Society of London B* 270:1461–66.

Frei, C., and C. Schar. 2001. Detection probability of trends in rare events: Theory and application to heave precipitation in the alpine region. *Journal of Climate* 14:1568–84.

Gunn, A., Shank, C., and B. McLean. 1991. The history, status and management of muskoxen on Banks Island. *Arctic* 44:188–95.

Haylock, M., and N. Nicholls. 2000. Trends in extreme rainfall indices for an updated high quality dataset for Australia, 1910–1998. *International Journal of Climatology* 20:1533–41.

Hulme, P. 2005. Adapting to climate change: Is there scope for ecological management in the face of a global threat? *Journal of Applied Ecology* 42:784–94.

Ibanez, I., J. Clarke, M. Dietze, K. Feeley, M. Hersh, S. Ladeau, A. McBride, N. Welch, and M. Wolosin. 2006. Predicting biodiversity change: Outside the climate envelope, beyond the species-area curve. *Ecology* 87:1896–1906.

Intergovenmental Panel on Climate Change (IPCC). 2001. *Climate Change 2001: The Scientific Basis*. Cambridge: Cambridge University Press.

Kohler, J., and R. Aanes. 2004. Effect of winter snow and ground-icing on a Svalbard reindeer population: Results of a simple snowpack model. *Arctic, Antarctic, and Alpine Research* 36:333–41.

Jentsch, A., J. Kreyling, and C. Beierkuhnlein. 2007. A new generation of climate change experiments: events, not trends. *Frontiers in Ecology and the Environment* 5:315–24.

Lovejoy, T. 2008. Climate change and biodiversity. *Revue Scientifique et Technique : Office International des Epizooties* 27:331–38.

Marshall, J., J. Blair, D. Peters, G. Okin, A. Rango, and M. Williams. 2008. Predicting and understanding ecosystem responses to climate change at continental scales. *Frontiers in Ecology and the Environment* 6:273–80.

McCabea, G. J., M. P. Clark, and L. E. Haya. 2007. Rain-on-snow events in the western United States. *Bulletin of the American Meteorological Society* 88:319–28.

Miller, F. L., and A. Gunn. 2003a. Catastrophic die-off of Peary caribou on the western Queen Elizabeth Islands, Canadian High Arctic. *Arctic* 56:381–90.

———. 2003b. Status, population fluctuations and ecological relationships of Peary caribou on the Queen Elizabeth Islands: Implications for their survival. *Rangifer* 14:213–26.

Nicholls, N. 2008. Australian climate and weather extremes: Past, present, and future. Report for the Department of Climate Change, Australian Government. Department of Climate Change, Commonwealth of Australia, Canberra.

Ohlemuller, R., B. Anderson, M. Araujo, S. Butchart, O. Kudrna, R. Ridgely, and C. Thomas. The coincidence of climatic and species rarity: high risk to small-range species from climate change. *Biology Letters* 4:568–72.

Parmesan, C. 2006. Ecological and evolutionary responses to recent climate change. *Annual Review of Ecology, Evolution, and Systematics* 37:637–69.

Picket, S., and P. White. 1985. Definitions: Patch dynamics, perturbation, and disturbance. In *The Ecology of Natural Disturbance and Patch Dynamics*, edited by S. Picket and P. White. California: Academic Press.

Post, E., M. C. Forchhammer, M. S. Bret-Harte, T. V. Callaghan, T. R. Christensen, B. Elberling, A. D. Fox, O. Gilg, D. S. Hik, T.T . Hoye, R. A. Ims, E. Jeppesen, D.R.Klein, J. Madsen, A. D. McGuire, S. Rysgaard, D. E. Schindler, I. Sterling, M.P. Tamstorf, N. J. C. Tyler, R. van der Wal, J. Welker, P. A. Wookey, N. M. Schmidt, and P. Aastrup. 2009. Ecological dynamics across the Arctic associated with recent climate change. *Science* 325:1355–58.

Pounds, J. A., M. R. Bustamante, L. A. Coloma, J. A. Consuegra, M. P. L. Fogden, P. N. Foster, E. La Marca, K. L. Masters, A. Merino-Viteri, R. Puschendorf, S. R. Ron, G. A. Sánchez-Azofeifa, C. J. Still, and B. E. Young. 2006. Widespread amphibian extinctions from epidemic disease driven by global warming. *Nature* 439:161–67.

Putkonen, J., and G. Roe. 2003. Rain-on-snow events impact soil temperatures and affect ungulate survival. *Geophysical Research Letters* 30:1188.

Running, S. 2006. Is global warming causing more, larger wildfires? *Science* 313: 927–28.

Schneider, S., and J. Lane. 2006. An overview of "dangerous" climate change. In *Avoiding Dangerous Climate Change*, edited by H. Schellnhuber, W. Cramer, N. Nakicenovic, T. Wigley, and G. Yohe, 7–23. Cambridge: Cambridge University Press.

Schwartz, M., L. Iverson, A. Prasad, S. Matthews, and R. O'Connor. 2006. Predicting extinctions as a result of climate change. *Ecology* 87:1611–15.

Shoo, L., S. Williams, and J. Hero. 2005. Detecting climate change induced range shifts: Where and how should we be looking? *Austral Ecology* 31:22–29.

Smith, R., and T. Smith. 2001. *Ecology and Field Biology*, 6th Edition. Benjamin Cummings; Addison Wesley Longman Inc. USA.

Tejeda-Cruz, C., and W. Sutherland. 2005. Cloud forest bird responses to unusually severe storm damage. *Biotropica* 37:88–95.

Tews, J., M. A. D. Ferguson, and L. Fahrig. 2007a. Modeling density dependence and climatic disturbances in caribou: a case study from the Bathurst Island complex, Canadian High Arctic. *Journal of Zoology* 272:209–17.

———. 2007b. Potential net effects of climate change on High Arctic Peary caribou: Lessons from a spatial-explicit simulation model. *Ecological Modelling* 207:85–98.

Trenberth, K. 2008. The impact of climate change and variability on heavy precipitation, floods and droughts. *Encyclopedia of Hydrological Sciences*, edited by M. Anderson. Chichester, UK: J. Wiley and Sons.

Webster, P., G. Holland, A. Curry, and H. Chang. 2005. Changes in tropical cyclone number, duration, and intensity in a warming environment. *Science* 309:1844–46.

Welbergen, J., S. Klose, N. Markus, and P. Eby. 2008. Climate change and the effects of temperature extremes on Australian flying-foxes. *Proceedings of the Royal Society* 275:419–25.

Promoting Resilience: Wildlife Management in the Face of Climate Change

14

Harvest Models for Changing Environments

Mark S. Boyce, Kyle Knopff, Joseph Northrup,
Justin Pitt, and Liv S. Vors

Climate change and land-use changes will continue to alter the distribution and abundance of fish and wildlife populations by influencing carrying capacity, seasonality, and vital rates of survival and fecundity. Accordingly, optimal harvest policies will change for those populations that are managed for human exploitation. Habitat fragmentation and climate change have caused population declines in many species (Mora et al. 2007, Brook et al. 2008). But responses will vary among species and among populations (Williams et al. 2008), and not all populations will be affected negatively by climate and landscape change. For example, we expect the carrying capacity for some species to decrease as habitats deteriorate (e.g., declines in caribou [*Rangifer tarandus*; Vors and Boyce 2009] and polar bears [*Ursus maritimus*; Stirling et al. 1999]) whereas other species will thrive under climate and landscape change (e.g., white-tailed deer [*Odocoileus virginianus*; Johnston and Schmitz 1997] and raccoons [*Procyon lotor*; Larivière 2004]). Generally, we expect vital rates to become more variable under climate change due to more extreme weather events (Boyce et al. 2006; IPCC 2001, 2007). Milder winters might result in less seasonal variation in some areas, and climate change will entail altered seasonal precipitation regimes in others (Rinke and Dethloff 2008).

Harvest regimes can be altered to adapt to changing environmental conditions, permitting sustainable harvests to be maintained, or even mitigating the negative consequences of environmental change in some cases. Harvest policies can include selective removals by age, size, or sex, and can involve management intervention of season, quota, bag limits, and gear (Getz and Haight 1989, Xu et al. 2005). Some harvest policies are deliberately density-dependent, which ensures population resilience in the face of climate change (Varley and Boyce 2006). More generally, harvesting can be adaptable and harvest models can be employed to identify how harvesting might be adjusted to accommodate future changes in climate and patterns of land use. In this chapter we review the theory behind harvest models, explore several relevant examples of their application with respect to environmental change, and provide direction for effective harvest management in a changing world.

Harvest Models

Sustainable harvest management depends fundamentally on density-dependent vital rates; that is, survival and reproduction are higher at reduced population size (Mendelssohn 1976, Boyce et al. 1999, Xu and Boyce 2009). The basic elements of harvesting can be modeled by removing a quota, Q, of animals from a population of N individuals growing according to the logistic model

$$dN\!\big/\!dt = rN\left(1 - \frac{N}{K}\right) - Q \qquad\qquad \text{[eq. 1]}$$

with a potential per capita growth rate of r, achieving an equilibrium carrying capacity of K without harvest. Here the population growth rate and associated potential yield is quadratic, as illustrated in figure 14.1, with an equilibrium population size of N^*. Maximum sustained yield, MSY, is achieved at population size $N^* = K/2$, but this is a dangerous target because if $Q > $ MSY, the population will surely decline, ultimately to extinction unless the quota is reduced. Thus it is crucial to monitor the population to ensure that it stabilizes, and typically managers attempt to manage for yields less than MSY with $N^* > K/2$ to accommodate uncertainty in

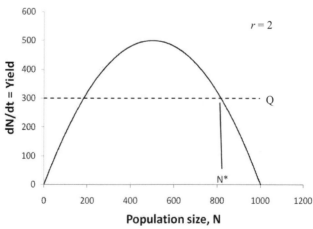

Figure 14.1. Quota harvesting of the logistic model, in which $r = 2$, $K = 1,000$, and $Q = 300$. For this quota of 300, sustainable yield occurs at equilibrium population size of N^* where the production curve intersects the quota-harvest line. See equation 1 for model definition.

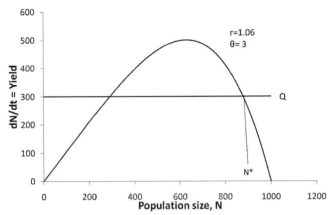

Figure 14.2. Nonlinear density dependence in quota harvest model, in which $r = 1.06$ and $\theta = 3$. Qualitatively the model is the same as in figure 14.1 for the logistic, except that equilibrium population size, N^*, is higher for $\theta > 1$. Yield occurs where the quota harvest line intersects the production curve.

population estimates, environmental stochasticity, and climate change. For example, a manager might target a Q of 80% or less of MSY to avoid overharvesting the population (Boyce 2000). Although another critical point occurs where the quota line intersects the yield curve for $N < K/2$, this is a saddle point and populations quickly converge on N^* if abundance should become larger, or to $N = 0$ if harvests are too large.

Many variations of this model can be constructed, depending on the biology of the population. For example, nonlinearity in density dependence is typical in vertebrates (Fowler 1981, 1987) and it might be represented by a nonlinear exponent, θ:

$$dN\!\big/\!dt = rN(1 - [\tfrac{N}{K}]^{\theta}) - Q \qquad \text{[eq. 2]}$$

This exponent has the consequence of altering the population size at which maximum sustainable yield can occur (figure 14.2). For example, for $\theta > 1$, as is typical for mammal populations (Fowler 1981, 1987), MSY will occur at $N^* > K/2$; if $\theta < 1$, MSY will be at $N^* < K/2$.

Likewise, one easily can model alternative harvesting regimens. A proportional harvest, h, of the population might be used instead of a quota:

$$dN\!\big/\!dt = rN\left[1 - \frac{N}{K}\right] - hN \qquad \text{[eq. 3]}$$

with a different graphical interpretation (figure 14.3). As for quota harvests of the logistic model, MSY would be achieved where the hN line intersects the maximum in the yield curve at $K/2$. Likewise, for harvests smaller than MSY, the population will equilibrate at $K/2 < N^* < K$. The caveat tied to such a proportional harvest is that the manager requires good estimates of abundance, so populations must be monitored closely. Proportional harvesting is sometimes equated with constant-effort harvesting (Ludwig 2001), which can be easier to implement using consistent harvest seasons and regulations. However, effort is seldom a simple linear function of yield, and the functional response of human hunters and fishers can be dynamic.

In many cases we expect that both climate change and landscape change will alter carrying capacity. And clearly a change in K will result in large consequences to the yield curve (figure 14.4). Smaller K will reduce MSY and it likewise can be expected to reduce the harvests that can be sustained safely, whether by using a quota or proportional harvesting policy. Conversely, if changes in the environment cause an increase in K, we can sustainably harvest many more animals from the population. Changes in r can be expected as well; thereby, we can envisage that the entire density-dependent function might be altered by future climate and landscape changes.

The consequences of environmental variability can be examined using harvest models. For example, seasonality can be incorporated into models by defining K a time-varying function:

$$K(t) = \bar{K} + K_a \cos(2\pi t/\tau) \qquad \text{[eq. 4]}$$

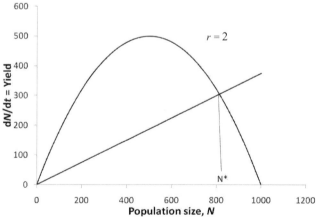

Figure 14.3. Proportional harvesting for $h = 0.3$. Again, yield occurs where the harvest line intersects the production curve at equilibrium population size, N*.

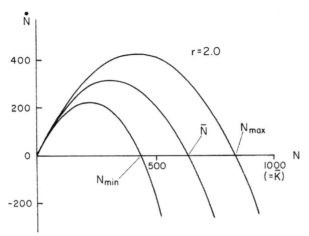

Figure 14.4. Time-varying $K(t)$ produces production curves that vary temporally with the smallest curve when $K(t)$ is at minimum, and the largest curve with greatest yield when $K(t)$ is at maximum (from Boyce and Daley 1980). Deviations of a certain magnitude will produce less of an increase when $N < K$, but a greater decrease if $N > K$.

with $K(t)$ oscillating through time, t, in a sinusoidal manner around an average $\bar{K}$ with amplitude K_a and period length τ. The consequence of increasing amplitude of seasonality in carrying capacity, K_a, is that average population size declines (Boyce and Daley 1980). In fact, more generally the greater the variability in $K(t)$, the lower the long-term population size, and the lower the sustainable yields (Xu et al. 2005). Similarly, $K(t)$ can be defined as a random variable to characterize stochastic variation in the environment. But the message is clear that harvesting levels should be set based on variable $K(t)$, to avoid the risk of overharvest.

Carefully monitored and regulated harvesting can be used to benefit conservation. For populations with a discrete harvesting season, and in which most natural mortality occurs in the non-harvest season, overcompensation can actually increase annual survival and abundance (Boyce et al. 1999)—a phenomenon termed the hydra effect (Abrams 2009). If harvest intensity increases with population size, harvesting can increase population resilience beyond that afforded by natural density-dependent processes (Varley and Boyce 2006).

Additional complexity in harvest models can be incorporated by detailing population growth into vital rates of survival and fecundity (Getz and Haight 1989). Selective harvesting of individuals with low reproductive value (e.g., males, young, old) can be used to increase sustainable yields. Most principles of harvesting are the same in structured population models as in models without age structure. For example, fluctuating vital rates usually result in reductions in average population size and reduced yields (Boyce et al. 2006).

Harvesting in Changing Environments

Changing climates and landscapes will enhance conditions for some species while eroding the environment for others. Caribou and polar bears are well-known examples for which climate warming and landscape change are reducing carrying capacity and thereby suggest lower sustainable harvests. Indeed, the Northwest Territories government has substantially reduced the quotas for caribou hunting, and the import of polar bears harvested in Canada by American hunters is prohibited by the US Endangered Species Act (but see below).

Harvest management for these two species presents unique complications. Restricting the harvests of arctic and migratory caribou herds has been controversial with both subsistence and recreational hunters, although these restrictions probably are necessary and appropriate given the magnitude of climate change in northern latitudes and their propensity to negatively influence caribou population growth. For nonmigratory woodland caribou further south, abundance has declined and hunting has been substantially reduced in most jurisdictions of Canada. In many areas industrial development has resulted in timber harvest which in turn creates early seral stage forests that are excellent habitat for moose (*Alces alces*) and white-tailed deer (Wittmer et al. 2007). Due to increased abundance of moose and deer, wolf (*Canis lupus*) populations respond numerically with concomitant increases in wolf predation on caribou, an ecological process that has been termed "apparent competition" (Holt 1987). In response, provincial governments have attempted to increase the harvests of wolves, deer, and moose, with mixed success (Robichaud and Boyce 2009).

Even though polar bear populations are declining rapidly in southern portions of their distribution in Canada, continued harvest might produce substantial conservation benefit by increasing the economic value of this species (Foote and Wenzel 2009, Freeman and Foote 2009). Polar bears are circumpolar predators that specialize on phocid seals that they hunt on sea ice. Other food sources are far less profitable, and bears will hunt seals at all times of year, provided that the ice persists (Stirling and McEwan 1975). The extent and duration of sea ice is therefore considered a critical limiting factor for polar bears (Derocher et al. 2004). The pace of climate change in polar regions is far more rapid than in other parts of the globe (Anisimov et al. 2007), and as the Arctic warms, the extent and intraannual duration of sea ice have been declining, resulting in the loss of preferred near-shore habitat for polar bears, a trend that is expected to continue over the next century (Durner et al. 2009). Although the topic remains contentious (Dyck et al. 2007, Stirling et al. 2008a, Dyck et al. 2008), a substantial body of research suggests that polar bear carrying capacity, and hence polar bear populations, will decline as the extent and duration of sea ice is reduced (Stirling et al. 1999, Derocher et al. 2004, Stirling and Parkinson 2006, Durner et al. 2009, Cherry et al. 2009).

Most polar bear harvest occurs in Canada, where the number of animals taken annually is tightly controlled by a quota system. Quotas are set using a combination of information gleaned from scientific studies and the traditional knowledge of indigenous people (Stirling and Parkinson 2006). Bears are harvested primarily for subsistence, with a small portion of the tags sold to nonindigenous trophy hunters (Freeman and Foote 2009). Although early estimates of polar bear numbers are unreliable, populations are thought to have increased from fewer than 10,000 animals in the early 1970s to between 20,000 and 25,000 today as a consequence of a unified circumpolar management regime, which has included harvest by hunting. In 2008 the United States listed polar bears as "threatened" under the Endangered Species Act in response to concern over potential declines in their numbers as a result of climate change. This triggered restrictions on polar bear products under the purview of the US Marine Mammal Protection Act. Although subsistence and trophy hunting continue in Canada, the US restrictions mean that US citizens can no longer import polar bears into their country, thus reducing their incentive to participate in the hunt and creating potentially deleterious consequences for Inuit communities that depend on revenues from US hunters (Slavik 2009).

Polar bears have poorer body condition (Stirling et al. 1999) and spend more time fasting (Cherry et al. 2009) in years when sea ice breaks up early, rendering seals inaccessible. Several polar bear populations have declined where early breakups have occurred, suggesting that population size has exceeded carrying capacity. That polar bears are nutritionally stressed and some populations are too large is further evidenced by bears dying of starvation and engaging in intraspecific predation and cannibalism (Amstrup et al. 2006, Stirling et al. 2008b). When populations exceed carrying capacity, harvests should take place to compensate. Bears may also be spending more time ashore in search of food (Stirling and Parkinson 2006), increasing conflict with local people, and reducing social carrying capacity (Tyrrell 2009). This can be circumvented somewhat if there are incentives, economic or otherwise, for local people to maintain carnivore populations (Foote and Wenzel 2009). Conservation hunting of surplus animals under carefully controlled and monitored quotas can maintain economic value for residents of arctic communities, and might improve short-term conservation prospects despite the decline in carrying capacity caused by climate change.

In contrast with caribou and polar bears, other species such as white-tailed deer, cougars (*Puma concolor*), and raccoons have been increasing in abundance and moving farther north in response to recent climate warming (Johnston and Schmitz 1997, Larivière 2004, Pitt et al. 2008, Anderson et al. 2009). As higher carrying capacies increase for these species, sustainable yields also increase, although in many areas there is little demand for hunting them.

Consequences of Climatic Variability

Climate models consistently predict increased climatic variation and more frequent extreme weather events (IPCC 2001, 2007; Carney et al. this volume; Tews et al. this volume). As wildlife habitat, vegetation, and the vital rates of many species are strongly influenced by climate (Stirling et al. 1999, Coulson et al. 2001, Wang et al. 2002, Keith et al. 2008), one potential effect of climate change is that wildlife populations might experience increased fluctuations in carrying capacity, vital rates, and population size (Boyce et al. 2006).

Examples for which the demographic consequences of fluctuating environments have been studied include elk (*Cervus elaphus*), Soay sheep (*Ovis aries*), alpine ibex (*Capra ibex*), and Svalbard reindeer (*R. tarandus platyrhynchus*). Populations of these species experience large fluctuations in carrying capacity which are linked to heavy winter precipitation that causes decreased forage availability or increased nutritional demands. Decreases in carrying capacity can lead to crashes in population size, especially when herds are very dense. When herds are less dense, they rebound quickly, partly because of the increased forage that results from higher than average precipitation (Coulson et al. 2001, Solberg et al. 2001, Wang et al. 2002; Jacobson et al. 2004, Berryman and Lima 2006, Lima and Berryman 2006).

Long-term studies have monitored each of these four species for more than 20 years (Clutton-Brock et al. 1992, Aanes et al. 2000, Sæther et al. 2002). Clearly, careful monitoring is crucial if we are to accurately model sustainable harvest in the face of climate change. Furthermore, it will be important to monitor key variables such as winter precipitation, vegetation cover, and composition along with population numbers. These variables are often the mechanisms driving population dynamics (Johnston and Schmitz 1997), and thus they may offer early warning signs of changes in carrying capacity or vital rates. Close monitoring will be of particular importance in high-latitude and high-altitude populations because climate change is predicted to be extreme in these regions (Parmesan 2006).

Elk, Soay sheep, alpine ibex, and Svalbard reindeer all experience large fluctuations in carrying capacity due to the combined pressures of harsh winters and high population densities. Harsh winters also are a common cause of mortality and population crashes in harvested populations—again, particularly at high densities (Singer et al. 1997, Nesslage and Porter 2001, DelGiudice et al. 2002). Harvest, however, might be used to maintain stable populations in a changing climate. Indeed, it has been suggested as a possible means of maintaining stable populations of white-tailed deer in some areas where populations have increased substantially (Cote et al. 2004).

The main difficulty in managing harvested populations will be to understand whether dynamics similar to those in the examples above are in play, and to set quotas accordingly. This requires accurate and appropriate harvest models. For example, restricting harvest in extreme winters could lead to large die-offs under certain scenarios. Harvest models must address different populations and segments of populations separately, because they will be affected differently by climate (Walther et al. 2002, Post et al. 2009). Some models cannot be extrapolated across large areas, and age- and sex-structured harvest models might be necessary to manage some wildlife populations. Indeed, such models were used to obtain more complete information on the influence of climate on Soay sheep, revealing differential effects of climate variation by age and sex, which led to disproportionate survival in some age and sex classes (Coulson et al. 2001).

Actual implementation mechanisms can be a crucial concern in establishing harvest policies. Quota harvests often are used to manage harvests of large mammals by issuing a certain number of hunting permits, with projected harvest anticipated by observations of hunter success. Proportional harvesting can be difficult to implement because it requires precise continuing estimates of abundance. An alternative is threshold harvesting, in which the hunt is terminated if estimates of abundance fall below a certain level. Although it can result in high variation in annual yield, Sæther et al. (2001) state that threshold harvesting for moose is a more sustainable harvesting strategy than proportional harvesting because harvest is reduced or absent during years with low population size.

Adaptive management is ideally suited to harvest management in changing environments. Indeed, waterfowl management in North America has followed this paradigm successfully for the past 30 years (Williams and Johnson 1995). Adaptive harvest management involves an iterative system of (1) prediction, ideally based on harvest models, (2) harvest manipulation to stimulate a population response, (3) monitoring to document the response to the harvest manipulation, and (4) evaluation of model predictions with results obtained "on the ground." Assuming the existence of some disparity between predicted and observed data, the harvest model should be adjusted accordingly and a new iteration initiated. This "learning by doing" process allows a gradual improvement in harvest models—which is almost certainly necessary in the face of environmental change, given that its pace is unlikely to be either constant or unidirectional.

Conclusions

Future environments on earth will change because human activities are altering global climates and increasing the intensity of land use. The response to these changes will vary among areas, among species, and among populations. Ecologists will need to use GIS tools and demographic investigations to evaluate whether the environmental changes influence carrying capacity or vital rates for species of concern. Clearly anticipating future change involves a great deal of uncertainty. Data required to estimate the population parameters needed for harvest models are often unavailable and are subject to uncertainty or error (e.g., climate change predictions from other models). Such uncertainty is always a concern when one attempts to use harvest models in managing quotas or other harvest regimes. Simulation modeling is useful for exploring such uncertainty and for adapting harvest policies to anticipate them.

Perhaps most important is the need for rigorous monitoring data, both for changes in the environment and for changes in the populations of harvested populations. Adjustments to harvesting can be justified on the basis of modeling results, but predictions made with models are only as good as the data used to parameterize the models.

ACKNOWLEDGMENTS
We thank the Alberta Conservation Association and the Natural Sciences and Engineering Research Council of Canada for support.

LITERATURE CITED
Aanes, R., B.-E. Sæther, and N. A. Oritsland. 2000. Fluctuations of an introduced population of Svalbard reindeer: The effects of density dependence and climatic variation. *Ecography* 23:437–43.

Abrams, P. A. 2009. When does greater mortality increase population size? The long history and diverse mechanisms underlying the hydra effect. *Ecology Letters* 12:462–74.

Amstrup, S. C., I. Stirling, and T. S. Smith. Recent observations of intraspecific predation and cannibalism among polar bears in the southern Beaufort Sea. *Polar Biology* 29:997–1002.

Anderson, C. R., Jr., F. Lindzey, K. H. Knopff, M. G. Jalkotzy, and M. S. Boyce. 2009. Cougar management in North America. In *Cougar: Ecology and Conservation*, edited by M. Hornocker, S. Negri, and A. Rabinowitz, 41–54. Chicago: University of Chicago Press.

Anisimov, O. A., D. G. Vaughan, T. V. Callaghan, et al. 2007. Polar regions (Arctic and Antarctic). In *Climate Change 2007: Impacts, Adaptation and Vulnerability. Fourth Assessment Report of the Intergovernmental Panel on Climate Change*, edited by M. L. Parry et al., 653–85. New York: Cambridge University Press.

Berryman, A., and M. Lima. 2006. Deciphering the effects of climate on animal populations: Diagnostic analysis provides new interpretation of Soay sheep dynamics. *American Naturalist* 168:784–95.

Boyce, M. S. 2000. Whaling models for conservation. In *Quantitative Methods for Conservation Biology*, edited by S. Ferson and M. Burgman, 109–26. New York: Springer-Verlag.

Boyce, M. S., and D. J. Daley. 1980. Population tracking of fluctuating environments and natural selection for tracking ability. *American Naturalist* 115:480–91.

Boyce, M. S., A. R. E. Sinclair, and G. C. White. 1999. Seasonal compensation of predation and harvesting. *Oikos* 87:419–26.

Boyce, M. S., C. V. Haridas, C. T. Lee, and NCEAS Stochastic Demography Working Group. 2006. Demography in an increasingly variable world. *Trends in Ecology and Evolution* 21:141–48.

Brook, B. W., N. S. Sodhi, and C. J. A. Bradshaw. 2008. Synergies among extinction drivers under global change. *Trends in Ecology and Evolution* 23:453–60.

Cherry, S. G., A. E. Derocher, I. Stirling, and E. S. Richardson. 2009. Fasting physiology of polar bears in relation to environmental change and breeding behavior in the Beaufort Sea. *Polar Biology* 32:383–91.

Clutton-Brock, T. H., O. F. Price, S. D. Albon, and P. A. Jewell. 1992. Early development and population fluctuations in Soay sheep. *Journal of Animal Ecology* 61:381–96.

Cote, S. D., T. P. Rooney, J. Tremblay, C. Dussault, and D. M.Waller. 2004. Ecological impacts of deer overabundance. *Annual Review of Ecology, Evolution and Systematics* 35:113–47.

Coulson, T., E. A. Catchpole, S. D. Albon, B. J. T. Morgan, J. M. Pemberton, T. H. Clutton-Brock, M. J. Crawley, and B. T. Grenfell. 2001. Age, sex, density, winter weather, and population crashes in Soay Sheep. *Science* 292:1528–531.

DelGiudice, D. G., M. R. Riggs, P. Joly, and W. Pan. 2002. Winter severity, survival, and cause-specific mortality of female white-tailed deer in north-central Minnesota. *Journal of Wildlife Management* 66:698–717.

Derocher, A. E., N. J. Lunn, and I. Stirling. 2004. Polar bears in a warming climate. *Integrated Comparative Biology* 44:163–76.

Durner, G. M., D. C. Douglas, R. M. Nielson, S. C. Amstrup, T. L. McDonald, I. Stirling, M. Mauritzen, E. W. Born, O. Wiig, E. DeWeaver, M. C. Serreze, S. E.

Belikov, M. M. Holland, J. Maslanik, J. Aars, D. A. Bailey, and A. E. Derocher. 2009. Predicting 21st-century polar bear habitat distribution from global climate models. *Ecological Monographs* 79:25–58.

Dyck, M. G., W. Soon, R. K. Baydack, D. R. Legates, S. Baliunas, T. F. Ball, and L. O. Hancock. 2007. Polar bears of western Hudson Bay and climate change: are warming spring air temperatures the "ultimate" survival control factor? *Ecological Complexity* 4:73–84.

———. 2008. Reply to response to Dyck et al. (2007) on polar bears and climate change in western Hudson Bay by Stirling et al. (2008). *Ecological Complexity* 5:289–302.

Foote, L., and G. W. Wenzel. 2009. Polar bear conservation hunting in Canada: economics, culture and unintended consequences. In *Inuit Polar Bears and Sustainable Use*. edited by M. M. R. Freeman and L. Foote, 13–24. Edmonton: CCI Press.

Fowler, C. W. 1981. Density dependence as related to life history strategy. *Ecology* 62:602–10.

———. 1987. A review of density dependence in populations of large mammals. *Current Mammalogy* 1:401–41.

Freeman, M. M. R., and L. Foote. 2009. *Inuit Polar Bears and Sustainable Use*. Edmonton: CCI Press.

Getz, W. M., and R. G. Haight. 1989. *Population Harvesting: Demographic Models of Fish, Forest, and Animal Resources*. Princeton, NJ: Princeton University Press.

Grosbois, V., O. Gimenez, J.-M. Gaillard, R. Pradel1, C. Barbraud, J. Clobert, A. P. Møller, and H. Weimerskirch. 2008. Assessing the impact of climate variation on survival in vertebrate populations. *Biological Reviews* 83:357–99.

Holt, R. 1977. Predation, apparent competition and structure of prey communities. *Theoretical Population Biology* 12:197–29.

IPCC. 2001. *Climate Change 2001: The Scientific Basis*, edited by J. T. Houghton, Y. Ding, D. J. Griggs, M. Noguer, P. J. van der Linden and D. Xiaosu. Cambridge: Cambridge University Press.

IPCC. 2007. *Climate Change 2007: Synthesis Report*, edited by R. K. Pachauri and A. Reisinger Geneva: IPCC.

Jacobson, A. R., A. Provenzale, A. von Hardenberg, B. Bassano, and M. Festa-Bianchet. 2004. Climate forcing and density dependence in a mountain ungulate population. *Ecology* 85:1598–610.

Jensen, A. L. 2002. Analysis of harvest and effort data for wild populations in fluctuating environments. *Ecological Modelling* 157:43–49.

Johnston, K. M., and O. J. Schmitz. 1997. Wildlife and climate change: Assessing the sensitivity of selected species to simulated doubling of atmospheric CO_2. *Global Change Biology* 3:531–44.

Keith, D. A., H. R. Akçakaya, W. Thuiller, G. F. Midgley, R. G. Pearson, S. J. Phillips, H. M. Regan, M. B. Araújo, and T. G. Rebelo. 2008. Predicting extinction risks under climate change: coupling stochastic population models with dynamic bioclimatic habitat models. *Biology Letters* 4:560–63.

Larivière, S. 2004. Range expansion of raccoons in the Canadian prairies: Review of hypotheses. *Wildlife Society Bulletin* 32:955–63.

Lima, M., and A. Berryman. 2006. Predicting nonlinear and non-additive effects of climate: The Alpine ibex revisited. *Climate Research* 32:129–35.

Ludwig, D. 2001. Can we exploit sustainably? In *Conservation of Exploited Species*, edited by J. D. Reynolds, G. M. Mace, K. H. Redford, and J. G. Robinson, 16–38. Cambridge: Cambridge University Press.

Mendelssohn, R. 1976. Optimization problems associated with a Leslie matrix. *American Naturalist* 110:339–49.

Mora, C., R. Metzger, A. Rollo, and R. A. Myers. 2007. Experimental simulations about the effects of overexploitation and habitat fragmentation on populations facing environmental warming. *Proceedings of the Royal Society B* 274:1023–28.

Nesslage, G. M., and W. F. Porter. 2001. A geostatistical analysis of deer harvest in the Adirondack Park, 1954–1997. *Wildlife Society Bulletin* 29:787–94.

Nielsen, S. E., G. B. Stenhouse, H. L. Beyer, F. Huettmann, and M. S. Boyce. 2008. Can natural disturbance-based forestry rescue a declining population of grizzly bears? *Biological Conservation* 141:2193–2207.

Parmesan, C. 2006. Ecological and evolutionary responses to recent climate change. *Annual Review of Ecology, Evolution and Systematic* 37:637–69.

Pitt, J. A., S. Larivière, and F. Messier. 2008. Survival and body condition of raccoons at the edge of their range. *Journal of Wildlife Management* 72:389–95.

Post, E., J. Brodie, M. Hebblewhite, A. D. Anders, J. A. K. Maier, and C. C. Wilmers. 2009. Global population dynamics and hot spots of response to climate change. *BioScience* 59:489–97.

Rinke, A., and K. Dethloff. 2008. Simulated circum-arctic climate changes by the end of the 21st century. *Global and Planetary Change* 62:173–86.

Robichaud, C., and M. S. Boyce 2009. Wolf control to protect the Little Smoky caribou herd. *Alberta Outdoorsmen* 16:12–15.

Saether, B.-E., S. Engen, and E. J. Solberg. 2001. Optimal harvest of age-structured populations of moose *Alces alces* in a fluctuating environment. *Wildlife Biology* 7:171–79.

Saether, B.-E., S. Engen, F. Filli, R. Aanes, W. Schroder, and R. Andersen. 2002. Stochastic population dynamics of an introduced Swiss population of the ibex. *Ecology* 83:3457–65.

Singer, F. J., A. Harting, K. K. Symonds, and M. B. Coughenour. 1997. Density dependence, compensation, and environmental effects on elk calf mortality in Yellowstone National Park. *Journal of Wildlife Management* 61:12–25.

Slavik, D. 2009. The economics and client opinions of polar bear conservation hunting in the Northwest Territories, Canada. In *Inuit Polar Bears and Sustainable Use*, edited by M. M. R. Freeman and L. Foote, 65–77. Edmonton: CCI Press.

Solberg, E. J., P. Jordhoy, O. Strand, R. Aanes, A. Loison, B.-E. Sæther, and J. D. C. Linell. 2001. Effects of density-dependence and climate on the dynamics of a Svalbard reindeer population. *Ecography* 24:441–51.

Stirling, I., N. J. Lunn, and J. Iacozza. 1999. Long-term trends in the population ecology of polar bears in the western Hudson Bay in relation to climate change. *Arctic* 53:294–306.

Stirling, I., and E. H. McEwan. 1975. The caloric value of whole ringed seals (*Phoca hispida*) in relation to polar bear (*Ursus maritimus*) ecology and hunting behaviour. *Canadian Journal of Zoology* 53:1021–27.

Stirling, I., and C. L. Parkinson. 2006. Possible effects of climate warming on selected populations of polar bears in the Canadian Arctic. *Arctic* 59:**261–75.**

Stirling, I., A. E. Derocher, W. A. Gough, and K. Rode. 2008a. Response to Dyck et al. (2007) on polar bars and climate change in western Hudson Bay. *Ecological Complexity* 5:193–201.

Tyrrell, M. 2009. West Hudson Bay polar bears: the Inuit perspective. In *Inuit Polar Bears and Sustainable Use*, edited by M. M. R. Freeman and L. Foote, 95–110. Edmonton: CCI Press.

Varley, N., and M. S. Boyce. 2006. Adaptive management for reintroductions: updating a wolf recovery model for Yellowstone National Park. *Ecological Modelling* 193:315–39.

Vors, L. S., and M. S. Boyce. 2009. Global declines of caribou and reindeer. *Global Change Biology* 15:2626–33.

Walther, G. R., E. Post, P. Convey, A. Menzel, C. Parmesan, T. J. C. Beebee, and J. M. Fromentin. 2002. Ecological responses to recent climate change. *Nature* 416:389–95.

Wang, G., N. T. Hobbs, F. J. Singer, D. S. Ojima, and B. C. Lubow. 2002. Impacts of climate change on elk population dynamics in Rocky Mountain National Park, Colorado, U.S.A. *Climatic Change* 54:205–223.

Williams, S. E., L. P. Shoo, J. L. Isaac, A. A. Hoffmann, and G. Langham. 2008. Towards an integrated framework for assessing the vulnerability of species to climate change. *PLoS Biology* 6:2621–26.

Wilmers, C. C., E. Post, and A. Hastings. 2007. The anatomy of predator–prey dynamics in a changing climate. *Journal of Animal Ecology* 76:1037–44.

Wittmer, H. U., B. N. McLellan, R. Serrouya, and C. D. Apps. 2007. Changes in landscape composition influence the decline of a threatened woodland caribou population. *Journal of Animal Ecology* 76:568–79.

Xu, C., and M. S. Boyce. 2009. Oil sardine (*Sardinella longiceps*) off the Malabar Coast: density dependence and environmental effects. *Fisheries Oceanography* 18:359–70.

Xu, C., M. S. Boyce, and D. J. Daley. 2005. Harvesting in seasonal environments. *Journal of Mathematical Biology* 50:663–82.

From Connect-the-Dots to Dynamic Networks: Maintaining and Enhancing Connectivity to Address Climate Change Impacts on Wildlife

Molly S. Cross, Jodi A. Hilty, Gary M. Tabor,
Joshua J. Lawler, Lisa J. Graumlich, and Joel Berger

Little doubt remains among scientists that humans are changing the global climate system, and that these changes will have far-reaching and fundamental impacts on ecosystems and biodiversity (Solomon et al. 2007). Even if atmospheric greenhouse gas concentrations were stabilized at year 2000 levels, average global temperature would continue to rise due to lags associated with greenhouse gas absorption by the oceans (Solomon et al. 2007). Therefore, prudence dictates the development of specific conservation and management strategies that help species persist in the face of rapidly changing climate.

During paleoclimatic changes, many plants and animals avoided extinction through a combination of adapting in place through phenotypic plasticity or genetic evolution and dispersing to areas that were more suitable (Graham et al. 1996, Davis and Shaw 2001, Davis et al. 2005). While in situ persistence did occur, paleoecological records indicate that many species' ranges shifted in response to past climatic changes. For example, many plant and animal species in North America and Europe experienced dramatic range shifts during the most recent glacial and interglacial periods (DeChaine and Martin 2004). Since the last glacial period, species have responded to warming largely individualistically rather than as a cohesive community, and movements have varied significantly in rate and direction (Hunter et al. 1988, Graham et al. 1996).

There is a growing body of evidence that 20th-century warming trends are already altering the distribution of species (e.g., see meta-analyses and reviews by Parmesan and Yohe 2003, Root 2003, Parmesan 2006). Range shifts have been detected at both poles and in northern temperate, tropical, marine, and montane regions. Species that have exhibited shifts include mammals, amphibians, birds, fish, marine invertebrates, plants, and pathogens (see Parmesan 2006 and Paremsan 2005 for detailed reviews of evidence of recent range shifts). In the Northern Hemisphere a meta-analysis demonstrated that range boundaries for 99 species of birds, butterflies, and alpine herbs have moved an average of 6.1 km/decade northward or 6.1 m/decade upward (Parmesan and Yohe 2003).

While paleoecological and recent information suggests some capacity for species to move in response to changing climate, the unprecedented rate of future climate change may mean that the movements and range shifts that occurred during previous climate changes may be more difficult to achieve in the future, especially for species with limited dispersal ability (Malcolm et al. 2002, Pearson and Dawson 2005). The expanding global human footprint will also inhibit species movement by continuing to accelerate habitat loss and fragmentation of the natural world (Sanderson et al. 2002). There is evidence that human land use already deters the movement of some species where they once moved freely (e.g., Berger 2004).

The designation of protected areas (i.e., parks, reserves, and refuges) has historically been the primary approach for conserving species and ecosystems of societal concern. However, the fixed boundaries of these protected areas present a problem as species' ranges shift with climate change. The loss and arrival of species could lead to significant turnover of species diversity and new species combinations (Burns et al. 2003). Reserves designed to protect particular species or communities may no longer serve their intended purpose as the climate changes.

For species that respond to climate change with range shifts, maintaining biodiversity will require conservation strategies that operate on scales much larger than individual parks or preserves. Maintaining and restoring connectivity across landscapes is an existing conservation tool that could reduce species extinction by facilitating range shifts among biota in response to long-term environmental change (Hunter et al. 1988, Halpin 1997, Shafer 1999, Noss 2001, Hannah et al. 2002, Hannah and Hansen 2005, Hulme 2005, Huntley 2007). Increasing connectivity between protected areas is the most commonly cited strategy for long-term biodiversity conservation (Heller and Zavaleta 2009), yet there have been few detailed discussions of exactly what connectivity requirements will look like as climate changes, and how to implement that strategy on the ground (e.g., Halpin 1997). We also lack an examination of the extent to which connectivity needs for responding to climate change are similar to or different from connectivity needs for other purposes.

In this chapter we discuss the definition of connectivity for addressing wildlife (i.e., terrestrial vertebrate) responses to climate change and compare it to connectivity under current conditions. We provide guidance on identifying, prioritizing, and protecting connectivity as a tool for facilitating wildlife conservation in light of climate change. Clarity on how to define and identify these connectivity needs will be important to determining how on-the-ground conservation practitioners and managers can implement this important strategy.

Defining Connectivity

Connectivity is a widely used term that can take on multiple meanings and serve many purposes. We define it as the measure of the extent to which organisms can move between habitat patches (Taylor et al. 1993). Connectivity is inherently a species-specific concept, and it is determined from the perspective of individual species' vagility and behavior (e.g., Hilty et al. 2006; Taylor et al. 2006). For that reason, a given landscape at a given time might be connected for some species, but not others. Species movement is critical for resource acquisition, seasonal migration, demographic and genetic dispersal, metapopulation dynamics, niche expansion, and predator/competitor avoidance. Connecting habitat fragments has been shown to result in increased species and population persistence, greater functional population sizes, more species diversity, and higher genetic exchange (Hilty et al. 2006). Many species are more prone to extinction in less connected and more isolated habitat fragments, including isolated protected areas (Newmark 1987; Parks and Harcourt 2002).

While there are clearly many benefits to fostering connectivity, there are also concerns about potential disadvantages, including the possibility that corridors might facilitate movement of invasive species or diseases, and increase the risk of predation (e.g., Simberloff et al. 1992; and see box 6.1 in Hilty et al. 2006 and table 1.1 in Crooks and Sanjayan 2006 for more extensive lists of potential disadvantages). It is also true that some degree of isolation is natural, and at times perhaps necessary, for species conservation. For example, intentional isolation of native westslope cutthroat trout (*Oncorhynchus clarkii lewisi*) populations in the western United States is a conservation tool aimed at preventing competitive displacement by nonnative brook trout (*Salvelinus fontinalis*; Peterson et al. 2008). While it is important to consider potential pitfalls in the context of connectivity conservation efforts, there is little empirical evidence of any negative effects of corridors on wildlife (Beier and Noss 1998). Conversely, mounting evidence supports their use in conservation, and the benefits of connectivity are likely to outweigh the negatives in the majority of cases (Beier and Noss 1998, Crooks and Sanjayan 2006, Hilty et al. 2006).

Connectivity can occur across a diverse range of landscape patterns that facilitate organism movement—from narrowly defined linear pathways to more diffuse landscape permeability. These ecological patterns can range from contiguous patches of landscape to discontinuous "stepping stones" (e.g., stopover points along bird migratory routes). Corridors (sometimes called linkages) are natural or human-designed linear features (e.g, a greenway, a mountain range, strips of riparian vegetation), and they represent one way of facilitating connectivity. There is likely to be connectivity value to many terrestrial and aquatic areas, although at any given moment some areas will have higher connectivity values than others. Connectivity areas do not necessarily preclude land from providing livelihoods to people (Knight et al. 1995). Traditional "working landscapes," such as large ranches in western North America and the seminomadic grazing systems of east Africa, have proven to be functional in maintaining biodiversity and connectivity (Reid et al. 2003, DeFries et al. 2005).

A challenge for conservation practitioners is determining how patterns that appear geospatially connected to human observers ("structural connectivity") represent actual species use ("functional connectivity"; Taylor et al. 2006). The goal of connectivity conservation efforts is to facilitate functional connectivity, but often there is limited knowledge of exactly why and how species move across the landscape. Therefore, it is difficult to pin-point where to focus connectivity conservation efforts, and what type of connectivity (e.g., corridors, linkages, stepping stones, or landscape permeability) would be most effective.

Connectivity for Climate Change

Enhancing connectivity in light of climate change requires designing landscapes that allow species to track shifting climate and habitat conditions through time. This differs from the traditional purpose of connectivity in three key ways. First, from a species-specific perspective, promoting range shifts requires connecting targets that are moving and changing. Areas of current habitat for a given species need to be connected with areas that are not currently habitat, but that likely will be in the future. The connectivity value of particular lands may also change as habitats change and species ranges shift. Second, enhancing connectivity to facilitate wildlife response to climate change requires considering the needs of a much wider array of species than is traditionally considered, since they all need to move in response to climate change and other wildlife depends on them. Finally, connecting landscapes to address climate change will often require consideration of broader spatial and temporal scales than does connectivity for other purposes.

Connecting Moving and Changing Targets

Connectivity is often conceived of as static lines connecting static core habitat areas. Identifying connectivity needs as climate changes will require anticipating how optimal climate and habitat conditions for spe-

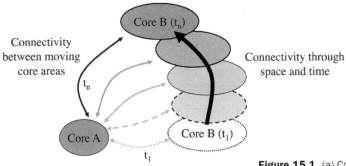

Connectivity between moving core areas

Connectivity through space and time

t_n

t_1

Core B (t_n)

Core A

Core B (t_1)

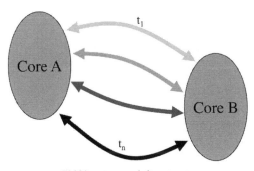

t_1

Core A

Core B

t_n

Shifting connectivity areas through space and time

Figure 15.1. (a) Core habitat area B shifts as climate changes, creating the need for connectivity through space and time between the original location of core area B at time $t = 1$ and its future location at $t = n$. Connectivity may also be needed between core areas B and A as core area B shifts through time. (b) Conditions that define an existing connectivity area may shift across the landscape through time $t = 1$ to time $t = n$, resulting in new routes for connectivity between two core habitat areas.

cies will shift and change across the landscape through time. While this dynamic concept of connectivity is acknowledged under current conditions due to human impact on the landscape as well as processes like hydrology and wildfire (e.g., Crooks and Sanjayan 2006, Taylor et al. 2006), climate change is likely to further increase this dynamism.

As core habitat areas shift in response to changing climate, the land in between the current and future locations of a core habitat area will become important to the survival of species because they will need to be able to track the movement of optimal climate and habitat conditions through space and time (figure 15.1a). For example, coastal marsh species will need to be able to migrate inland as sea level rise inundates areas along the current coastline. At each incremental time step into the future, coastal marsh species will need to be able to move into and inhabit new areas. For some species and habitats, these changes might entail gradual movement across the landscape, whereas other species may experience abrupt or discontinuous shifts.

Climate change will also shape other forms of connectivity, such as migrations and dispersal. New connections between shifting core habitat areas and other core areas (which may or may not be shifting themselves) might become important at different times (figure 15.1a). For example, core A in figure 15.1a might represent a particular species' winter habitat and core B its summer habitat, with annual migrations that occur along fairly regular paths between them. If vegetation and other factors that make up the species' summer habitat are sensitive to climate change and shift across the landscape, it may alter the distance and direction of that species' annual migration route.

Even if climate change does not affect the location of core habitat areas, it could alter the suitability of certain corridors or linkages for species that require particular conditions (e.g., vegetation structure or food resources) for connectivity. Those shifts may result in newly identified connectivity areas within a region that is not currently deemed important (figure 15.1b). It also may increase the suitability of some corridors for particular species (e.g., generalists, invasives, or drought-tolerant organisms). An increase in the ability of these species to move across the landscape could have significant consequences for individual species and for ecosystems more broadly.

Climate change may increase the need for or importance of connectivity in some areas. While there is a general premise that species and ecosystem distributions will shift poleward and/or up in elevation as climate changes, there is also evidence that the spatial extent of some systems will contract rather than shift (Lawler et al. 2010). The survival of some species could be threatened as a core habitat area large enough to support a viable population shrinks to a size that is no longer sufficient. In that case, species may need to disperse to or establish genetic exchange with other core areas, thereby creating a necessity for connectivity where it previously has not existed (figure 15.2). For species that now persist through meta-population dynamics, connectivity is already a requirement for species survival, and shrinkage of core habitat areas will only make that connectivity more important. A related concern involves contiguous patches of habitat that become fragmented, leading to isolated smaller patches of habitat that require connectivity between them to sustain viable populations (figure 15.2).

Climate-induced changes in ecological processes may also modify connectivity. For example, as climate change increases wildfire frequency and severity, large swaths of forest may transition to a more permanent early successional stage, characterized by a high density of short, young saplings (figure 15.3a). Those new conditions may render those forested

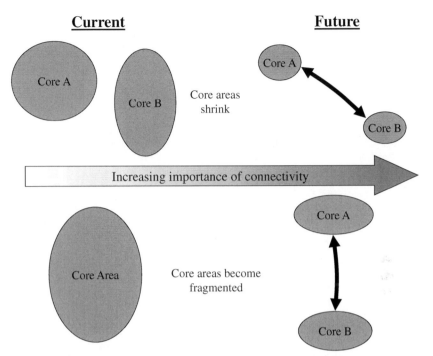

Figure 15.2. Climate change may increase the importance of connectivity between smaller core habitat areas as they shrink (*top*) or become fragmented (*bottom*).

(A) Increased wildfire frequency and severity

(B) Decreased stream flows

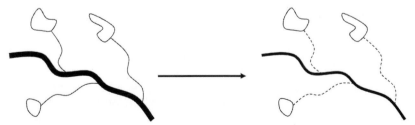

Figure 15.3. Climate-induced changes in ecological processes such as wildfire (*A*) or stream flows (*B*) may modify connectivity or conditions within connectivity areas.

areas unusable by species that depend on mature forest conditions (e.g., for food resources or physical protection from aerial predators), or impassable by species that have trouble physically moving through dense post-fire, early-succession forests. Climate change impacts on stream flows represent another example, whereby decreased flows may diminish or completely cut off connectivity between lakes, tributary streams, and a river's main stem (figure 15.3b).

Anticipating future habitat or ecosystem conditions and species distributions can be done using a variety of techniques, from statistically based climate envelope models for specific species (e.g., Pearson and Dawson 200,; Thuiller 2004, Lawler et al. 2006) to dynamic models depicting the distribution of broad vegetation categories (e.g., Cramer et al. 2001, Bachelet et al. 2003) and models of ecological processes such as wildfire (e.g., Arora and Boer 2005) and hydrology (e.g., Liang et al. 1994). Determining appropriate methodologies will depend on the extent to which a species uses particular core habitat areas and connectivity zones because of physical features (e.g., topography, geology, the distance between core habitats), biological characteristics (e.g., presence of a particular plant or animal food source, vegetation community type, vegetation structure), and/or human land use (e.g., avoidance of human population centers or infrastructure). Climate change will impact biological factors more directly than most physical features, whereas human land use may be either independent of or related to changes in climate (e.g., agricultural patterns or practices that shift in response to altered climate conditions).

Connectivity for a Broader Suite of Species

While some connectivity conservation efforts have focused on facilitating connectivity for less vagile organisms (e.g., amphibians), connectivity is mostly considered necessary for relatively wide-ranging wildlife species that require more space than is typically found within single protected areas to persist. As climate conditions shift across the landscape in the future, more organisms will need to move in ways that are not deemed necessary under current conditions. A challenge is how to accommodate the movements of so many species, each of which may have differing, or even conflicting, temporal and spatial connectivity needs.

The location and importance of connectivity for many wildlife species will depend on how other organisms on which they depend (e.g., plants, invertebrates, other vertebrates) respond to climate change. Some wildlife movements are determined by the presence or absence of other species (e.g., forage, prey, predators, competitors, or pollinators), so understanding how climate change could affect species interactions and behaviors will be crucial to determining overall impacts on some wildlife species' connectivity needs. One way in which species interactions may be altered is through climate-induced changes in the timing of seasonal events (e.g., bud burst, migrations, and emergence from hibernation; Inouye et al. 2000, Root et al. 2005, Parmesan 2006). These phenological shifts could create species interaction mismatches and modify connectivity needs. For example, as individuals are forced to search outside of their current range for food that is not available at traditional places and points in time, the timing and location of seasonal migration routes currently seen as static may be altered.

Wildlife and the species on which they depend can have very different movement abilities, and therefore different connectivity requirements. Species vary significantly in their ability to disperse (e.g., some can travel long distances through wind-dispersed seeds or fly over nonhabitat patches, while others are relatively or completely sessile) and their ability to tolerate different types of land use and levels of fragmentation. Modeled estimates of the distance between current and future optimal climate conditions also vary for different species (figure 15.4). Combining information on species' movement abilities with estimates of distances between current and future optimal conditions can help in estimating which species are more likely to benefit from connectivity as a conservation strategy.

An additional consideration is not just how far wildlife species will have to move, but whether they will have suitable habitat to colonize once they get there. Even if suitable climate conditions can be found in places relatively near a species' current range, appropriate vegetation, soils, or other habitat factors may be unavailable in those new areas. Many animal distributions and movements are driven by interactions with vegetation, so understanding plant responses to climate change may be a key limiting factor in determining how those species will respond to changing climate, and what role connectivity might play. The ability of plant populations to adapt and move in response to past and future changes in climate is complicated and remains uncertain (Clark et al. 1998, Davis and Shaw 2001, Malcolm et al. 2002, Jump and Peñuelas 2005, McLachlan et al. 2005, Pearson 2006), thus making it difficult for us to predict subsequent impacts on vegetation-dependent wildlife.

Temporal and Spatial Scales

What constitutes a connected landscape for a given species in a changing climate will depend on the spatial and temporal scale considered. For example, over the next 30 to 50 years, the location and condition of

(A)   (B)

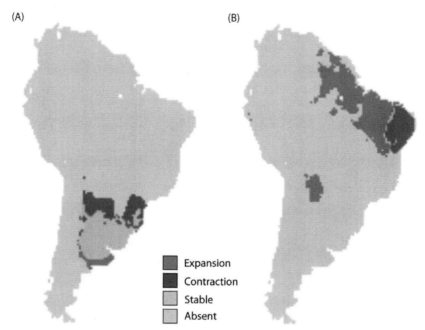

Expansion
Contraction
Stable
Absent

Figure 15.4. Projected climate-driven range shifts for two amphibians, *Leptodactylus gracilis* (A) and *Proceratophrys cristiceps* (B), illustrating how the distance between current and future optimal climate conditions may vary across species. The projected shifts are for an averaged time period between 2071 and 2099, and are based on climate simulated by the Hadley Centre UKMO-HadCM3 model run for the SRES A2 emissions scenario for the Fourth Assessment Report of the Intergovernmental Panel on Climate Change (adapted from Lawler et al. 2009).

habitats for particular species may be only slightly shifted or altered. By 2100, however, the magnitude of climate change may lead to a species or its habitat moving entirely out of a region such that increasing connectivity is no longer a viable solution for maintaining that species in that location (Hannah and Hansen 2005). The greater uncertainty associated with longer-term projections of future climate conditions, along with the emergence of novel climate conditions and species assemblages (Williams et al. 2007), will make connectivity needs over the next century especially difficult to predict.

Spatial scale also needs to be explicitly considered when discussing species' connectivity needs under changing climate. Our understanding of how a species' range might shift at a coarse scale and across large geographic areas may be different from how that species moves across the landscape at a finer resolution. This could be due to the relative importance of various factors (e.g., climate, topography, land cover, human land use) in determining species ranges at different spatial scales, or due to fine-scale idiosyncrasies in how climate will change or species will move.

Connectivity at a fine spatial scale may be especially important in the near term, whereas coarse-scale connectivity over large areas may be more important over longer timeframes. For example, as climate change negatively impacts boreal forest isolates at the southern extent of the boreal ecotone in the northeastern United States, like the Adirondack State Park, the near-term focus (e.g., over the next few decades) may be on how to facilitate movements of boreal species over several kilometers between their current locations and those places within the park where boreal forests are most likely to persist. In the long term (over the next 50 to 100 years), boreal forests may move completely out of the Adirondack State Park and the focus of connectivity efforts for those boreal species will be on how to help them move several hundreds of kilometers north into the boreal forests of Canada. Connectivity conservation efforts within the park in the longer term are likely to be less about maintaining boreal species, and more about allowing the in-migration of new species to form new assemblages and ecological communities.

Another spatial consideration is how connectivity needed to help species move in mountainous areas may differ from connectivity needs in flatter areas. Predicted velocities of temperature change (the pace at which temperature is expected to change in km per year) are lowest for topographically diverse landscapes like mountainous biomes, and fastest for flatter biomes such as flooded grasslands, mangroves, and deserts (Loarie et al. 2009). Heterogeneous mountain systems are also more likely than relatively flat regions to house potential climatic refugia, due to diverse combinations of slope, aspect, and other topographic features. Therefore, species in mountainous biomes may not have to move as far as species in more flat biomes to encounter suitable climate conditions in the future. The challenge facing species in mountainous areas is that while elevation shifts may allow them to reach more suitable climates more quickly, upslope shifts are limited since the amount of available areas decreases and eventually disappears at the tops of mountains.

Identifying Areas for Connectivity in a Changing Climate

Designing connected landscapes to address climate change involves understanding how connectivity facilitates movement. Recognizing the potential for, and location of, future connectivity is important for conserving biodiversity as climate changes. Connectivity needs may be fully captured by current conceptualizations, or there may be new needs as climate changes in the future. While many current aspects of connectivity will continue to be important as climate changes, if we focus only on them, we might miss other important future needs.

Strategies for identifying connectivity priorities in light of changing climate are likely to rely on species-specific approaches as well as relatively coarse approaches that consider wildlife species or biodiversity in the aggregate. These approaches have tradeoffs in terms of the nature and scale of connectivity being addressed, the amount of information they require, and their ability to address climate change. Ideally, a combination of these approaches will be applied whenever possible.

Species-Specific Approaches

The inherently species-specific nature of connectivity suggests that identifying specific corridors or broader areas for allowing wildlife movement will require some consideration of the changing needs of particular species. Figure 15.5 outlines a series of broad steps towards identifying priority areas for species-specific connectivity conservation efforts in response to climate change. Many researchers recommend that these efforts concentrate on a suite of focal species that together represent the connectivity and habitat needs of a larger number of native species and ecological processes (e.g., Beier et al. 2008; also see box 1 in figure 15.5). These focal species could be selected on the basis of factors including their current importance for connectivity conservation, the sensitivity of their current

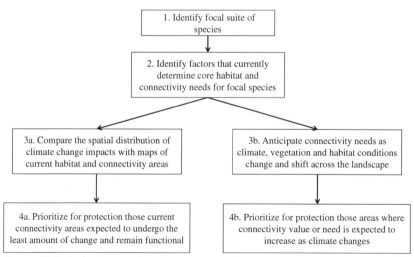

Figure 15.5. A species-specific approach to identifying connectivity areas important for facilitating wildlife responses to climate change

habitat and connectivity areas to climate change, or a general understanding of their current or future dispersal or other movement needs.

There are two important components of identifying and promoting the type and location of connectivity needed to facilitate focal species movements in light of climate change, both of which will be important. The first component involves understanding the potential impact of climate change on currently identified zones of connectivity, and highlighting the ones expected to change the least and which are therefore most likely to remain functional (box 4a in figure 15.5). The other component anticipates how climate and vegetation or ecosystem conditions will shift in the future, and identifies the areas in which connectivity will be necessary for species to relocate (box 4b in figure 15.5). Together, these approaches will ensure that current connectivity is valued and maintained without ignoring the need for new connectivity in less obvious areas.

The primary challenge in implementing the first component is that we lack a strong understanding of current connectivity priorities for wildlife in most places (box 2 in figure 15.5), and the priority connectivity areas that have been identified often are not formally protected. Several empirical and modeling approaches can help identify the location of important connectivity areas for particular species and ecological systems (e.g., Beier and Noss 1998, Bunn et al. 2000, Moilanen and Nieminen 2002, Tewksbury et al. 2002, Williams et al. 2005, Chetkiewicz et al. 2006, Beier et al. 2008, McRae et al. 2008, Phillips et al. 2008, Cushman et al. 2009, Schwartz et al. 2009, Carroll et al. 2010). But while connectivity needs have been studied extensively for some species in some places (e.g., the South Coast Missing Linkages project described in Beier et al. 2006), many areas lack any such identification. State fish and wildlife agencies in the western United States are currently working to compile priority connectivity maps (Western Governors' Association 2008). Additional efforts to understand the factors that determine current connectivity needs will improve our ability to project how climate change may alter those needs.

As mentioned above, there are many techniques for anticipating the impact of climate change on habitat and connectivity areas (boxes 3a and 3b in figure 15.5). Comparisons of the spatial distribution of climate changes and associated impacts with maps of current habitat and connectivity areas can help us identify areas expected to undergo the least change, which are therefore likely to provide connectivity benefits under both current and future climate conditions (boxes 3a and 4a in figure 15.5). Rose and Burton (2009) propose a method for mapping the persistence of climate space through time as a way of prioritizing areas for protection given the temporal connectivity needs of species. Climate change impact information can also be used to identify any additional areas whose connectivity value may increase (boxes 3b and 4b in figure 15.5). Identifying such areas requires anticipating how climate, vegetation, and other habitat conditions may change and shift across the landscape, and understanding the other connectivity needs of a species or group (box 3b in figure 15.5). This challenge is largely being addressed in two ways: (1) by relying on general rules of thumb about how species will move in response to climate change (e.g., poleward or upslope in elevation), and (2) by using spatial models to project future climate, habitat, and species distributions.

Using models to predict where species and their habitats may be distributed in the future is appealing, because we anticipate that not all species will adhere to the generalized rules of thumb. For example, increasing drought will likely cause species to move toward relatively cool or moist microclimate refugia, or to seek out areas with persistent water availability. Movement to these areas may or may not be in line with poleward and elevational movements, so focusing on the rules of thumb may not be sufficient.

Spatial models that attempt to project future climate, habitat, and species distributions have recently been used to identify areas of connectivity for facilitating species shifts in response to climate change (Williams et al. 2005, Phillips et al. 2008, Vos et al. 2008, Carroll et al. 2010). But while model projections can be used to consider more complicated consequences of climate change, they all have uncertainties associated with both the climate projections they take as inputs and the biotic models themselves (Pearson et al. 2006, Beier and Brost 2010). The challenge is to determine how to integrate spatial models into connectivity conservation efforts in a way that increases the chance of effectively allowing species movements as climate changes, even given model uncertainties.

For areas where uncertainty about the future consequences of climate change is fairly high, one strategy could be to identify priority connectivity zones under current conditions and see how well they would allow for wildlife movement poleward and upward in elevation, or towards more well-watered sites. If currently identified connectivity cannot adequately capture those movements, then areas that augment the movements could be added as conservation priorities. This would provide a good first approximation of connectivity needs under climate change. Because not all species will follow those general rules, another strategy would be to look at the best available research and expert opinion on wildlife responses to climate change in a region, and broadly outline the type and location of connectivity that would be needed to facilitate those responses. A third strategy would be to apply more complex modeling approaches that explicitly link climate change and connectivity models to project potential future needs (e.g., Williams et al. 2005, Phillips et al. 2008, Vos et al. 2008, Carroll et al. 2010).

Focusing on Wildlife and Biodiversity in the Aggregate

In some cases it may be preferable or necessary to take a coarser approach to targeting connectivity conservation. If there is not sufficient information about the current or future habitat or connectivity needs for a species, it may not be possible to take a species-specific approach. There also may be a preference for strategies that aim to capture a wide range of potential connectivity needs, given the uncertainties involved in knowing exactly how species and their habitats will be affected by climate change. Some of these coarser strategies might include protecting landscape features that encompass latitudinal gradients, elevation, or other physical gradients, and managing the matrix to promote permeability of movement for wildlife and other species.

Although not all species will adhere to general rules of thumb about species distributions shifting poleward and up in elevation, the fact that many species are already following those patterns suggests that protecting areas that encompass north-south gradients and elevation gradients may be reasonable for increasing connectivity. As mentioned earlier, it might be appropriate in situations where uncertainty about future climate change is fairly high. Another coarse-filter approach involves protecting and connecting areas that encompass diverse physical environments (Hunter et al. 1988). For example, Beier and Brost (2010) recommend protecting landscape units that include diverse "land facets" (recurring land units that represent similar topographic and soil conditions). They argue that wildlife corridors that contain continuous areas of different land facets should also contain continuous areas of vegetation, even though the specific species assemblages may change through time. Therefore, protecting a diversity of land facets should facilitate the persistence of diverse vegetation and wildlife species, and complement conservation strategies that are based on species-specific responses to climate change (Beier and Brost 2010).

Yet another way to maxime the number of species that can move in response to climate change is to think beyond corridors and stepping stones, and to consider managing the matrix of lands between protected areas in ways that promote permeability and species movement (Da Fonseca et al. 2005, Franklin and Lindenmayer 2009). Because 83% of the earth's land surface (excluding Antarctica) is directly affected by human infrastructure and land use activities (Sanderson et al. 2002), and because we cannot predict exactly how future vegetation and ecosystem conditions may shift and change, these human-dominated matrix lands (e.g., agricultural areas, private forested lands, ranchlands) present an important opportunity for securing the kind of landscape-scale connectivity that will be necessary for future movements of plant and animal species.

One Tool in the Conservation Toolbox

While connectivity will obviously help some species respond to climate change, it will not be a panacea. Relatively immobile or poorly dispersed species may not benefit as much as relatively vagile species, and may require more aggressive management interventions, such as translocation to future suitable areas (Hoegh-Guldberg et al. 2008, Richardson et al. 2009, Willis et al. 2009). Climate change may present less of a threat to generalist species that are broadly distributed in a diverse array of habitat patches. However, species that are more limited by climate or are isolated in relatively few patches will be less able to rapidly relocate to survive climate change, especially if landscape fragmentation prevents their movement (Opdam and Wascher 2004).

The rate and magnitude of shifts in climate and ecosystems and the future existence of suitable climate and habitat somewhere on the landscape will ultimately determine the usefulness of maintaining or enhancing connectivity for wildlife conservation. Such efforts will also be futile unless sufficient investments are made in establishing core habitat areas towards which species can move. Some argue that focusing conservation on increasing the size, quality, and number of protected areas is better than attempting to identify and protect connectivity, which is fraught with uncertainties (Hodgson et al. 2009). Others point out that the distances that species will need to move in response to climate change are likely to be much greater than what can be accommodated within individual protected areas (Krosby et al. 2010). For these reasons, connectivity conservation efforts need to be applied in the context of broader landscape conservation work in which protected areas, connectivity, and land management approaches are considered simultaneously.

A Way Forward

Climate change underscores the importance of rethinking the old paradigm of connectivity as "connecting the dots" between disjunct core habitats or protected areas, where both the dots and connecting lines are static concepts in space and time. We need to consider connectivity

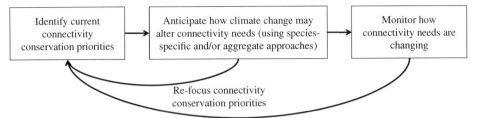

Figure 15.6. An iterative process for setting connectivity conservation priorities that anticipates how climate change may alter connectivity needs and incorporates monitoring information on how connectivity needs actually change through time

to be dynamic and in some cases transitory. This paradigm shift challenges us to design monitoring and modeling efforts that anticipate when and where current connectivity may fail, and when and where new connectivity may form over short (e.g., the next few decades) and long (e.g., the next century) time scales. One answer to this challenge likely lies in an iterative process of identifying current connectivity needs, anticipating how those needs may change in the future, monitoring how those needs are changing, and refocusing conservation attentions as necessary (figure 15.6).

Although the available tools for modeling future conditions have limitations, they are useful for visualizing future scenarios that can form the basis for anticipating future connectivity needs using the best available science. Coupling those modeling efforts with strategic monitoring of species' habitat conditions, distributions, and connectivity needs will allow us to track changes happening on the ground, and to compare and test model projections. Monitoring and research targeted at improving our understanding of functional connectivity under current conditions will elucidate where the most important connections are, and why they are important. This information will in turn strengthen the accuracy and applicability of connectivity models. An iterative process would allow for increasing levels of sophistication in the use of general rules of thumb, expert opinion, and efforts that explicitly link climate change impact and connectivity models. It would also allow for the inclusion of updated ecological information, climate projections, and modeling techniques as they become available.

The good news is that by building a more dynamic paradigm for how we think about connectivity needs as climate changes, we can focus on developing creative, dynamic solutions. Land trusts may be able to devise conservation plans that reflect transitory stages and therefore do not necessarily require conservation in perpetuity. Restoration projects on currently disturbed or degraded land could be managed specifically to maximize the land's value for connectivity, both now and in the future. For example, the removal of human infrastructure and some restriction of human access to a critical wildlife corridor in Banff National Park in Canada resulted in a significant increase in that corridor's use by wolves (Duke et al. 2001). Low-intensity land uses may also be managed to augment connectivity of undisturbed natural ecosystems in order to provide broader landscape permeability (Hannah and Hansen 2005).

Wildlife species are increasingly challenged to adapt and move in response to changing climatic conditions. As humans, we also need to adapt by designing conservation strategies that allow species to cope with this challenge. Restoring and maintaining connectivity between core protected areas in space and through time will undoubtedly be an important tool. Implementing this strategy through an iterative process whereby the best available science is used to examine both current and changing connectivity needs will increase the likelihood of benefit for as many species as possible. If we do not act quickly to develop and implement a dynamic approach to connectivity, the rapid encroachment of the human footprint will preclude many conservation options.

ACKNOWLEDGMENTS
We thank Jedediah F. Brodie, Eric S. Post, Daniel F. Doak, Malcolm Hunter Jr., and Kevin Crooks for valuable comments on an earlier version of this manuscript.

LITERATURE CITED
Arora, V. K. and G. J. Boer. 2005. Fire as an interactive component of dynamic vegetation models. *Journal of Geophysical Research* 110:G02008.1–G02008.20.
Bachelet, D., R. P. Neilson, T. Hickler, R. J. Drapek, J. M. Lenihan, M. T. Sykes, B. Smith, S. Sitch, and K. Thonicke. 2003. Simulating past and future dynamics of natural ecosystems in the United States. *Global Biogeochemical Cycles* 17: 14.1–14.21.
Beier, P. and B. Brost. 2010. Use of land facets to plan for climate change: Conserving the arenas, not the actors. *Conservation Biology* 24:701–10.
Beier, P., D. R. Majka, and W. D. Spencer. 2008. Forks in the road: Choices in procedures for designing wildland linkages. *Conservation Biology* 22:836–51.
Beier, P. and R. F. Noss. 1998. Do habitat corridors provide connectivity? *Conservation Biology* 12:1241–52.
Beier, P., K. L. Penrod, C. Luke, W. D. Spencer, and C. Cabanero. 2006. South Coast Missing Linkages: Restoring connectivity to wildlands in the largest metropolitan area in the USA. In K. R. Crooks and M. Sanjayan, eds., *Connectivity Conservation*. Cambridge University Press, Cambridge.
Berger, J. 2004. The last mile: How to sustain long-distance migration in mammals. *Conservation Biology* 18:320–31.
Bunn, A. G., D. L. Urban, and T. H. Keitt. 2000. Landscape connectivity: A

conservation application of graph theory. *Journal of Environmental Management* 59:265–78.

Burns, C. E., K. M. Johnston, and O. J. Schmitz. 2003. Global climate change and mammalian species diversity in US national parks. *Proceedings of the National Academy of Sciences, USA* 100:11474–77.

Carroll, C., J. R. Dunk, and A. Moilanen. 2010. Optimizing resiliency of reserve networks to climate change: Multispecies conservation planning in the Pacific Northwest, USA. *Global Change Biology* 16:891–904.

Chetkiewicz, C.-L. B., C. C. St. Clair, and M. S. Boyce. 2006. Corridors for conservation: Integrating pattern and process. *Annual Review of Ecology, Evolution, and Systematics* 37:317–42.

Clark, J. S., C. Fastie, G. Hurtt, S. T. Jackson, C. Johnson, G. A. King, M. Lewis, J. Lynch, S. Pacala, C. Prentice, E. W. Schupp, T. Webb, III, and P. Wyckoff. 1998. Reid's paradox of rapid plant migration. *BioScience* 48:13–24.

Cramer, W., A. Bondeau, F. I. Woodward, I. C. Prentice, R. A. Betts, V. Brovkin, P. M. Cox, V. Fisher, J. A. Foley, A. D. Friend, C. Kucharik, M. R. Lomas, N. Ramankutty, S. Sitch, B. Smith, A. White, and C. Young-Molling. 2001. Global response of terrestrial ecosystem structure and function to CO_2 and climate change: Results from six dynamic global vegetation models. *Global Change Biology* 7:357–73.

Crooks, K. R. and M. Sanjayan. 2006. Connectivity conservation: Maintaining connections for nature. In *Connectivity Conservation*, edited by K. R. Crooks and M. Sanjayan. Cambrdige: Cambridge University Press.

Cushman, S. A., K. S. McKelvey, and M. K. Schwartz. 2009. Use of empirically derived source-destination models to map regional conservation corridors. *Conservation Biology* 23:368–76.

Da Fonseca, G. A. B., W. Sechrest, and J. Oglethorpe. 2005. Managing the matrix. In *Climate Change and Biodiversity*, edited by T. E. Lovejoy and L. Hannah. New Haven: Yale University Press.

Davis, M. B. and R. G. Shaw. 2001. Range shifts and adaptive responses to Quaternary climate change. *Science* 292:673–79.

Davis, M. B., R. G. Shaw, and J. R. Etterson. 2005. Evolutionary responses to changing climate. *Ecology* 86:1704–14.

DeChaine, E. G. and A. P. Martin. 2004. Historic cycles of fragmentation and expansion in *Parnassius smintheus* (Papilionidae) inferred using mitochondrial DNA. *Evolution* 58:113–27.

DeFries, R., A. Hansen, A. C. Newton, and M. C. Hansen. 2005. Increasing isolation of protected areas in tropical forests over the past twenty years. *Ecological Applications* 15:19–26.

Duke, D. L., M. Hebblewhite, P. C. Paquet, C. Callaghan, and M. Percy. 2001. Restoration of a large carnivore corridor in Banff National Park, Alberta. In *Large Mammal Restoration: Ecological and Sociological Challenges in the 21st Century*, edited by D. S. Maehr, R. F. Noss, and J. F. Larkin. New York: Island Press.

Franklin, J. F. and D. Lindenmayer. 2009. Importance of matrix habitats in maintaining biological diversity. *Proceedings of the National Academy of Science of the United States of America* 106:349–50.

Graham, R. W., E. L. Lundelius, M. A. Graham, E. K. Schroeder, R. S. Toomey, E. Anderson, A. D. Barnosky, J. A. Burns, C. S. Churcher, D. K. Grayson, R. D. Guthrie, C. R. Harington, G. T. Jefferson, L. D. Martin, H. G. McDonald,

R. E. Morlan, H. A. Semken, S. D. Webb, L. Werdelin, and M. C. Wilson. 1996. Spatial response of mammals to late quaternary environmental fluctuations. *Science* 272:1601–6.

Halpin, P. N. 1997. Global climate change and natural-area protection: Management responses and research directions. *Ecological Applications* 7:828–43.

Hannah, L. and L. Hansen. 2005. Designing landscapes and seascapes for change. In *Climate Change and Biodiversity*, edited by T. E. Lovejoy and L. Hannah. New Haven: Yale University Press.

Hannah, L., G. F. Midgley, T. Lovejoy, W. J. Bond, M. Bush, J. C. Lovett, D. Scott, and F. I. Woodward. 2002. Conservation of biodiversity in a changing climate. *Conservation Biology* 16:264–68.

Heller, N. E. and E. S. Zavaleta. 2009. Biodiversity management in the face of climate change: a review of 22 years of recommendations. *Biological Conservation* 142:14–32.

Hilty, J., W. Lidicker Jr., and A. Merenlender. 2006. *Corridor Ecology: The Science and Practice of Linking Landscapes for Biodiversity Conservation*. Washington: Island Press.

Hodgson, J. A., C. D. Thomas, B. A. Wintle, and A. Moilanen. 2009. Climate change, connectivity and conservation decision making: Back to basics. *Journal of Applied Ecology* 46:964–69.

Hoegh-Guldberg, O., L. Hughes, S. McIntyre, D. B. Lindenmayer, C. Parmesan, H. P. Possingham, and C. D. Thomas. 2008. Assisted colonization and rapid climate change. *Science* 321:345–46.

Hulme, P. E. 2005. Adapting to climate change: Is there scope for ecological management in the face of a global threat? *Journal of Applied Ecology* 42:784–94.

Hunter, M. L., Jr., G. L. Jacobson, Jr., and T. Webb, III. 1988. Paleoecology and the coarse-filter approach to maintaining biological diversity. *Conservation Biology* 2:375–85.

Huntley, B. 2007. Climatic change and the conservation of European biodiversity: Towards the development of adaptation strategies. Discussion paper prepared for the 27th meeting of the Standing Committee, Convention on the Conservation of European Wildlife and Natural Habitats, Council of Europe , Strasbourg, 26–29 November 2007.

Inouye, D. W., B. Barr, K. B. Armitage, and B. D. Inouye. 2000. Climate change is affecting altitudinal migrants and hibernating species. *Proceedings of the National Academy of Sciences, USA* 97:1630–33.

Jump, A. S. and J. Peñuelas. 2005. Running to stand still: Adaptation and the response of plants to rapid climate change. *Ecology Letters* 8:1010–20.

Knight, R. L., G. N. Wallace, and W. E. Riebsame. 1995. Ranching the view: Subdivisions versus agriculture. *Conservation Biology* 9:459–61.

Krosby, M., J. Tewksbury, N.M. Haddad, and J. Hoekstra. 2010. Ecological connectivity for a changing climate. *Conservation Biology* 24:1686–89.

Lawler, J. J., S. L. Shafer, B. A. Bancroft, and A. R. Blaustein. 2010. Projected climate impacts for the amphibians of the western hemisphere. *Conservation Biology* 24: 38–50.

Lawler, J. J., S. L. Shafer, D. White, P. Kareiva, E. P. Maurer, A. R. Blaustein, and P. J. Bartlein. 2009. Projected climate-induced faunal change in the Western Hemisphere. *Ecology* 90:588–97.

Lawler, J. J., D. White, R. P. Neilson, and A. R. Blaustein. 2006. Predicting climate-induced range shifts: model differences and model reliability. *Global Change Biology* 12:1568–84.

Liang, X., D. P. Lettenmaier, E. F. Wood, and S. J. Burges. 1994. A simple hydrologically based model of land surface water and energy fluxes for general circulation models. *Journal of Geophysical Research* 99:14415–28.

Loarie, S. R., P. B. Duffy, H. Hamilton, G. P. Asner, C. B. Field, and D. D. Ackerly. 2009. The velocity of climate change. *Nature* 462:1052–55.

Malcolm, J. R., A. Markham, R. P. Neilson, and M. Garaci. 2002. Estimated migration rates under scenarios of global climate change. *Journal of Biogeography* 29:835–49.

McLachlan, J. S., J. S. Clark, and P. S. Manos. 2005. Molecular indicators of tree migration capacity under rapid climate change. *Ecology* 86:2088–98.

McRae, B. H., B. G. Dickson, T. H. Keitt, and V. B. Shah. 2008. Using circuit theory to model connectivity in ecology, evolution, and conservation. *Ecology* 89:2712–24.

Moilanen, A. and M. Nieminen. 2002. Simple connectivity measures in spatial ecology. *Ecology* 83:1131–45.

Newmark, W. D. 1987. A land-bridge island perspective on mammalian extinctions in western North American parks. *Nature* 325:430–32.

Noss, R. F. 2001. Beyond Kyoto: Forest management in a time of rapid climate change. *Conservation Biology* 15:578–90.

Opdam, P. and D. Wascher. 2004. Climate change meets habitat fragmentation: Linking landscape and biogeographical scale levels in research and conservation. *Biological Conservation* 117:285–97.

Parks, S. A. and A. H. Harcourt. 2002. Reserve size, local human density, and mammalian extinctions in US protected areas. *Conservation Biology* 16:800–808.

Parmesan, C. 2005. Biotic responses: Range and abundance changes. In *Climate Change and Biodiversity*, edited by T. E. Lovejoy and L. Hannah. New Haven: Yale University Press.

———. 2006. Ecological and evolutionary responses to recent climate change. *Annual Review of Ecology, Evolution, and Systematics* 37:637–69.

Parmesan, C. and G. Yohe. 2003. A globally coherent fingerprint of climate change impacts across natural systems. *Nature* 421:37–42.

Pearson, R. G. 2006. Climate change and the migration capacity of species. *Trends in Ecology & Evolution* 21:111–13.

Pearson, R. G. and T. P. Dawson. 2003. Predicting the impacts of climate change on the distribution of species: Are bioclimate envelope models useful? *Global Ecology and Biogeography* 12:361–71.

———. 2005. Long-distance plant dispersal and habitat fragmentation: identifying conservation targets for spatial landscape planning under climate change. *Biological Conservation* 123:389–401.

Pearson, R. G., W. Thuiller, M. B. Araujo, E. Martinez-Meyer, L. Brotons, C.McClean, L. Miles, P. Segurado, T. P. Dawson, and D. C. Lees. 2006. Model-based uncertainty in species range prediction. *Journal of Biogeography* 33:1704–11.

Peterson, D. P., B. E. Rieman, J. B. Dunham, K. D. Fausch, and M. K. Young. 2008. Analysis of trade-offs between threats of invasion by nonnative brook trout (*Salvelinus fontinalis*) and intentional isolation for native westslope cutthroat trout (*Oncorhynchus clarkii lewisi*). *Canadian Journal of Fisheries and Aquatic Sciences* 65:557–73.

Phillips, S. J., P. Williams, G. Midgley, and A. Archer. 2008. Optimizing dispersal corridors for the Cape Proteaceae using network flow. *Ecological Applications* 18:1200–1211.

Reid, R. S., M. Rainy, J. Ogutu, R. L. Kruska, K. Kimani, M. Nyabenge, M. McCartney, M. Kshatriya, J. Worden, L. Ng'ang'a, J. Owuor, J. Kinoti, E. Njuguna, C. J. Wilson, and R. Lamprey. 2003. People, Wildlife and Livestock in the Mara Ecosystem. Report, Mara Count 2002, International Livestock Research Institute, Nairobi.

Richardson, D. M., J. J. Hellmann, J. S. McLachlan, D. F. Sax, M. W. Schwartz, P. Gonzalez, E. J. Brennan, A. Camacho, T. L. Root, O. E. Sala, S. H. Schneider, D. M. Ashe, J. R. Clark, R. Early, J. R. Etterson, E. D. Fielder, J. L. Gill, B. A. Minteer, S. Polasky, H. D. Safford, A. R. Thompson, and M. Vellend. 2009. Multidimensional evaluation of managed relocation. *Proceedings of the National Academy of Science of the United States of America* 106:9721–24.

Root, T. L., D. P. MacMynowski, M. D. Mastrandrea, and S. H. Schneider. 2005. Human-modified temperatures induce species changes: Joint attribution. *Proceedings of the National Academy of Science of the United States of America* 102:7465–69.

Root, T. P., K. R. Hall, S. H. Schneider, C. Rosenzweig, and J. A. Pounds. 2003. Fingerprints of global warming on wild animals and plants. *Nature* 421:57–60.

Sanderson, E. W., M. Jaiteh, M. A. Levy, K. H. Redford, A. V. Wannebo, and G. Woolmer. 2002. The human footprint and the last of the wild. *BioScience* 52:891–904.

Schwartz, M. K., J. P. Copeland, N. J. Anderson, J. R. Squires, R. M. Inman, K. S. McKelvey, K. L. Pilgrim, L. P. Waits, and S. A. Cushman. 2009. Wolverine gene flow across a narrow climatic niche. *Ecology* 90:3222–32.

Shafer, C. L. 1999. National park and reserve planning to protect biological diversity: Some basic elements. *Landscape and Urban Planning* 44:123–53.

Simberloff, D., J. A. Farr, J. Cox, and D. W. Mehlman. 1992. Movement corridors: Conservation bargains or poor investments? *Conservation Biology* 6:493–504.

Solomon, S., D. Qin, M. Manning, Z. Chen, M. Marquis, K. B. Averyt, M. Tignor, and H. L. Miller, eds. 2007. *Climate Change 2007: The Physical Science Basis. Contribution of Working Group I to the Fourth Assessment Report of the Intergovernmental Panel on Climate Change.* Cambridge and New York: Cambridge University Press.

Taylor, P. D., L. Fahrig, K. Henein, and G. Merriam. 1993. Connectivity Is a vital element of landscape structure. *Oikos* 68:571–73.

Taylor, P. D., L. Fahrig, and K. A. With. 2006. Landscape connectivity: A return to basics. In *Connectivity Conservation*, edited by K. R. Crooks and M. Sanjayan. Cambridge: Cambridge University Press.

Tewksbury, J. J., D. J. Levey, N. M. Haddad, S. Sargent, J. L. Orrock, A. Weldon, B. J. Danielson, J. Brinkerhoff, E. I. Damschen, and P. Townsend. 2002. Corridors affect plants, animals, and their interactions in fragmented landscapes. *Proceedings of the National Academy of Sciences of the United States of America* 99:12923–26.

Thuiller, W. 2004. Patterns and uncertainties of species' range shifts under climate change. *Global Change Biology* 10:2020–27.

Vos, C.C., P. Berry, P. Opdam, H. Baveco, B. Nijhof, J. O'Hanley, C. Bell, and H. Kuipers. 2008. Adapting landscapes to climate change: Examples of climate-proof ecosystem networks and priority adaptation zones. *Journal of Applied Ecology* 45:1722–31.

Western Governors' Association. 2008. Wildlife corridors initiative report. Accessed at www.westgov.org/wga/meetings/am2008/wildlife08.pdf.

Williams, J. W., S. T. Jackson, and J. E. Kutzbacht. 2007. Projected distributions of novel and disappearing climates by 2100 AD. *Proceedings of the National Academy of Sciences of the United States of America* 104:5738–42.

Williams, P., L. Hannah, S. Andelman, G. Midgley, M. Araujo, G. Hughes, L. Manne, E. Martinez-Meyer, and R. Pearson. 2005. Planning for climate change: Identifying minimum-dispersal corridors for the Cape Proteaceae. *Conservation Biology* 19:1063–74.

Willis, S. G., J. K. Hill, C. D. Thomas, D. B. Roy, R. Fox, D. S. Blakeley, and B. Huntley. 2009. Assisted colonization in a changing climate: A test-study using two U.K. butterflies. *Conservation Letters* 2:46–52.

Restoring Predators as a Hedge against Climate Change

Christopher C. Wilmers, Chris T. Darimont, and Mark Hebblewhite

Climate change and the loss of ecologically relevant large terrestrial carnivore populations are two important challenges currently facing conservation practitioners (Ray et al. 2005, Sutherland et al. 2009). Little attention is given, however, to how these two problems might be related, and how mitigation efforts for each might be united. Here we explain how predators cannot only influence the cause of climate change (atmospheric carbon) but also influence—directly and indirectly—climate impacts on their prey and on entire ecological communities. We draw on an emerging body of work to conceive ways in which the restoration of large carnivores might provide opportunities for both carnivore conservation and mitigation of climate change.

Mitigation of Atmospheric Carbon

Hairston et al. (1960) hypothesized that the world is "green" because predators hold herbivores in check, thus allowing plants to thrive. Fretwell (1977) generalized this idea by suggesting that the parity of a food chain determines whether plants will be primarily limited in their biomass by resources or herbivory. He hypothesized that plants in odd-numbered food chains are limited by resources because predators hold herbivores in check, whereas plants in even-numbered food chains are limited by herbivory. As most large mammal food webs in terrestrial systems are comprised of three links, theoretically plants in ecosystems with large mammalian predators should be, at least in part, released from the pressures of herbivory and should thus become more resource-limited. While this general phenomenon has been well documented in aquatic systems (Strong 1992), its broad applicability to terrestrial systems remains a topic of active research (Pace et al. 1999).

This is a challenging question to address because of the difficulty in experimenting with large mammals and the long time scales over which terrestrial systems cycle. Additionally, some herbivores are able to escape predator limitation through either migration or large body size (Sinclair 2003). Recent studies on wolves (*Canis lupus*) and pumas (*Puma con-*

color) in North America (Berger et al. 2001, Ripple and Beschta 2004a, Beschta 2005, Fortin et al. 2005, Hebblewhite et al. 2005, Ripple and Beschta 2006, Ripple and Beschta 2008) and on jaguars (*Panthera onca*) in the tropics (Terborgh et al. 2001) suggest, however, that the indirect effects of large predators on plant biomass can be substantial if predators are at ecologically relevant densities (Berger and Smith 2005). Most of the above-ground carbon in the biosphere is contained in the tissue of plants. If predators indirectly influence the biomass of plants, what is their potential influence on the ability of plants to sequester atmospheric carbon?

Few mammalian-initiated trophic cascades are as well understood as the one generated by sea otters on kelp forests in nearshore environments (Estes and Palmisano 1974). Sea otters along the North American west coast prey on sea urchins, which in turn feed on kelp. In areas with sea otters, kelp is plentiful. In areas without them, kelp is rare. Recent work by Wilmers et al. (in press) shows that kelp forests throughout the North American range of otters contain 1×10^{10} kg more carbon in total than they would without sea otters. This amounts to an approximately 11% reduction in atmospheric carbon present in the three-dimensional column projected above the otter range, or 43% of the increase in atmospheric carbon since preindustrial times. This substantial amount of carbon would currently fetch in excess of $700 million on the European carbon market.

While the mechanism by which sea otters impact kelp densities is the same as that operating in terrestrial ecosystems (namely, trait and density impacts on prey) the outcome is more extreme. The near-total depletion of plants by herbivores in aquatic environments is made possible by, among other causes, the lack of cellulose (Power 1992), which is indigestible by most herbivores. In terrestrial systems, cellulose comprises much plant matter, so the indirect effects of predators on plant biomass are less than in aquatic systems. Yet it is in cellulose that much of the above-ground biosphere's carbon resides. Coniferous forests and grasslands, for instance, hold nearly 100 and 10 times as much carbon per square meter of earth's surface respectively than do kelp forests. Thus, extrapolating the impact of sea otters to terrestrial ecosystems implies only a 1% to 10% impact of predators on terrestrial plant biomass. While studies attempting to quantify the indirect effects of large terrestrial carnivores on plant biomass at the ecosystem scale have not been conducted to our knowledge, indirect evidence suggests that these influences can be important. For instance, the disappearance of top predators from forest fragments in Venezuela led to a dramatic rise in herbivore numbers and a consequent decline in plant biomass (Terborgh et al. 2001). Similarly, the absence of wolves in eastern North America has led to a fivefold increase in white-tailed deer placing increased pressure on vegetation (Crete 1999). If predators were restored to ecologically effective densities throughout all ice-free portions of the earth's surface, and if they exerted a 1% to 10% impact on plant biomass, they would indirectly help sequester 23 gigatons of atmospheric carbon. This represents roughly 15% of the expected increase in atmospheric carbon over the next 50 years, which is equivalent to one of the seven carbon reduction "wedges" needed to stabilize global CO_2 (Pacala and Socolow 2004); it would currently fetch some $1.6 trillion on the European Carbon Exchange.

This simple calculation illustrates the potential importance of large predators to the global carbon cycle. In practice, the effects of predator repatriation on plant carbon stocks in many places would not be practicable, as a sizeable portion of the earth's land surface has been converted to agriculture and is thereby unsuitable for either native herbivores or predators. Moreover, in many areas large ungulates themselves are being (or have already been) destroyed through overhunting by humans, so there is no need for control of their populations by predator reintroduction. Still, many opportunities, in addition to the sea otter example, might exist for mitigating atmospheric carbon levels by restoring large predators to ecologically effective densities.

Predators as Buffers

Ecologically effective predator populations not only can directly offset atmospheric carbon increases, but also can partially buffer those systems from existing or impending changes in climate. While climate may directly impact dynamics at all levels of a food web, studies of the effects of climate on large mammal food webs have primarily focused on bottom-up pathways (but see Post et al. 1999). That is, changes in climate that impact the timing, growth, and composition of plants in turn influence the quality and quantity of forage for herbivores. Herbivores that are resource-limited, therefore, might be impacted more by a changing climate than are herbivores that are predator-limited. Additionally, the feedback of herbivores on plants and other components of an ecosystem might be driven more by climate in the absence of predators. As we discuss below, these effects might play out over both ecological and evolutionary time scales.

Ecological Time Sscales

Over ecological time scales, predators can act as keystone species by exerting strong top-down control over community dynamics (Power et al.

1996). This maintains biodiversity by suppressing competitively dominant prey (Paine 1966) and possibly by enhancing spatial heterogeneity in prey resource use (Mech 1977). A substantial body of ecological theory now supports the hypothesis that more biodiverse systems are more resilient to outside perturbations (McCann 2000). We hypothesize, then, that ecosystems with keystone species at ecologically functional densities will be more resilient to climate change. Below we explore some of the different mechanisms that operate on ecological time scales and are imposed by predators that might contribute to this role.

Population Dynamics of Prey

The population dynamics of large herbivores that are resource-limited (as in the absence of carnivores) are more likely to be influenced by variation in climate than are the population dynamics of those that are predator-limited (Wilmers et al. 2006). Mild years lead to the buildup of prey populations that then decline rapidly during climatically harsh years. This can lead to overcompensating density dependence and boom-bust cycles in herbivore populations. Populations with boom-bust dynamics are more at risk of extinction than are stable populations of equivalent mean population size, because they risk stochastic extinction during the bust phase of the cycle, and also because they have lower effective population sizes, which makes them less adaptable to novel changes in the environment, such as new diseases.

Examination of ungulate dynamics supports this idea. Reindeer introduced to Saint Matthew Island in the Bering Sea in 1944 in the absence of predators grew to a population of 6,000 individuals in 1960 before crashing to fewer than 42 individuals after food supply shrank dramatically when climate conditions worsened (Klein 1968). The Soay sheep population on Saint Kilda Island, Scotland, displays similar dynamics, though on a shorter time scale. Here the population experiences boom-bust cycles every two to four years as density interacts with climatic conditions to create high-variance population cycles (Coulson et al. 2001). In Greenland, recent analyses have shown that predatorless caribou populations are subject to trophic mismatches between the timing of parturition and phenology of vegetation (Post et al. 2008). In years when the mismatch is greatest, caribou calf production drops as much as fourfold.

A recent analysis of moose population dynamics on Isle Royale, Michigan, suggests that when predators occur at ecologically relevant densities, climate-induced fluctuations in prey populations are dampened. By taking advantage of a disease outbreak in wolves, Wilmers et al. (2006)

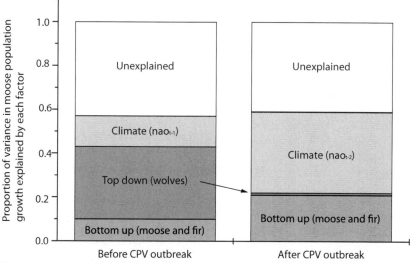

Figure 16.1. Role of canine parvovirus (CPV) outbreak on trophic factors affecting moose population dynamics. Hatched areas represent the variance R_x^2 in moose population growth rate explained by each variable. Before the outbreak of CPV, known biotic factors regulating moose population dynamics are primarily top-down (3:1 ratio) while after outbreak they are primarily bottom-up (28:1). Climate is a small factor governing moose population dynamics when moose are controlled by wolves, and a large factor after wolf populations become decimated by CPV. The arrow indicates the proportion of variation explained by top-down control in the post-CPV period. Redrawn from Wilmers et al. 2006.

showed that when wolves were unaffected by disease and were thus dynamically coupled to moose, they regulated the moose populations. When wolf populations crashed due to the introduction of canine parvovirus (CPV) to the island, however, the moose population mimicked the dynamics of the Saint Matthews caribou herd; they grew to record high numbers before plummeting to very low numbers during a severe winter (figure 16.1).

Ungulate populations in ecosystems where wolves have been extirpated, such as elk in Rocky Mountain and Yellowstone national parks before wolf reintroduction, are often limited by interactions between density and winter severity. Severe winters restrict access to forage and create a food bottleneck that limits the size of ungulate populations. As the climate warms and winters become increasingly mild, ungulate populations in wolf-free systems are predicted to increase in size (Wang et al. 2002, Creel and Creel 2009). This is likely to lead to overgrazing and boom-bust cycles in ungulate population dynamics, as less frequent severe winters coincide with larger herds and lead to large winter die-offs.

Recent theoretical work has elucidated the mechanisms by which predators, large herbivore life histories, and climate interact to influence ungulate population dynamics. More than 30 years ago, Eberhardt (1977) hypothesized that the vital rates of large vertebrates in different life-history stages change in a predictable sequence as density increases. Juvenile survival declines first, followed by adult fecundity and finally by adult survivorship. By incorporating these features into a stochastic population model, Wilmers et al. (2007b) showed that the level of yearly variation in climatic conditions can have profound impacts on moose population dynamics on Isle Royale. If a good year during which the population grows is followed by a bad year, then density dependence is experienced only in juvenile survivorship and the population declines by a small amount. If a few good years in a row occur before a bad year, then the population can grow to larger densities such that when a bad year strikes, density dependence is experienced in both juvenile survivorship and adult fecundity, thus resulting in a larger population decline. The extreme population crashes occur when many good years in a row are followed by a bad year. This allows the population to grow so large that when a bad year strikes, density dependence is experienced by individuals in all life-history stages, and the population crashes. Paradoxically, then, a higher frequency of good years leads to more dramatic boom-bust cycles, with the population in bust years reaching much smaller sizes than it would given a lower frequency of good years.

Climate can be thought of as a factor that ratchets the negative effects of density dependence up a life-history gradient from declines in juvenile survival through adult fecundity and finally to adult survival. The stronger this ratchet, the less stable the population. If climate becomes increasingly mild (i.e., favorable to herbivores) but more variable over time, as many climate models predict (Boyce et al. 2006), it suggests a strengthening of this ratchet over time.

Large predators can act in opposition to this climate ratchet (figure 16.2). By killing prey and modifying herbivore behavior through fear so that they eat less, predators lower the population growth rates of large herbivores during good years (Wilmers et al. 2007a). This means that when a bad year arrives, the population density is smaller than it would have been without predators, and the consequent decline in population size is smaller. Hence, as in Isle Royale prior to the outbreak of CPV in the wolf population, the model suggests that predators buffer prey popula-

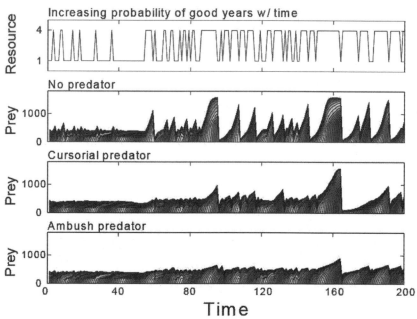

Figure 16.2. Climate change scenario depicting the influence of predation and the increasing frequency of favorable environmental conditions on the population fluctuations of an age-structured prey. As the frequency of favorable years increases, population fluctuations increase in magnitude because favorable conditions allow the population to grow to large densities which, in a poor year, leads to density dependence in multiple vital rates, and finally to a crash. Predators dampen the magnitude of these crashes because they retard the population's growth in good years. A climate change scenario depicting decreasing frequency of favorable conditions through time can be visualized by reading the figure right to left.

tions from the effects of climate, thus resulting in less variable prey population dynamics.

Under certain circumstances, predators might destabilize prey communities. Most famously this has been shown to occur in Fennoscandia rodent populations above 60 degrees latitude (Jedrzerjeski and Jedrzerjewska 1996, Kausrud et al. 2008). This is thought to arise because of inherent time lags associated with reproduction in a specialist predator feeding on a single prey. Below 60 degrees of latitude, the added presence of generalist predators stabilizes prey dynamics. This suggests that low-diversity ecosystems with single specialist predators will be less buffered from climate change than those with more diverse multiple-generalist predators. This example also emphasizes the importance of species diversity in promoting resilience to climate change effects (Kausrud et al. 2008).

Facilitative Effects

Another way in which predators might buffer the impact of climate change on ecological time scales is their influence on community dynamics, via their facilitative effects on scavengers. Wolves, for instance, have been documented to provide winter carrion to more than 40 species of vertebrate scavengers (Paquet et al. 1996, Wilmers et al. 2003b, Selva and Fortuna 2007) and more than 50 species of beetles (Sikes 1998, Wilmers unpublished data). For some species, this predator-derived subsidy can increase overwinter survival and reproduction (Wilmers et al. 2003a). In Yellowstone, the reintroduction of wolves dramatically changed the winter availability of carrion. Before wolf reintroduction, carrion availability was primarily a function of winter severity. During mild winters or at the beginning of winter, very little carrion was available. In contrast, during severe winters or at the end of winter, so much carrion would be available that vertebrate scavengers would be saturated; in the late spring, excess carcasses not exploited by this guild were consumed primarily by invertebrates. After wolf reintroduction, the dominant source of ungulate (in particular, elk) mortality shifted from winter severity to predation by wolves. The effect on carrion availability was that it became more predictable and less variable within and between years (Wilmers et al. 2003a).

The shift in the prime determinant of carrion availability away from winter severity and towards predation by wolves also meant that climate change would have less influence on carrion availability with wolves than without them. Winters in Yellowstone, as in many places across the globe, have been shortening (Wilmers and Getz 2005). This narrowing of the winter season would imply less carrion over a shorter time period in the absence of wolves. With wolves, however, carrion availability is primarily a function of wolf predation, so even as winters grow shorter, carrion is predicted to be available over roughly the same window of time (Wilmers and Getz 2005). Late-winter carrion is also reduced with the presence of wolves in the system (due to a predicted decrease in wolf kill rate), but to a much lesser extent, and over a longer time scale. This conceivably allows scavengers time to adapt to a changing environment over a time scale commensurate with natural processes (figure 16.3).

Predator-mediated carrion supply can also favor some scavenger species over others. The more pulsed and hence abundant a resource is, the more it favors "recruitment specialists" over "competitive dominants" (Wilmers et al. 2003b). Carrion is first consumed by dominant species (e.g., in Yellowstone, coyotes are dominant over eagles and ravens). But if the carrion is so plentiful that dominant species cannot consume it all, recruitment specialists will consume the rest. In Yellowstone, coyotes are

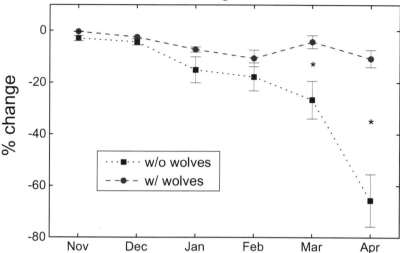

Figure 16.3. Percentage reduction (± SE) in winter carrion available to scavengers due to climate change from 1950 to 2000 under scenarios with and without wolves. The symbol * denotes a significant difference between the two scenarios.

limited to feeding on carrion within a few miles of their home range. If they are saturated with carcasses, bald eagles and ravens, which can recruit from many miles away, will build up their numbers and consume much of the carrion. The presence of wolves, by acting in opposition to the climate-induced pulsing of carrion, will therefore favor competitively dominant species over recruitment specialists.

By extending the time horizon over which winter carrion is available, wolf presence favors species, such as grizzly bears, that have strong seasonal patterns in resource use (figure 16.4). Recent work suggests that as temperatures rise, the buffering effect of wolves on scavengers grows stronger, and in particular on those with strong seasonal use in resources (Wilmers and Post 2006).

Structural Effects

Optimal foraging theory predicts that to maximize fitness, individuals will trade benefits associated with acquiring resources against the costs of acquiring those resources, which often take the form of increased predation risk (Sih 1987, Abrams 1991, Post et al. 2009). Several consequences of this perceived trade-off have been observed in large herbivores. Mech (1977) and later Lewis and Murray (1993) showed that boundaries between wolf pack territories serve as refuges for herbivore prey. Recently, investigators have shown that elk avoid areas of high predation risk near creeks and rivers (Ripple and Beschta 2004a, Fortin et al. 2005).

Avoidance by large herbivores of areas with high predation risk (Mech 1977, Fortin et al. 2005) can increase spatial heterogeneity in plant composition and biomass. For instance, in Yellowstone and Banff national parks, cascading behavioral interactions appear to be influencing willow stand dynamics (Beyer et al. 2007), but not aspen (Hebblewhite et al. 2005, Kauffman et al. 2010). This increased heterogeneity can increase the resilience of communities in the face of environmental perturbations associated with interannual fluctuations in climate (Allen-Diaz and

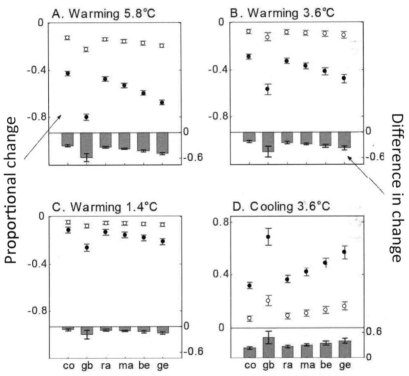

Figure 16.4. Changes in carrion consumption by the six most common scavenger species under (A) maximum, (B) mid-level, and (C) minimum warming, and (D) moderate cooling. Top panels display proportional change in mean carrion abundance (± SE) to coyotes (co), grizzly bears (gb), ravens (ra), magpies (ma), bald eagles (be), and golden eagles (ge) from 2000 to 2100 under various climate change scenarios with and without wolves. Bottom panels display differences in availability of carrion to each scavenger species from 2000 to 2100 under scenarios with and without wolves. These graphs illustrate the magnitude of wolf buffering against changes in ENSO.

Jackson 2000). An evenly grazed meadow, for instance, is more likely to respond uniformly to changes in precipitation or temperature than one with different vegetation heights and the associated differences in shade and soil moisture levels.

This effect is likely to be pronounced in riparian areas, and it can influence other species. The reemergence of wolves and cougars in areas of western North America has led to the return of riparian woody vegetation such as willow and cottonwood in some areas (Beschta 2003, Ripple and Beschta 2004b, Hebblewhite et al. 2005, Ripple and Beschta 2006, Beyer et al. 2007). Areas where riparian woody vegetation is not returning may be indications that the system has entered an alternative stable state, requiring the restoration of beaver to provide higher levels of fine sediment and shallower stream incision profiles necessary for high rates of willow growth (Wolf et al. 2007). Increased riparian woody vegetation provides shade for creeks, keeping water temperatures cool during the hot summer months. This is particularly important for native trout (e.g., *Salvelinus* spp.) that need cool water to survive through summer. While warmer ambient temperatures will raise the temperature of any body of water, shaded areas protected from direct sunlight will warm more slowly. Additionally, by stabilizing stream banks, providing leaf litter to streams, and providing woody debris important for pool formation, increased riparian woody vegetation likely increases habitat quality for native trout (Harig and Fausch 2002).

By shaping the spatial ecology of prey, predators indirectly buffer the base of the food web against climate change. At both population and individual levels and at multiple scales, foragers must constantly evaluate and respond to a shifting mosaic of benefits and costs of foraging in particular areas. And importantly, the trade-offs are not static. For example, predation risk can be spatially and temporally dynamic (Lima and Zollner 1996). Consequently, predator-mediated foraging behavior restricts herbivores from staying in one place (Forester et al. 2007), thus allowing plants to recover from browsing or grazing pressure. This in turn potentially provides them with resources to deal better with climate stressors.

Evolutionary Time

Predators might also drive diversity in foraging behaviors by herbivores over evolutionary time scales. Darimont et al. (2007) used stable isotope analysis to demonstrate that individuals within a black-tailed deer (*Odocoileus hemionus*) population in a coastal rain forest differed considerably in their use of different forest stand types under the risk of predation. Deer were killed by wolves in all areas, but the probability of mortality increased in types of forest where the protein content of forage plants was higher—consistent with the hypothesis that foragers trade forage quality against the risk of predation. Moreover, individuals that specialized in any one stand type were more likely to be killed. This selection against specialization and individuals' differential responses to risk-reward trade-offs under the risk of predation together explained the observed diversity in foraging behaviors. This is important in the context of climate change because species with generalist life-history strategies are more likely to adapt to a changing climate than are specialists.

Management Implications

As we outline above, there are myriad reasons why managers might consider restoring predators to ecologically effective densities as an adaptive response to climate change. While large predators have recolonized certain areas and been reintroduced to others, this action will generally require a sea change in attitude towards predator management. Currently, predators are often managed for very low densities. This occurs for a number of reasons, chief among which in North America is to control depredation on livestock, and because of a perceived or real notion that large predators substantially suppress game populations.

The management of large game populations and the quota systems that go with it developed over the last century in an era of extremely low predator abundance. In North America, the eradication of wolves, bears, and pumas over large expanses of land meant that wildlife managers could essentially ignore predator effects in setting hunting quotas on deer and elk. Now that the importance of predators to proper ecological function is becoming well accepted in other domains, managers should be further motivated to act, given realized or impending climate change. This could take the form of adjusting hunting quotas (of predators or their prey) or control efforts to levels that can support ecologically effective densities of predators. Moreover, restoration efforts could reintroduce native predators to previously occupied portions of their historic range.

While human hunting might provide an adequate substitute for large carnivores in some situations, the influence of human hunting on ecosystems often has a very different impact than that of large carnivores (Berger 2005). Many of the indirect effects of predators on community structure, for instance, are medaited by behavior rather than by density. In Yellowstone, for instance, elk were managed for very low densities until the park service adopted a policy of natural regulation in the early 1970s (Houston 1982). During that period of low elk population density, however, riparian vegetation did not recover. It was not until reintroduced wolves altered the "forage quality–predation risk tradeoff" that elk began to avoid valley bottoms, potentially allowing willow and cottonwood to come back. Furthermore, a recent review of the relative importance of trait-mediated interactions (TMIs) and density-mediated interactions (DMIs) showed that TMIs are as strong as DMIs on prey demographics and much stronger on cascading interactions. Density effects attenuate through food chains, while trait effects remain strong (Bolnick and Preisser 2005, Preisser et al. 2005).

Implementing climate change adaptation strategies will require new models that explicitly take predators into account. Conversely, strategies to conserve carnivores should now consider their potential role as hedges against climate change. Indeed, economic incentives or payoffs in the carbon markets of tomorrow might influence carnivore conservation and management. While the legal framework for buying and selling carbon from different sources is still rapidly evolving, it is not inconceivable that the potential carbon sink provided by predators such as sea otters could be sold to fund restoration to ecologically effective densities.

Harvest models that consider predators are needed to implement more effective game management, as discussed above. Predictive models of community response to climate change would also benefit by including strongly interacting species (Soule et al. 2005) such as top predators in order to improve the accuracy of their predictions. Recent empirical work has shown that ignoring species interactions can reverse the predicted response of community composition to climate change (Suttle et al. 2007). Areas with and without top predators may respond very differently, and these differences should be considered in predicting the ecological impacts of increased atmospheric carbon.

This emerging insight into the important role that terrestrial predators serve in buffering ecosystems from global climate change comes at a time when their own future is uncertain. Predators are declining more rapidly than any other food web group. The 2003 IUCN red list of threatened species lists 125 carnivores as threatened with extinction, and carnivores not on the list have, for the most part, experienced dramatic contractions of their range (Laliberte and Ripple 2003). Accordingly, repatriating predators to their historic ranges has enormous potential not only to provide well-known ecological services, but also to improve ecosystem resilience to climate change and drive down atmospheric carbon levels.

ACKNOWLEDGMENTS
This work was supported by NSF grants #0963022, #0713994, and #0729707 to C. Wilmers, and by funding from the University of Montana to M. Hebblewhite.

LITERATURE CITED

Abrams, P. A. 1991. Life history and the relationship between food availability and foraging effort. *Ecology* 72:1242–52.

Allen-Diaz, B. and R. D. Jackson. 2000. Grazing effects on spring ecosystem vegetation of California's hardwood rangelands. *Journal of Range Management* 53:215–20.

Berger, J. 2005. Hunting by carnivores and humans: Does functional redundancy occur and does it matter? In *Large Carnivores and the Conservation of Biodiversity*, edited by J. C. Ray, K. H. Redford, R. S. Steneck, and J. Berger, 316–37. Washington: Island Press.

Berger, J., and D. W. Smith. 2005. Restoring functionality in Yellowstone with recovering carnivores: gains and uncertainties. In *Large Carnivores and the Conservation of Biodiversity*. edited by J. C. Ray, K. H. Redford, R. S. Steneck, and J. Berger, 316–37. Washington: Island Press.

Berger, J., P. B. Stacey, L. Bellis, and M. P. Johnson. 2001. A mammalian predator-prey imbalance: Grizzly bear and wolf extinction affect avian neotropical migrants. *Ecological Applications* 11:947–60.

Beschta, R. L. 2003. Cottonwoods, elk, and wolves in the Lamar Valley of Yellowstone National Park. *Ecological Applications* 13:1295–1309.

———. 2005. Reduced cottonwood recruitment following extirpation of wolves in Yellowstone's northern range. *Ecology* 86:391–403.

Beyer, H. L., E. H. Merrill, N. Varley, and M. S. Boyce. 2007. Willow on yellowstone's northern range: Evidence for a trophic cascade? *Ecological Applications* 17: 1563–71.

Bolnick, D. I. and E. L. Preisser. 2005. Resource competition modifies the strength of trait-mediated predator-prey interactions: A meta-analysis. *Ecology* 86:2771–79.

Boyce, M. S., C. V. Haridas, C. T. Lee, and N. s. d. w. group. 2006. Demography in an increasingly variable world. *Trends in Ecology and Evolution* 21:141–49.

Coulson, T., E. A. Catchpole, S. D. Albon, B. J. T. Morgan, J. M. Pemberton, T. H. Clutton-Brock, M. J. Crawley, and B. T. Grenfell. 2001. Age, sex, density, winter weather, and population crashes in Soay sheep. *Science* 292:1528–31.

Creel, S. and M. Creel. 2009. Density dependence and climate effects in Rocky Mountain elk: An application of regression with instrumental variables for population time series with sampling error. *Journal of Animal Ecology* 78:1291–97.

Crete, M. 1999. The distribution of deer biomass in North America supports the hypothesis of exploitation ecosystems. *Ecology Letters* 2:223–27.

Darimont, C. T., P. C. Paquet, and T. E. Reimchen. 2007. Stable isotopic niche predicts fitness of prey in a wolf-deer system. *Biological Journal of the Linnean Society* 90:125–37.

Eberhardt, L. L. 1977. Optimal policies for the conservation of large mammals, with special reference to marine ecosystems. *Environmental Conservation* 4:205–12.

Estes, J. A., and J. F. Palmisano. 1974. Sea otters: their role in structuring nearshore communities. *Science* 185:1058–60.

Forester, J. D., A. R. Ives, M. G. Turner, D. P. Anderson, D. Fortin, H. L. Beyer, D. W. Smith, and M. S. Boyce. 2007. State-space models link elk movement patterns to landscape characteristics in Yellowstone National Park. *Ecological Monographs* 77:285–99.

Fortin, D., H. L. Beyer, M. S. Boyce, D. W. Smith, T. Duchesne, and J. S. Mao.

2005. Wolves influence elk movements: Behavior shapes a trophic cascade in Yellowstone National Park. *Ecology* 86:1320–30.

Fretwell, S. D. 1977. Regulation of plant communities by food-chains exploiting them. *Perspectives in Biology and Medicine* 20:169–85.

Hairston, N. G., F. E. Smith, and L. B. Slobodkin. 1960. Community Structure, Population Control, and Competition. *American Naturalist* 94:421–25.

Harig, A. L. and K. D. Fausch. 2002. Minimum habitat requirements for establishing translocated cutthroat trout populations. *Ecological Applications* 12:535–51.

Hebblewhite, M., C. A. White, C. G. Nietvelt, J. A. McKenzie, T. E. Hurd, J. M. Fryxell, S. E. Bayley, and P. C. Paquet. 2005. Human activity mediates a trophic cascade caused by wolves. *Ecology* 86:2135–44.

Houston, D. B. 1982. *The Northern Yellowstone Elk: Ecology and Management.* New York: Macmillan.

Jedrzerjeski, W. and B. Jedrzerjewska. 1996. Rodent cycles in relation to biomass and productivity of ground vegetation and predation in the Palearctic. *Acta Theriologica* 41:1–34.

Kauffman, M. J., J. F. Brodie, and E. S. Jules. 2010. Are wolves saving Yellowstone's aspen? A landscape-level test of a behaviorally mediated trophic cascade. *Ecology* 91:2742–55.

Kausrud, K. L., A. Mysterud, H. Steen, J. O. Vik, E. Ostbye, B. Cazelles, E. Framstad, A. M. Eikeset, I. Mysterud, T. Solhoy, and N. C. Stenseth. 2008. Linking climate change to lemming cycles. *Nature* 456:93–U93.

Klein, D. R. 1968. Introduction Increase and Crash of Reindeer on St Matthew Island. *Journal of Wildlife Management* 32:350.

Laliberte, A. S. and W. J. Ripple. 2003. Wildlife encounters by Lewis and Clark: A spatial analysis of interactions between native Americans and wildlife. *Bioscience* 53:994–1003.

Lewis, M. A. and J. D. Murray. 1993. Modelling territoriality and wolf-deer interactions. *Nature* 366:738–40.

Lima, S. L. and P. A. Zollner. 1996. Anti-predatory vigilance and the limits to collective detection: Visual and spatial separation between foragers. *Behavioral Ecology and Sociobiology* 38:355–63.

McCann, K. S. 2000. The diversity-stability debate. *Nature* 405:228–33.

Mech, D. 1977. Wolf-pack buffer zones as prey resevoirs. *Science* 198:320–21.

Pacala, S. and R. Socolow. 2004. Stabilization wedges: Solving the climate problem for the next 50 years with current technologies. *Science* 305:968–72.

Pace, M. L., J. J. Cole, S. R. Carpenter, and J. F. Kitchell. 1999. Trophic cascades revealed in diverse ecosystems. *Trends in Ecology & Evolution* 14:483–88.

Paine, R. T. 1966. Food web complexity and species diversity. *American Naturalist* 100:65–75.

Paquet, P. C., J. Wierzchowski, and C. Callaghan. 1996. Summary report on the effects of human activity on gray wolves in the Bow River Valley, Banff National Park, Alberta. In *A Cumulative Effects Assessment and Futures Outlook for the Banff Bow Valley*, edited by J. Green, C. Pacas, S. Bayley, and L. Cornwell. Ottawa: Department of Canadian Heritage.

Post, E., J. Brodie, M. Hebblewhite, A. D. Anders, J. A. K. Maier, and C. C. Wilmers. 2009. Global population dynamics and hot spots of response to climate change. *Bioscience* 59:489–97.

Post, E., C. Pedersen, C. C. Wilmers, and M. C. Forchhammer. 2008. Warming, plant phenology and the spatial dimension of trophic mismatch for large herbivores. *Proceedings of the Royal Society B: Biological Sciences* 275:2005–13.

Post, E., R. O. Peterson, N. C. Stenseth, and B. E. McLaren. 1999. Ecosystem consequences of wolf behavioural response to climate. *Nature* 401:905–7.

Power, M. E. 1992. Top-down and bottom-up forces in food webs: Do plants have primacy? *Ecology* 73:733–46.

Power, M. E., D. Tilman, J. A. Estes, B. A. Menge, W. J. Bond, L. S. Mills, G. Daily, J. C. Castilla, J. Lubchenco, and R. T. Paine. 1996. Challenges in the quest for Keystones: Identifying keystone species is difficult-but essential to understanding how loss of species with affect ecosystems. *Bioscience* 46:609–20.

Preisser, E. L., D. I. Bolnick, and M. F. Benard. 2005. Scared to death? The effects of intimidation and consumption in predator-prey interactions. *Ecology* 86:501–9.

Ray, J. C., K. H. Redford, R. S. Steneck, and J. Berger, eds. 2005. *Large Carnivores and the Conservation of Biodiversity*. Washington: Island Press.

Ripple, W. J., and R. L. Beschta. 2004a. Wolves and the ecology of fear: Can predation risk structure ecosystems? *Bioscience* 54:755–66.

———. 2004b. Wolves, elk, willows, and trophic cascades in the upper Gallatin Range of Southwestern Montana, USA. *Forest Ecology and Management* 200:162–81.

———. 2006. Linking a cougar decline, trophic cascade, and catastrophic regime shift in Zion National Park. *Biological Conservation* 133:397–408.

———. 2008. Trophic cascades involving cougar, mule deer, and black oaks in Yosemite National Park. *Biological Conservation* 141:1249–56.

Selva, N. and M. A. Fortuna. 2007. The nested structure of a scavenger community. *Proceedings of the Royal Society B: Biological Sciences* 274:1101–8.

Sih, A. 1987. Predators and prey lifestyles: An evolutionary and ecological overview. In W. C. Kerfoot and A. Sih, eds., *Predation*. Hanover, NH: University Press of New England.

Sikes, D. S. 1998. Hidden biodiversity: The benefits of large rotting carcasses to beetles and other species. *Yellowstone Science* 6:10–14.

Sinclair, A. R. E. 2003. Mammal population regulation, keystone processes and ecosystem dynamics. *Philosophical Transactions of the Royal Society of London Series B: Biological Sciences* 358:1729–40.

Soule, M. E., J. A. Estes, B. Miller, and D. L. Honnold. 2005. Stongly interacting species: Conservation policy, managment, and ethics. *Bioscience* 55: 168–176.

Strong, D. R. 1992. Are trophic cascades all wet? Differentiation and donor-control in speciose ecosystems. *Ecology* 73:747–54.

Sutherland, W. J., W. M. Adams, R. B. Aronson, R. Aveling, T. M. Blackburn, S. Broad, G. Ceballos, I. M. Cote, R. M. Cowling, G. A. B. Da Fonseca, E. Dinerstein, P. J. Ferraro, E. Fleishman, C. Gascon, M. Hunter, J. Hutton, P. Kareiva, A. Kuria, D. W. MacDonald, K. MacKinnon, F. J. Madgwick, M. B. Mascia, J. McNeely, E. J. Milner-Gulland, S. Moon, C. G. Morley, S. Nelson, D. Osborn, M. Pai, E. C. M. Parsons, L. S. Peck, H. Possingham, S. V. Prior, A. S. Pullin, M. R. W. Rands, J. Ranganathan, K. H. Redford, J. P. Rodriguez, F. Seymour, J. Sobel, N. S. Sodhi, A. Stott, K. Vance-Borland, and A. R. Watkinson. 2009. One hundred questions of importance to the conservation of global biological diversity. *Conservation Biology* 23:557–67.

Suttle, K. B., M. A. Thomsen, and M. E. Power. 2007. Species interactions reverse grassland responses to changing climate. *Science* 315:640–42.

Terborgh, J., L. Lopez, P. Nunez, M. Rao, G. Shahabuddin, G. Orihuela, M. Riveros, R. Ascanio, G. H. Adler, T. D. Lambert, and L. Balbas. 2001. Ecological meltdown in predator-free forest fragments. *Science* 294:1923–26.

Wang, G. M., N. T. Hobbs, F. J. Singer, D. S. Ojima, and B. C. Lubow. 2002. Impacts of climate changes on elk population dynamics in Rocky Mountain National Park, Colorado, USA. *Climatic Change* 54:205–23.

Wilmers, C. C., R. L. Crabtree, D. Smith, K. M. Murphy, and W. M. Getz. 2003a. Trophic facilitation by introduced top predators: gray wolf subsidies to scavengers in Yellowstone National Park. *Journal of Animal Ecology* 72:909–16.

Wilmers, C. C., J. A. Estes, K. L. Laidre, M. Edwards, B. Konar. In press. Do trophic cascades affect the storage and flux of atmospheric carbon? An analysis for sea otters and kelp forests. *Frontiers in Ecology and the Environment.*

Wilmers, C. C. and W. M. Getz. 2005. Gray wolves as climate change buffers in Yellowstone. *PLoS Biology* 3:571–76.

Wilmers, C. C. and E. Post. 2006. Predicting the influence of wolf-provided carrion on scavenger community dynamics under climate change scenarios. *Global Change Biology* 12:403–9.

Wilmers, C. C., E. Post, and A. Hastings. 2007a. The anatomy of predator-prey dynamics in a changing climate. *Journal of Animal Ecology* 76:1037–44.

———. 2007b. A perfect storm: The combined effects on population fluctuations of autocorrelated environmental noise, age structure, and density dependence. *American Naturalist* 169:673–83.

Wilmers, C. C., E. Post, R. O. Peterson, and J. A. Vucetich. 2006. Predator disease out-break modulates top-down, bottom-up and climatic effects on herbivore population dynamics. *Ecology Letters* 9:383–89.

Wilmers, C. C., D. R. Stahler, R. L. Crabtree, D. W. Smith, and W. M. Getz. 2003b. Resource dispersion and consumer dominance: scavenging at wolf- and hunter-killed carcasses in Greater Yellowstone, USA. *Ecology Letters* 6:996–1003.

Wolf, E. C., D. J. Cooper, and N. T. Hobbs. 2007. Hydrologic regime and herbivory stabilize an alternative state in Yellowstone National Park. *Ecological Applications* 17:1572–87.

17

Assisted Colonization of Wildlife Species at Risk from Climate Change

Viorel D. Popescu and Malcolm L. Hunter Jr.

Rapid climate change is regarded as one of the top threats to biodiversity, potentially causing declines and extinctions of many species (Walther et al. 2002, Parmesan and Yohe 2003, Parmesan 2006) especially those that cannot shift their geographic ranges, and disrupting community-level interactions (Berg et al. 2010, Traill et al. 2010). Evolutionary adaptations to climate change may occur in some species (Skelly et al. 2007, Austin et al. this volume), but range shifts are likely to be a key response for many, and specific regions are predicted to experience high species turnover (Lawler et al. 2009). Shifting geographic ranges may depend upon the availability of contiguous suitable habitat for dispersal movements. The resilience of populations to climate change is affected by human-induced stressors, such as habitat alteration and fragmentation, invasive species, and overexploitation, all of which act synergistically (Warren et al. 2001, Travis 2003, Opdam and Wascher 2004) and have the potential to preclude natural range shifts. Thus, less vagile species occurring as isolated populations facing significant barriers in the required direction of colonization (poleward or towards higher altitudes) may need assistance to reach habitat that matches their climate requirements. In this context, moving species at greatest threat into regions outside their current or historical range may help overcome or reverse the loss of resilience and prevent populations from becoming functionally extinct. This process is known as *assisted colonization* (Hunter 2007, Hoegh-Guldberg et al. 2008), *assisted migration* (McLachlan et al. 2007), or *managed relocation* (Richardson et al. 2009), although there is still debate about a unified terminology for this conservation strategy (Seddon 2010). In this chapter we use the term *assisted colonization* (AC) because zoologists use migration to refer to cyclic (e.g., seasonal) movements, while most relocations are within a species' historical geographic range.

The concept of AC as a response to climate change is relatively new, and it has become highly controversial given the potential ecological and social implications. Plant biologists first started speculating on the potential of introducing species with a narrow climate envelope to novel sites (Kutner and Morse 1996), but the debate did not become prominent until McLachlan et al. (2007) evaluated AC as a tool to conserve species that are threatened by climate change. They started the dialogue, later extended by Hunter (2007) and Hoegh-Guldberg et al. (2008), that framed the various issues surrounding AC. Inevitably, this "hot topic" captured the attention of the scientific and conservation community, and even that of Christian theologians (Southgate et al. 2008). The pros and cons have been discussed further by others (Chapron and Samelius 2008, Davidson and Simkanin 2008, Huang 2008, Mueller and Hellmann 2008, Ricciardi and Simberloff 2009, Sax et al. 2009, Sandler 2010, Seddon 2010, Dawson et al. 2011, Loss et al. 2011). McLachlan et al. (2007) outline three conflicting perspectives or attitudes on the topic of AC that scientists, policy makers, and managers may or may not have already embraced: (1) aggressive AC, (2) avoidance of AC, and (3) constrained AC. These perceptions are based on interactions between perceived risks of AC and confidence in our understanding of how ecological systems work. Advocates for AC have high confidence in our ability to translocate species efficiently, and believe that doing nothing is riskier than carrying out AC. At the other end of the spectrum, critics of AC believe that uncertainty surrounding such endeavors (e.g., climate change projections, translocation strategies) or collateral negative outcomes (e.g., potential for invasiveness or other unforeseen species interactions that may adversely affect the relocated species and those onsite) outweigh the risks of no action. The intermediate option, constrained AC, supports using strong scientific evidence for a carefully planned AC strategy that might alleviate the effects of climate change on some species under imminent threat.

We are not advocating one option over any others. Rather, our objectives are to (1) provide an in-depth examination of the benefits and shortcomings of AC, and (2) offer insight into designing effective AC projects using lessons from species translocations. We will focus on terrestrial vertebrates, but use the broader literature as needed.

A Framework for Evaluating the Benefits and Costs of Assisted Colonization
Potential Benefits
The primary motivation for AC is to save a species from becoming extinct because it cannot adapt to rapid climate change. Although climate change has driven many extinctions in Earth's history (Raup 1992), human intervention to minimize extinctions during the current period of rapid climate change seems justified for at least two reasons. First, hu-

man activities are largely responsible for current climate change. Second, humans have increased the vulnerability of many species by reducing them to small, fragmented populations, and the interplay between land use patterns and climate change detrimentally affects their resilience and long-term viability (Travis 2003).

The arguments for saving species from extinction are well known to conservationists. They include diverse forms of instrumental value (e.g., food, medicine, the important roles they may play in ecosystems), potential value (instrumental value that is yet to be realized), and intrinsic value (the idea that species have value independent of their usefulness to humans or any other species; Sandler 2010). Some of the most likely candidates for AC may have rather constrained instrumental value simply because they are likely to be rare species. However, this rarity does not diminish their intrinsic value.

Although avoiding extinction has been cited as the main benefit of AC, some people might argue for AC of a potential dominant or keystone species to maintain the ecological integrity of the "receiving" ecosystem (Kreyling et al. 2011). Imagine that a reef-forming mollusk or coral species was disappearing in response to climate change, and that this was causing the degeneration of the whole reef ecosystem. If a congeneric species with a tolerance for warmer water were available, it would be tempting to use AC to replace the declining species. Saving species from extinction and saving ecosystems from profound degradation are important benefits, but AC comes with some significant risks and costs.

Potential Costs

Arrayed against the benefits of AC are a series of drawbacks, both ecological and socioeconomical. Foremost is the high potential for invasiveness by species moved outside their historical range (Davidson and Simkanin 2008, Ricciardi and Simberloff 2009; see also Fordham et al., this volume, for implications of climate change for invasive species). Considering both a species' invasive potential and the questionable resilience of the ecosystem into which it is to be introduced, the likelihood for significant unintended consequences could be considerable. Mueller and Hellmann (2008) provide an assessment of the potential for invasiveness by introduced species using intracontinental examples from North America as a proxy for species potentially to be used in AC. Species that are invasive within their own continent seem less prevalent than intercontinental invasives, but there are taxa such as fish and crustaceans that still pose a high intracontinental invasiveness risk (Mueller and Hellmann 2008). The lowest risk to native species is posed by highly specialized endemics

(Thomas 2011); carefully selected ecosystems could host many endemics at risk from climate change, even creating new ecological communities. For aquatic species, invasiveness potential can be minimized if species are translocated within the same major river basin and/or to physically isolated water bodies (Olden et al. 2010).

Huang (2008) also questions the feasibility of using AC to help rare species, which are likely to be the main candidates for such efforts. He argues that rare species are understudied, and thus their biology and habitat requirements are not well understood. Because careful AC requires understanding the species' ability to adapt to the new environment as well as its invasiveness potential, only a few thoroughly studied rare species would be candidates. This argument might be turned upside down: reviewing a species for AC could catalyze detailed studies that produce valuable knowledge regardless of whether or not the colonization project is executed. Sometimes, species at risk from climate change might already be at risk for other reasons such as poaching or diseases, so these threats must be eliminated in the receiving areas used for translocation (Dodd and Seigel 1991, Fischer and Lindenmayer 2000). Similarly, it is critical to determine whether climate change is one of the contributing factors driving a species' decline before deciding on AC. There is some risk that people will blame climate change to justify moving an endangered species, when they are primarily motivated to get it out of the way because it constrains economic development (Hunter 2007).

Spreading disease is another risk associated with moving species (Cunningham 1996). Emerging infectious diseases are threats to biodiversity, and translocation of organisms (whether intentional or accidental) is a main factor in transmission (Daszak et al. 2000). Although it may be difficult to distinguish between infection and disease (Scott 1988), the transmission of both infections (e.g., micro- and macroparasites present in at least one member of the host population; Scott 1988) and diseases (clinical conditions that can be observed and measured, and which can characterize a single individual or an entire population; Scott 1988) can severely compromise AC efforts. The risk of transmitting infections goes both ways between native and introduced species (Cunningham 1996). Native parasites can severely compromise a small founding population, while parasites in the introduced individuals can have major deleterious effects on the native community (Scott 1988). Cunningham (1996) reviewed the disease threats associated with translocations and recommended several means for minimizing them. However, his overall recommendation was to avoid introducing exotic species to new habitats.

Perhaps the greatest barrier to AC is economic. How do we undertake AC when resources for conservation are always scarce, and may well become even more scarce? The logistical costs can be high; they are largely dependent on the species in question, the depth of existing knowledge, the need for monitoring, the source of individuals to be relocated, and many other variables (Haight et al. 2000). Fischer and Lindenmayer (2000) reviewed animal translocation projects and found that their costs varied widely among taxa (with the caveat that only 6 out of 180 studies reported costs): for example, $1 million per year for the reintroduction of the California condor (*Gymnogyps californianus*), €70,000 per individual released for bearded vultures (*Gypaetus barbatus*) in the central Alps (Schaub et al. 2009), $22,000 per surviving golden lion tamarin (*Leontopithecus r. rosalia*), and $6,700,000 for two reintroductions of gray wolves (*Canis lupus*) in Yellowstone National Park over a period of eight years. Of course, most of these species are large vertebrates with large home ranges, and their relocations involved captive breeding and rehabilitation of individuals. Costs are likely to be lower for species that have small home ranges, are of wild origin, and/or require minimum periods in captivity (e.g., €20,000, excluding salaries, for two reintroductions of the lacertid lizard *Psammodromus algirus* in Spain; Santos et al. 2009).

Fischer and Lindenmayer (2000) also assessed the success of translocations through time (until the late 1990s, mostly for birds and mammals). Although the number of translocations increased through time, the proportion of successes decreased (from 35% to 50% in the 1970s to 17% in the 1990s), thus suggesting that over a long period of time we did not get better at doing this. The same conclusion was reached by Dodd and Seigel (1991) for translocations of amphibians and reptiles. However, a recent review of only amphibians and reptiles (Germano and Bishop 2009) found that in recent years (since 1990) the success rate of translocations doubled in comparison to the findings cited by Dodd and Seigel (1991), from 19% to 41%. From a taxonomic perspective, reviews of translocations prior to 1990 found that birds were more successful than mammals (Wolf et al. 1996), which in turn were more successful than herpetofauna (Dodd and Seigel 1991). Among birds, species with broader ecological niches and large brains relative to body size were more likely to establish viable populations (Vall-llosera and Sol 2009). The uncertainty of a successful outcome is definitely a concern for AC initiatives, and careful planning needs to be undertaken with an eye to aborting the project if the risks and costs are likely to exceed benefits (see the section on designing a successful project below).

Which Species to Move and Where?
Candidate Species

The footprint of climate change in global ecosystems is quantifiable, but not all species or ecosystems are affected equally. The negative effects discussed throughout this book (e.g., Matthews et al., this volume; Owen-Smith, this volume; Tews et al., this volume) are balanced by possible range expansions (e.g., amphibians and reptiles in Europe; Araújo et al. 2006), or in-situ adaptation (Skelly et al. 2007). These and other issues mean that relatively few species are likely to be candidates for AC.

Hunter (2007) identified three main considerations for evaluating candidate species: (1) likelihood of extinction from climate change, (2) vagility, and (3) ecological role. As mentioned in the previous section, pinpointing climate change as one of the causes for decline is difficult, and thorough studies on other possible causes must be conducted prior to selecting a species. Obviously, species that are less vagile or which exist only as isolated populations (e.g., mountaintop or cave biotas) are among the main candidates. Species whose ecological roles are generic (e.g., aquatic detritivorous invertebrates) might be more likely to "blend in" ecologically with the native species of the colonized area, thus making them less risky candidates for AC. As mentioned above, one of the incentives for AC might be the loss of a keystone or dominant climate-threatened species and the need to replace it with another species capable of performing the same ecological functions. While such replacement might indeed help a suite of other species, it is likely to have a profound impact on ecosystems and pose high risks for translocations, as exemplified by many invasive species (e.g., beaver [*Castor canadensis*] in Tierra del Fuego; Martinez Pastur et al. 2006). At the opposite end of the spectrum, Thomas (2011) argues that narrow endemics are ideal candidates for assisted colonization, as long as the receiving localities do not host endemics themselves and relocations occur within the same broad geographic range.

From a practical perspective, Griffith et al. (1989) found that moving endangered species even in high-quality habitat had a low success rate, and recommended that action be taken before translocation becomes the last conservation option for the species. Hence, we must be aware that even with strong financial and social support, some endangered species might not be movable at all.

The decision framework for AC provided by Hoegh-Guldberg et al. (2008; figure 17.1) advances the discussion initiated by Hunter (2007). Since it is self-explanatory we will not discuss it further, except to note

Decisions

Options

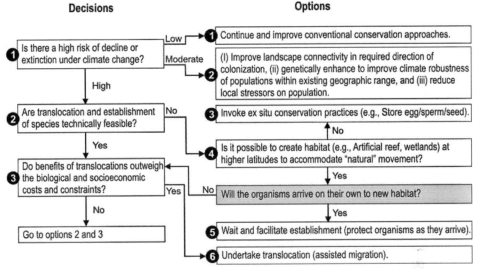

Figure 17.1. Decision framework for assisted colonization. Redrawn from Hoegh-Guldberg et al. 2008; used with permission from AAAS.

that, in their view, AC is considered almost the last resort for saving species, followed only by ex-situ conservation.

Complementary to this tree approach, which relies mainly on ecological traits of species and technical aspects of translocations, Richardson et al. (2009) developed a multidimensional decision-making framework that accounts for differences in societal values while simultaneously addressing the costs and benefits of AC. The ecological and social criteria are divided into four categories: (1) focal impact (impacts of conducting or not conducting AC), (2) collateral impact (impacts on the recipient ecosystems), (3) practical feasibility, and (4) acceptability (as a social criterion only). One of the main implications of this multidimensional approach is that, given the inherent complexities of the decision-making process, AC should be considered an option in the conservation portfolio to be used whenever feasible and necessary, rather than as a measure of last resort (Seddon 2010, Dawson et al. 2011).

Candidate Sites

Identifying sites that are potential candidates for assisted colonization might also pose challenges for managers. Hunter (2007) outlined four potential issues: (1) disturbance history, (2) geographic isolation, (3) relation to the species' "long-term" range, and (4) current biological diversity. First, given the uncertainties surrounding the introduction of species in new environments, sites that are already disturbed and altered might be preferred over undisturbed sites (e.g., an ecosystem restoration site versus a pristine ecological reserve). Similarly, geographically well-connected areas where changes in species composition have been common (e.g., continents) are generally preferable to isolated areas (e.g., islands harboring endemic species). Sometimes, however, isolated remnants of formerly connected habitat might be preferred in order to limit the spread of unintended consequences. Furthermore, moving species within their long-term distribution boundaries might be more acceptable than introducing them entirely outside their previous range, as it might limit the potential for invasiveness (Mueller and Hellmann 2008, Olden et al. 2010, Kreyling et al. 2011). Also, there is debate about the relationship between the recipient ecosystem's species richness and its capacity to cope with introduced or invasive species. Traditionally, species-rich systems have been thought more resilient to invasion, and small-scale experiments on plants support this assertion (Fargione and Tilman 2005). However, other studies encompassing a variety of ecosystems found that nonnative plant species were likely to become established in species-rich systems because such systems have attributes that both maintain native species richness and elevate nonnative species richness (Stohlgren et al. 2002). In contrast, high-productivity native systems were likely to limit the potential for invasiveness (Cleland et al. 2004). Experiments by Dunstan and Johnson (2004) showed that species richness in a marine epibenthic community was not a good predictor of invasiveness potential, which depended on the dynamics of the local community and the properties of particular species. These contrasting views are probably tied to the scale of analysis, with a negative relation between richness and invasive species at a local scale, and a positive relation at the broader scale (across ecosystems), because some sites are more prone to species establishment than others (Levine and D'Antonio 1999, Levine 2000). Thus, when performing AC, it is of interest to identify the "invasiveness history" for that particular type of ecosystem, or for similar ecosystems in cases where local data is missing. Decisions on where to move species should therefore consider ecosystems with the fewest known invasive species.

Assisted colonization efforts are thought to involve mostly single-species translocations, but that might be difficult to achieve since an array of parasites and microorganisms are associated with virtually every species. In some taxa, such as plants, intentional introductions of the associated communities might be considered (e.g., moving soil to maintain mycorrhizal associations). Introduced species are likely to perform better in new environments if they can become free of natural enemies, such as parasites (Torchin et al. 2003). In poleward expansions, some plant species performed better when inoculated with rhizosphere from the projected expansion range, than when inoculated with rhizosphere from their original range, potentially because they were released from enemies borne by their native soil (van Grunsven et al. 2010). However, this release also tends to enhance the species' invasiveness potential (Torchin et al. 2003).

There are issues of time lag and spatial incoherence between current and future climate conditions that make identifying suitable translocation areas and timing the translocation difficult, especially given the uncertainties surrounding various climate change models. Small-scale experiments can be performed to investigate the likelihood of successful introductions and to improve translocation technology (Olden et al. 2010). One option is to translocate species to locations outside their current ranges that clearly fall within the suitable climate space (Kreyling et al. 2011). Such translocation experiments with two butterfly species in the United Kingdom, *Melanargia galathea* and *Thymelicus sylvestris* (Willis et al. 2008) were successful in terms of both population growth and range expansion. The researchers first modeled the butterflies' climate space using bioclimatic variables, and then selected locations for introductions 35 to 65 km beyond the northern limit of their range, an area that fell within the predicted suitable climate space. A similar bioclimatic modeling approach revealed potential locations in the United Kingdom for the establishment of two regionally extinct butterfly species (*Aporia crataegi* and *Polyommatus semiargus*), but further habitat studies needed to be performed before performing AC (Carroll et al. 2009). Another alternative would be to undertake introductions within the same range (moving the species from a "warm" site, such as a low-elevation valley, to a "colder" site, such as a higher elevation). Such actions have already been undertaken with trees (red oak, *Quercus rubra*, in Baxter State Park, Maine; C. Redelsheimer, personal communication).

Multiple locations need to be considered, because the success of translocations may be directly related to the number of locations, and hence the number of newly founded populations (Griffith et al. 1989). Three reviews of animal translocations (Griffith et al. 1989, Wolf et al. 1996, Miller et al. 1999) associated the success of translocations with the release of individuals in optimal habitat. Even in optimal habitat, subsequent supportive measures (e.g., supplemental feeding) are likely to be needed to support the new populations (Fischer and Lindenmayer 2000).

Establishing new stepping-stone populations in areas within species' current ranges could help them shift toward higher latitudes, while providing experimental evidence for the potential success of the introduction (Hoegh-Guldberg et al. 2008; option 2(1) in figure 17.1).

Designing an Effective Assisted Colonization Project

Evidence from translocation research emphasizes that AC should be employed before the species becomes endangered or is on the brink of extinction; after a thorough assessment of candidate species and introduction sites, decisions should be made before the populations collapse. If AC is judged to be a viable option, managers will be faced with a broad array of issues. Some relate to basic ecology and biology, such as number of individuals, number of locations, timing and number of releases, demography and genetic diversity of the founding population, and habitat selection. On the other hand, attention to economic (e.g., initial introduction costs, funding for monitoring), administrative (e.g., appropriate policies and coherent management objectives), and social issues (e.g., acceptance, cultural and moral attitudes, private property rights) is also essential in guaranteeing the ultimate success of translocation efforts.

Assisted colonization can be viewed within a broader spectrum of conservation translocation, a fairly common management practice (Seddon 2010). IUCN guidelines (IUCN 1998) indicate that a species should be translocated outside its historical range only if there is no suitable habitat left within it. This provision allows for the establishment of populations (usually threatened species, sometimes extinct in the wild) in new areas where human threat is low.

Although translocations have been a rather common practice in species management, they have often proven unsuccessful (Armstrong and Seddon 2008). Most such efforts focus on reintroduction, defined as the "intentional movement of an organism into a part of its native range from which it has disappeared or become extirpated in historic times" (IUCN 1987), and restocking, defined as the "movement of individuals to build up an existing population" (IUCN 1987). "Pure" introductions of species, which are "movements of organisms outside their historically known native ranges" (IUCN 1987), are less common, generally due to the greater uncertainty about their potential effects on native biota and ecosystems.

Nonetheless, we can build upon advances in reintroduction biology and restoration ecology (Lipsey and Child 2007) to provide guidelines for successful assisted colonization projects. Armstrong and Seddon (2007) propose 10 key questions about populations, metapopulations, and ecosystems that need to be addressed before considering moving species (figure 17.2).

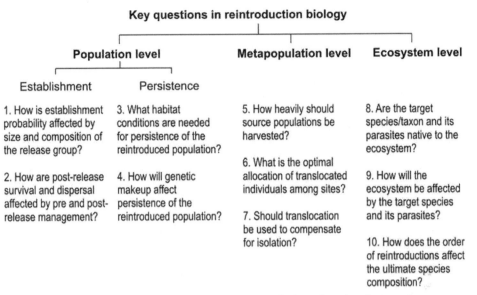

Key questions in reintroduction biology

Population level   **Metapopulation level**   **Ecosystem level**

Establishment   **Persistence**

1. How is establishment probability affected by size and composition of the release group?

2. How are post-release survival and dispersal affected by pre and post-release management?

3. What habitat conditions are needed for persistence of the reintroduced population?

4. How will genetic makeup affect persistence of the reintroduced population?

5. How heavily should source populations be harvested?

6. What is the optimal allocation of translocated individuals among sites?

7. Should translocation be used to compensate for isolation?

8. Are the target species/taxon and its parasites native to the ecosystem?

9. How will the ecosystem be affected by the target species and its parasites?

10. How does the order of reintroductions affect the ultimate species composition?

Figure 17.2. Key questions in reintroduction biology. Redrawn from Armstrong and Seddon 2007; used with permission from Elsevier.

Animal species targeted for translocation have ranged from invertebrates and amphibians to birds and large mammals. The translocation literature is vast, and a detailed discussion of translocation issues is beyond the scope of this chapter. We also refer the readers to the wealth of published data on plant translocations (see Falk et al. 1996), which offer further insight into translocation policy and technology. Here, we will focus on a few key ecological and socioeconomic problems that may apply to AC.

Ecological Perspectives

A primary concern is the number of animals to be introduced into the new range. Initial introductions are often followed by supplemental releases intended to augment the initial population because managers assume low survival of the initial population and decreasing genetic variation due to small population sizes. However, subsequent releases are costly, and most of them are performed without a good understanding of the actual population dynamics after translocation (Armstrong and Ewen 2001). Optimizing translocation strategies with limited financial resources is critical (Haight et al. 2000), so follow-up releases should be backed up by population viability analyses (PVA) based on data gathered after the release. Moreover, Griffith et al. (1989) and Wolf et al. (1996)

found no association between the probability of success and the number of releases (using data mostly from bird and mammal translocations), thus suggesting that this parameter is specific to species and to management goals. For example, Armstrong and Ewen (2001) found that in the case of New Zealand robins (*Petroica australis*), supplemental releases one year after the initial translocation had little benefit in increasing the likelihood of population persistence. On the contrary, those later releases had a negative effect on the overall population dynamics, and the costs involved (e.g., disease transmission, social stress, cost of labor) outweighed the benefits (reduction of Allee effect, increase in genetic variation). For the bearded vulture (*Gypaetus barbatus*), successive yearly releases for more than two decades (1986–2006) were necessary to establish a self-sustaining breeding population in the Alps (Schaub et al. 2009). A PVA indicated that although continuing releases would benefit the population, no releases were necessary after 2006 from a demographic standpoint, and thus the project could shift towards translocating vultures into other areas of Europe.

It is also critical to consider the impact of removal on the source population. The individuals selected for introduction can come from either captive breeding (as is often the case for endangered populations) or wild origins. Griffith et al. (1989) and Fischer and Lindenmayer (2000) found that success was more likely when wild populations were used as the source than when captive-bred individuals were used. If captive populations are the only source of animals, the goal should be to minimize the number of generations in captivity to avoid genetic adaptations to captive environments, which can be deleterious when populations are to be established in the wild (Frankham 2008). Also, caution needs to be used if the intent is combine wild and captive-bred populations. For the Atlantic salmon (*Salmo salar*) interbreeding between wild individuals and maladapted captive-bred individuals led to depressed recruitment and disrupted the wild population's capacity to adapt to warmer conditions; this relation is likely also to apply to other poikilothermic species (McGinnity et al. 2009). Disease in captive populations is a frequent issue due to higher transmission rates in high-density situations (Scott 1988), so performing AC using only captive-bred animals may be even more problematic than when a combination of captive-bred and wild animals are used.

The geographical location of the source population is a consideration because different populations have different local adaptations. Because of interest in climate warming much attention has focused on the high-latitude populations at the expanding edge of species ranges, but "rear-edge" stable populations (i.e., those located at lower latitudes) of a climate-driven species are also of interest because they are likely to have specific adaptations that allow them to cope with climate oscillations (Hampe and Petit 2005). Such local adaptations need to be investigated and perhaps exploited, particularly because some such populations might exhibit higher fitness (Kinnison et al. 2008). On the other hand, many potential candidates for assisted colonization are likely to have a small number of unstable populations, and dilemmas related to mixing individuals from different populations might not be relevant.

The demography of the species will affect the success of an AC project. The initial size of the founder population also needs to be considered, but this number will vary by taxon. For example, Germano and Bishop (2009) reported that translocations of amphibians involving more than 1,000 individuals were likely to be most successful, while for reptiles the outcome was independent of the number released. Fischer and Lindenmayer (2000) found greater probability of success when more than 100 individuals (mostly in birds and mammals) were released. Griffith et al. (1989) associated a higher probability of success with releasing 80 to 120 individuals in birds, and 20 to 40 individuals in large mammal translocations; releasing more individuals did not add any additional benefit. Generally, sufficient numbers of animals with an appropriate social structure need to be released because of potential Allee effects. For social carnivores, Miller et al. (1999) emphasize the importance of the age and sex ratios of the individuals released, which have to mimic natural populations.

The translocation of species does not necessarily entail the use of young, sexually mature individuals. Propagules, such as eggs or larvae, might be just as effective in establishing new populations (e.g., in invertebrates, fish, and amphibians). For aquatic-breeding amphibians, Semlitsch (2002) suggests that the release of 10,000 to 50,000 eggs over several consecutive years would establish an adult breeding population of 100 individuals. Generation time is an important issue to consider when translocating species. Species with short generation time are more likely to be successful and have lower translocation cost than species that reach sexual maturity late in life. For example, the successful reintroduction of the lacertid lizard *Psammodromus algirus* was attributed to short generation time, as well as to the absence of learned behaviors (Santos et al. 2009). However, as mentioned above, species with short generation times are more prone to genetic adaptation to captive environments, so minimizing time spent in captivity is important (Frankham 2008). If this is not possible (i.e., if a suitable location for translocation does not exist), an alternative could be storage of sperm and ova.

Behavior, especially parental care, is likely to play an important role in translocations. Species that require little or no parental care (most invertebrates, fish, amphibians, reptiles, and a few birds) have management requirements different from those of species with prolonged parental care and altricial offspring (e.g., most primates, carnivores, marsupials, and passerines). Other behavioral traits such as migration, dispersal, territoriality, and homing behavior are also important. In many cases, translocation failures have been the direct result of individuals leaving the introduction areas (Miller et al. 1999, Fischer and Lindenmayer 2000, Germano and Bishop 2009).

Finally, AC projects must account for genetic concerns. Some candidate species for assisted colonization are likely to be characterized by small population sizes, uncertain genetic makeup, and a small number of distinct populations. These populations are prone to inbreeding and genetic drift, leading to reduction in fitness and local extinction. Founding populations resulting from assisted colonization are also likely to be small, so the same general management rules apply, and a key goal is to increase population size fast. It might seem obvious that to prevent problems endemic to small populations, one would want to enhance the genetic variability of a founding population by introducing individuals from various sources. On the other hand, if different populations have evolved different traits and local adaptations (Hampe and Petit 2005), mixing them might lead to loss of useful divergence among ecotypes, and lower offspring fitness (e.g., through introgression) in the resulting population (Storfer 1999, Tallmon et al. 2004). Furthermore, moving species to sites already inhabited by other species closely related to them may potentially lead to hybridization, thus affecting the genetic makeup of both resident and introduced species. Moving species to different latitudes raises the issue of sudden changes in photoperiod and seasonal patterns that may be problematic. For example, it could result in breeding during the wrong period (Storfer 1999). What then is the ideal genetic makeup of a founding population? The answer is not simple, and it is likely to depend very much on landscape context and species.

Social and Economic Perspectives

Legal protection must follow any intentional introduction of species at risk from climate change (Chapron and Samelius 2008). Species likely to be subject to assisted colonization are probably already on protected lists in their place of origin, but perhaps not at their destination site. Therefore, if species are translocated across borders, legislative and political support must be coordinated. A good example of such legislation is the European Union's Birds and Habitats Directives, which form the core of

conservation policy in Europe. If conservation regulations were specifically to list species at risk from climate change, it would facilitate communication among managers responsible for translocations across borders.
Akçakaya et al. (2006) asserted that the current IUCN Red List criteria
(IUCN 2001) were not developed specifically for climate-change threatened species, and that shorter-lived species would be disadvantaged. The
rationale behind the assertion is that the population status assessments
conducted over 10 years or three generations (as recommended by IUCN)
would not likely reflect the potential effects of climate change, which are
not usually apparent on temporal scales of less than 50 years. Akçakaya
et al. (2006) propose that new information, such as predicted shifts in
geographic ranges, landscape context, life-history parameters, and interaction with other threats, be incorporated in the IUCN criteria for
assigning climate-threatened species to various Red List categories (see
Foden et al. 2008 for a thorough evaluation of the protection status of
bird, amphibian, and coral species threatened by climate change). Areas
selected for translocation also need to be placed under protection before
the actual movement of species occurs (Chapron and Samelius 2008). Legal and policy coordination must also occur across taxa; for example, the
US Endangered Species Act impedes the AC of many endangered animal
species, while the same regulations do not apply to endangered plants
(Shirey and Lamberti 2010).

Social acceptability is one of the critical issues in intentional introduction of climate-threatened species (Richardson et al. 2009), and it
draws upon people's beliefs and preferences pertaining to certain wildlife species (Decker and Purdy 1988). Because the social acceptability of
endangered species varies greatly between and within taxonomic groups
(Czech et al. 1998), deeming an AC candidate as "experimental" rather
than offering it full protection status might facilitate the social acceptance
of its transfer, especially for species and/or regions over which conflicts
between endangered species and private property rights have already occurred. Given these high social and political controversies, Minteer and
Collins (2010) suggest that a more comprehensive social-policy response
to climate change that includes a dynamic and pragmatic approach to
ecological ethics is needed if AC is to be considered a viable conservation
tool.

The key economic issue in translocations is that they can be extremely
costly. Haight et al. (2000) used robust optimization under scenarios involving both hard and soft releases and various founder population sizes,
both with and without monitoring, to evaluate how best to allocate funds
for translocations when future funding and population growth are uncer-

tain. Some of the findings are relevant to assisted colonization, and describe a tradeoff between size of founding population and risk of failure: the larger the population, the smaller the risk. For assisted colonization, which is likely to deal with small populations of endangered species, it is important to ensure appropriate protection and monitoring to enhance population growth. The survival of the released animals is extremely important, so release methods with a low risk of failure are desirable regardless of cost (Haight et al. 2000). Monitoring of the introduced individuals also poses a dilemma; as the proportion of a fixed budget spent for monitoring increases, the performance gain from monitoring decreases (Haight et al. 2000).

Summary

Overall, assisted colonization is likely to be a reasonable solution to a conservation dilemma under certain circumstances. As with many other conservation strategies, the focus of AC is to increase the resilience of populations faced with multiple threats. Climate change–threatened populations that cannot shift their geographic ranges naturally may benefit from AC helping them to overcome biogeographic and man-made barriers, and establishing new viable and resilient populations within their climatic envelope. AC efforts have already been undertaken, and further investigations of the feasibility of this conservation tool are certainly warranted. AC must rely on a careful consideration of costs and benefits, and it must involve a transparent decision-making process that synthesizes ecological information (on both the focal species and the recipient system) with societal values. However, given the complexities of performing AC, we suggest that only rarely will the benefits outweigh the limitations. Thus, taking proactive steps that maximize species' chances of adapting to climate change without being translocated is likely to be more effective.

Based on the costs and benefits of AC outlined in this chapter, future AC projects should address the following steps, not necessarily in chronological order:

1. *Identify the candidate species.* Detailed studies must assess whether a species is threatened mainly by climate change or by habitat loss and other human-induced pressures. This requirement is not necessarily a drawback, because the scientific evidence needed to accept or reject an AC proposal will certainly increase our understanding of poorly-known species. After a list of species clearly under threat from climate change is produced, the list of candidates for AC is likely to shrink after the deletion of species

that either might become invasive or are sufficiently mobile to benefit from restored habitat connectivity.

2. *Identify recipient sites*. The main problem hindering otherwise feasible AC efforts is the lack of suitable sites for introductions. Given the uncertainty surrounding the effects of candidate species on recipient systems, disturbed sites are preferable to pristine sites, and intracontinental translocations to intercontinental ones. Also, introductions within the long-term range boundaries of the species are preferred to relocations across extreme distances. The identification of potential sites should also take into account the sites' invasiveness history, the known relations between species richness and invasive species, and the dispersal potential of the candidate species. There are also uncertainties surrounding climate models and their ability to predict where and when to move species (locations likely to fit a species' climate requirements and the time period over which climate change will take place). An alternative approach would be to use bioclimatic envelope models and identify suitable locations outside but near current ranges to perform initial introductions, followed by monitoring of the introduced population's growth and expansion. However, given the complexity of species' potential responses to climate warming, caution must be used because bioclimatic envelope models tend to oversimplify the species' climatic requirements, fail to account for potential interactions with other threats, and may have little or no predictive power (Brodie et al. this volume).

3. *Assess technical and economic feasibility*. Given the overall poor success of animal translocations, any future effort must be based on a thorough understanding of the species' life-history traits and habitat requirements. The translocation literature has identified many technical and funding attributes (i.e., monitoring costs, starting population size, subsequent releases, frequency of monitoring, etc.) and taxa-specific features for successfully moving species.

4. *Assess political support and social acceptance*. Political and social issues might make some AC efforts impossible even when they seem technically feasible (Minteer and Collins 2010). Carrying out AC ultimately depends on society's perception of the instrumental and intrinsic value of candidate species, and efforts that do not account for the dynamic nature of the public's perception are likely to fail (Sandler 2010). Policy support for AC would involve both preparing for a species' arrival at a recipient site by giving the

species protected status (although this is a debatable issue) and by
providing funding.

ACKNOWLEDGMENTS
We thank Jedediah F. Brodie, Eric S. Post, Daniel F. Doak, and an anonymous reviewer
for helpful comments on earlier versions of this manuscript.

LITERATURE CITED
Akçakaya, H. R., S. H. M. Butchart, G. M. Mace, S. N. Stuart, and C. Hilton-Taylor.
2006. Use and misuse of the IUCN Red List Criteria in projecting climate change
impacts on biodiversity. *Global Change Biology* 12:2037–43.
Araújo, M. B., W. Thuiller, and R. G. Pearson. 2006. Climate warming and the decline
of amphibians and reptiles in Europe. *Journal of Biogeography* 33:1712–28.
Armstrong, D. P. and J. G. Ewen. 2001. Assessing the value of follow-up
translocations: a case study using New Zealand robins. *Biological Conservation*
101:239–47.
Armstrong, D. P. and P. J. Seddon. 2008. Directions in reintroduction biology. *Trends
in Ecology and Evolution* 23:20–25.
Berg, M. P., E. T. Kiers, G. Driessen, M. van der Heijden, B. W. Kooi, F. Kuenen,
M.Liefting, H. A. Verhoef, and J. Ellers. 2010. Adapt or disperse: Understanding
species persistence in a changing world. *Global Change Biology* 16:587–98.
Carroll, M. J., B. J. Anderson, T. M. Brereton, S. J. Knight, O. Kudrna, and C. D.
Thomas. 2009. Climate change and translocations: The potential to reestablish
two regionally-extinct butterfly species in Britain. *Biological Conservation* 142:
2114–21.
Chapron, G. and G. Samelius. 2008. Where species go, legal protections must follow.
Science 322:1049b–50.
Cleland, E. E., M. D. Smith, S. J. Andelman, C. Bowles, K. M. Carney, M. C. Horner-
Devine, J. M. Drake, S. M. Emery, J. M. Gramling, and D. B. Vandermast. 2004.
Invasion in space and time: Non-native species richness and relative abundance
respond to interannual variation in productivity and diversity. *Ecology Letters*
7:947–57.
Cunningham, A. A. 1996. Disease risks of wildlife translocations. *Conservation
Biology* 10:349–53.
Czech, B., P. R. Krausman, and R. Borkhataria. 1998. Social construction, political
power, and the allocation of benefits to endangered species. *Conservation Biology*
12:1103–12.
Daszak, P., A. A. Cunningham, and A. D. Hyatt. 2000. Emerging infectious diseases of
wildlife: Threats to biodiversity and human health. *Science* 287:443–49.
Davidson, I. and C. Simkanin. 2008. Skeptical of assisted colonization. *Science* 322:
1048b–49.
Dawson, T. P., S. T. Jackson, J. I. House, I. C. Prentice, and G. M. Mace. 2011.
Beyond predictions: Biodiversity conservation in a changing climate. *Science*
332:53–58.
Decker, D. J., and K. G. Purdy. 1988. Toward a concept of wildlife acceptance capacity
in wildlife management. *Wildlife Society Bulletin* 16:53–57.
Dodd, C. K., Jr., and R. A. Seigel. 1991. Relocation, repatriation, and translocation

of amphibians and reptiles: Are they conservation strategies that work? *Herpetologica* 47:336–50.

Dunstan, P. and C. Johnson. 2004. Invasion rates increase with species richness in a marine epibenthic community by two mechanisms. *Oecologia* 138:285–92.

Falk, D. A., C. I. Millar, and M. Olwell. 1996. *Restoring Diversity: Strategies for Reintroduction of Endangered Plants*. Washington: Island Press.

Fargione, J. E., and D. Tilman. 2005. Diversity decreases invasion via both sampling and complementarity effects. *Ecology Letters* 8:604–11.

Fischer, J., and D. B. Lindenmayer. 2000. An assessment of the published results of animal relocations. *Biological Conservation* 96:1–11.

Foden, W., G. Mace, J.-C. Vié, A. Angulo, S. Butchart, L. DeVantier, H. Dublin, A. Gutsche, S. Stuart, and E. Turak. 2008. Species susceptibility to climate change impacts. In *The 2008 Review of The IUCN Red List of Threatened Species*, edited by J.-C. Vié, C. Hilton-Taylor, and S. N. Stuart. Gland, Switzerland: IUCN.

Frankham, R. 2008. Genetic adaptation to captivity in species conservation programs. *Molecular Ecology* 17:325–33.

Germano, J. M. and P. J. Bishop. 2009. Suitability of amphibians and reptiles for translocation. *Conservation Biology* 23:7–15.

Griffith, B., J. M. Scott, J. W. Carpenter, and C. Reed. 1989. Translocation as a species conservation tool: Status and strategy. *Science* 245:477–80.

Haight, R. G., K. Ralls, and A. M. Starfield. 2000. Designing species translocation strategies when population growth and future funding are uncertain. *Conservation Biology* 14:1298–1307.

Hampe, A. and R. J. Petit. 2005. Conserving biodiversity under climate change: The rear edge matters. *Ecology Letters* 8:461–67.

Hoegh-Guldberg, O., L. Hughes, S. McIntyre, D. B. Lindenmayer, C. Parmesan, H. P. Possingham, and C. D. Thomas. 2008. Assisted colonization and rapid climate change. *Science* 321:345–46.

Huang, D. 2008. Assisted colonization won't help rare species. *Science* 322: 1049a.

Hunter, M. L. 2007. Climate change and moving species: Furthering the debate on assisted colonization. *Conservation Biology* 21:1356–58.

IUCN. 1987. *IUCN Position Statement on the Translocation of Living Organisms: Introductions, Re-introductions, and Re-stocking*. Prepared by the Species Survival Commission in collaboration with the Commission on Ecology and the Commission on Environmental Policy, Law and Administration. Accessed at www.iucnsscrsg.org.

———. 1998. *Guidelines for Re-introductions*. Prepared by the IUCN/SSC Reintroduction Specialist Group. Accessed at www. iucnsscrsg.org.

———. 2001. *IUCN Red List Categories and Criteria: Version 3.1*. Gland, Switzerland: IUCN.

Kinnison, M. T., M. J. Unwin, and T. P. Quinn. 2008. Eco-evolutionary vs. habitat contributions to invasion in salmon: Experimental evaluation in the wild. *Molecular Ecology* 17:405–14.

Kreyling, J., T. Bittner, A. Jaeschke, A. Jentsch, M.J. Steinbauer, D. Thiel, and C. Beierkuhnlein. 2011. Assisted colonization: A question of focal units and recipient localities. *Restoration Ecology* 19:433–40.

Kutner, L. S., and L. E. Morse. 1996. Reintroduction in a changing climate. In

Restoring Diversity: Strategies for Reintroduction of Endangered Plants, edited by
D. A. Falk, C. I. Millar, and M. Olwell, 23–48. Washington: Island Press.

Lawler, J. J., S. L. Shafer, D. White, P. Kareiva, E. P. Maurer, A. R. Blaustein, and
P. J. Bartlein. 2009. Projected climate-induced faunal change in the Western
Hemisphere. *Ecology* 90:588–97.

Levine, J. M. 2000. Species diversity and biological invasions: Relating local process
to community pattern. *Science* 288:852–54.

Levine, J. M. and C. M. D'Antonio. 1999. Elton revisited: A review of evidence linking
diversity and invasibility. *Oikos* 87:15–26.

Lipsey, M. K. and M. F. Child. 2007. Combining the fields of reintroduction biology
and restoration ecology. *Conservation Biology* 21:1387–90.

Loss, S. R., L. A. Terwilliger, and A. C. Peterson. 2011. Assisted colonization:
Integrating conservation strategies in the face of climate change. *Biological
Conservation* 144:92–100.

Martinez Pastur, G., M. V. Lencinas, J. Escobar, P. Quiroga, L. Malmierca, and
M. Lizarralde. 2006. Understorey succession in Nothofagus forests in Tierra
del Fuego (Argentina) affected by *Castor canadensis*. *Applied Vegetation Science*
9:143–54.

McGinnity, P., E. Jennings, E. deEyto, N. Allott, P. Samuelsson, G. Rogan, K. Whelan,
and T. Cross. 2009. Impact of naturally spawning captive-bred Atlantic salmon on
wild populations: depressed recruitment and increased risk of climate-mediated
extinction. *Proceedings of the Royal Society B: Biological Sciences* 276:3601–10.

McLachlan, J. S., J. J. Hellmann, and M. W. Schwartz. 2007. A framework for
debate of assisted migration in an era of climate change. *Conservation Biology*
21:297–302.

Miller, B., K. Ralls, R. P. Reading, J. M. Scott, and J. Estes. 1999. Biological and
technical considerations of carnivore translocation: A review. *Animal Conservation*
2:59–68.

Minteer, B. A., and J. P. Collins. 2010. Move it or lose it? The ecological ethics of
relocating species under climate change. *Ecological Applications* 20:1801–4.

Mueller, J. M., and J. J. Hellmann. 2008. An assessment of invasion risk from assisted
migration. *Conservation Biology* 22:562–67.

Olden, J. D., M. J. Kennard, J. J. Lawler, and N. L. Poff. 2010. Challenges and
opportunities in implementing managed relocation for conservation of freshwater
species. *Conservation Biology* 25:40–47.

Opdam, P. and D. Wascher. 2004. Climate change meets habitat fragmentation:
Linking landscape and biogeographical scale levels in research and conservation.
Biological Conservation 117:285–97.

Parmesan, C. 2006. Ecological and evolutionary responses to recent climate change.
Annual Review of Ecology, Evolution, and Systematics 37:637–69.

Parmesan, C. and G. Yohe. 2003. A globally coherent fingerprint of climate change
impacts across natural systems. *Nature* 421:37–42.

Raup, D.M. 1992. *Extinction: Bad Genes or Bad Luck?* W.W. Norton, New York.

Ricciardi, A., and D. Simberloff. 2009. Assisted colonization is not a viable
conservation strategy. *Trends in Ecology and Evolution* 24:248–53.

Richardson, D. M., J. J. Hellmann, J. S. McLachlan, D. F. Sax, M. W. Schwartz,
P. Gonzalez, E. J. Brennan, A. Camacho, T. L. Root, O. E. Sala, S. H. Schneider,
D. M. Ashe, J. R. Clark, R. Early, J. R. Etterson, E. D. Fielder, J. L. Gill, B. A.

Minteer, S. Polasky, H. D. Safford, A. R. Thompson, and M. Vellend. 2009. Multidimensional evaluation of managed relocation. *Proceedings of the National Academy of Sciences* 106:9721–24.

Sandler, R. 2010. The value of species and the ethical foundations of assisted colonization. *Conservation Biology* 24:424–431.

Santos, T., J. Pérez-Tris, R. Carbonell, J. L. Tellería, and J. A. Díaz. 2009. Monitoring the performance of wild-born and introduced lizards in a fragmented landscape: Implications for ex situ conservation programmes. *Biological Conservation* 142: 2923–30.

Sax, D.F., K.F. Smith, and A.R. Thompson. 2009. Managed relocation: a nuanced evaluation is needed. *Trends in Ecology and Evolution* 24:472–73.

Schaub, M., R. Zink, H. Beissmann, F. Sarrazin, and R. Arlettaz. 2009. When to end releases in reintroduction programmes: Demographic rates and population viability analysis of bearded vultures in the Alps. *Journal of Applied Ecology* 46:92–100.

Scott, M. E. 1988. The impact of infection and disease on animal populations: Implications for conservation biology. *Conservation Biology* 2:40–56.

Seddon, P. J. 2010. From reintroduction to assisted colonization: Moving along the conservation translocation spectrum. *Restoration Ecology* 18:796–802.

Semlitsch, R. D. 2002. Critical elements for biologically based recovery plans of aquatic-breeding amphibians. *Conservation Biology* 16:619–29.

Shirey, P. D., and G. A. Lamberti. 2010. Assisted colonization under the U.S. Endangered Species Act. *Conservation Letters* 3:45–52.

Skelly, D. K., L. N. Joseph, H. P. Possingham, L. K. Freidenburg, T. J. Farrugia, M. T. Kinnison, and A. P. Hendry. 2007. Evolutionary responses to climate change. *Conservation Biology* 21:1353–55.

Southgate, C., C. Hunt, and D. G. Horrell. 2008. Ascesis and assisted migration: Responses to the effects of climate change on animal species. *European Journal of Science and Theology* 4:99–111.

Stohlgren, T. J., G. W. Chong, L. D. Schell, K. A. Rimar, Y. Otsuki, M. Lee, M. A. Kalkhan, and C. A. Villa. 2002. Assessing vulnerability to invasion by nonnative plant species at multiple spatial scales. *Environmental Management* 29:566–77.

Storfer, A. 1999. Gene flow and endangered species translocations: A topic revisited. *Biological Conservation* 87:173–80.

Tallmon, D. A., G. Luikart, and R. S. Waples. 2004. The alluring simplicity and complex reality of genetic rescue. *Trends in Ecology and Evolution* 19:489–96.

Thomas, C. D. 2011. Translocation of species, climate change, and the end of trying to recreate past ecological communities. *Trends in Ecology and Evolution* 26:216–21.

Torchin, M. E., K. D. Lafferty, A. P. Dobson, V. J. McKenzie, and A. M. Kuris. 2003. Introduced species and their missing parasites. *Nature* 421:628–30.

Traill, L. W., M. L. M. Lim, N. S. Sodhi, and C. J. A. Bradshaw. 2010. Mechanisms driving change: Altered species interactions and ecosystem function through global warming. *Journal of Animal Ecology* 79:937–47.

Travis, J. M. J. 2003. Climate change and habitat destruction: A deadly anthropogenic cocktail. *Proceedings of the Royal Society B: Biological Sciences* 270:467–73.

Vall-llosera, M., and D. Sol. 2009. A global risk assessment for the success of bird introductions. *Journal of Applied Ecology* 46:787–95.

Van Grunsven, R. H. A., W. H. van der Putten, T. M. Bezemer, F. Berendse, and E. M. Veenendaal. 2010. Plant-soil interactions in the expansion and native range of a poleward shifting plant species. *Global Change Biology* 16:380–85.

Walther, G.-R., E. Post, P. Convey, A. Menzel, C. Parmesan, T. J. C. Beebee, J.-M. Fromentin, O. Hoegh-Guldberg, and F. Bairlein. 2002. Ecological responses to recent climate change. *Nature* 416:389–95.

Warren, M. S., J. K. Hill, J. A. Thomas, J. Asher, R. Fox, B. Huntley, D. B. Roy, M. G. Telfer, S. Jeffcoate, P. Harding, G. Jeffcoate, S. G. Willis, J. N. Greatorex-Davies, D. Moss, and C. D. Thomas. 2001. Rapid responses of British butterflies to opposing forces of climate and habitat change. *Nature* 414:65–69.

Willis, S. G., J. K. Hill, C. D. Thomas, D. B. Roy, R. Fox, D. S. Blakeley, and B. Huntley. 2009. Assisted colonization in a changing climate: A test-study using two U.K. butterflies. *Conservation Letters* 2:46–52.

Wolf, C. M., B. Griffith, C. Reed, and S. A. Temple. 1996. Avian and mammalian translocations: Update and reanalysis of 1987 survey data. *Conservation Biology* 10:1142–54.

18

The Integration of Forest Science and Climate Change Policy to Safeguard Biodiversity in a Changing Climate

Nicholas Blay and Michael Dombeck

Climate change is arguably the most significant global challenge of modern times, threatening to have a major influence on all aspects of human society and natural systems. Addressing this issue requires urgent formulation of comprehensive and effective policy responses both nationally and internationally. The anthropogenic influences on earth's climate system threaten to have severe impact on the environment when coupled with more immediate and tangible environmental problems such as air and water pollution, overconsumption of resources associated with population growth, loss of biodiversity, and deforestation and land degradation. So far, efforts to combat climate change through national policies and international agreements have been largely ineffective and have focused primarily on the industrial and energy sectors (Streck et. al 2008). Over the past few years, the potential of the agriculture, forestry, and other land use sectors (AFOLU) to reduce emissions and mitigate the effects of climate change has been realized and has come to the fore-front in domestic and international climate negotiations. In fact, land use changes, particularly deforestation, account for approximately 20% of all global carbon emissions, and are the primary source of greenhouse gas (GHG) emissions in many developing countries such as Brazil and Indonesia (Houghton 2005). The prospect of the AFOLU sector, particularly the forestry sector, to influence climate change policy promises to be instrumental in more ways than one.

The goals of this chapter are (1) to provide an overview of the importance of forest systems for biodiversity, the role they play in the earth's biochemical cycles, and their influence on human society; (2) to examine some of the threats climate change poses to forest systems and assess some of the effects that are already occurring; and (3) to suggest how forest ecosystems could be incorporated into a climate policy focused around carbon sequestration and emission reductions at the national level. Ultimately, incorporating forest systems into both national and international policies to combat climate change will not only help minimize the effects of climate change, but could also have tremendous benefits for human society and biodiversity worldwide. Thus, comprehensive policies that address the needs of forest ecosystems and biodiversity can also be seen as part of a global climate change solution.

The Importance of Forest Ecosystems

The world's forests are predominantly divided into three subgroups based on region: (1.) boreal forests that extend across Canada, Scandinavia, and Russia; (2.) tropical rainforests, including forested areas in Central and South America, Southeast Asia, Australasia, Madagascar, and equatorial Africa; and (3.) temperate forests, including those found in the United States, northeast Asia, and west central Europe. Forests systems account for approximately 30% of global land area and contain some of the greatest concentrations and assemblages of species of any terrestrial ecosystem, harboring up to 80% of all terrestrial biodiversity (World Bank 2010).

Forest systems not only are storehouses for the majority of earth's terrestrial biodiversity, but also serve important functions in human society. Forests provide ecosystem services such as water storage, flood control, erosion control, improvement of water quality, renewable energy sources, and timber products. For example, US forests supply water to approximately 124 million Americans, or nearly 40% of the nation's population (USFS 2007). Aside from these important services, forests offer ample opportunity for enjoyment and recreation such as hiking, camping, fishing, and wildlife viewing, and are important for their aesthetic and spiritual value as well.

Forests play an important role in the planet's biochemical cycles, notably in the vast amount of gas exchange with the atmosphere; they are often described as the lungs of the planet. They are one of the earth's largest terrestrial stores of carbon, and they play a major role in regulating the planet's climate. An estimated 638 gigatons (Gt) of carbon, 283 Gt in forest biomass alone, is stored within forest systems (including soils, leaf litter, and deadwood). This is equivalent to approximately 50% more carbon than is in the atmosphere (Streck et al. 2008). Since climate plays such a critical role within forest systems, an altered climate poses a major threat to the sustainability of these systems. The distribution and unique assemblages of plants and animals that make up a forest ecosystem are dictated in large part by their climatic tolerances. Any change in atmospheric components will affect local climates predominantly through the alteration of precipitation patterns, temperatures, and soil conditions; thus, changes in the critical elements that define an ecosystem may affect the number and types of species found within that system (Streck et al.

2008). In addition, climatic changes may weaken or alter slow-adapting systems, making them more susceptible, in the case of forests, to more immediate stressors such as insect outbreaks, fires, disease (cf. Paull and Johnson this volume), fragmentation (cf. Cross et al. this volume), and invasive species (cf. Fordham et al. this volume), all of which threaten to destroy the integrity of forest ecosystems (Intergovernmental Panel on Climate Change 2007).

US Policy Recommendations for Protecting Biodiversity in a Changing Climate

The United States is poised to be a world leader in the development and implementation of climate policies, especially within the forestry sector, for two primary reasons. First, forests cover approximately one-third of the United States, or approximately 304 million ha. This makes the United States the fourth largest forest estate in the world, exceeded only by Russia, Brazil, and Canada. The US Forest Service (USFS) manages approximately 77 million ha, or about 25% of the nation's forests. Of these, 60 million ha are designated as national forests, which include tropical, boreal, and temperate forest systems (USFS 2001). The USFS employs approximately 550 scientists in 67 locations, making the agency one of the world's largest forest research entities, if not the largest (USFS 2007). Forests managed by the USFS are unique in that they are publicly owned and highly contiguous, with mandated management plans. These characteristics make US national forests ideal prototypes for the development and implementation of forest management policies, with the potential to be modeled around the world for alleviating the effects of climate change.

Of the remaining 75% of forested land, approximately 50% is privately owned, leaving another 25% to be managed by federal, state, county, or municipal agencies (USFS 2001). Approximately 80% of forested land in the east is privately owned, as opposed to approximately 30% of forests in the west (figure 18.1). This large expanse of forested land will be important in the implementation of any US climate change policy, the prospective details of which are outlined below.

To effectively address the threat of climate change to the earth's biodiversity, particularly that associated with forest systems, three critical goals must be achieved:

1. The effects of climate change on natural systems, particularly forests, must be understood and prioritized when developing management policies.

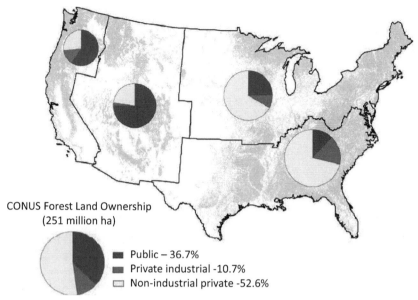

CONUS Forest Land Ownership
(251 million ha)

- Public – 36.7%
- Private industrial -10.7%
- Non-industrial private -52.6%

Figure 18.1. US forest ownership by region. US Forest Service 2007.

2. A national climate policy must be created in the United States.
3. The potential of forests to assist in climate mitigation must be realized and incorporated into any national and international climate change policy.

Forest Management Policy

The effects of climate change on forest systems have become better understood over the past several decades. Since forests play a key role in the global carbon cycle, any effective climate policy adopted by the United States must have forest management as a key component. To do this, the nation's largest forest management agency must have the tools, resources, management guidelines, and framework in place to do so.

Unfortunately, the USFS and other federal agencies have not addressed climate change in a timely manner, largely due to lack of concern by the executive and legislative branches of government over the past three decades, as well as a lack of unified public pressure. Also, national public policy shifts are typically slow, lagging behind need by several years (Dombeck 2009). In fact, policy shifts related to forest management have often come about as the result of social pressure due to a fear or realization of an imminent threat to human well-being, or as the direct consequence of a catastrophic event or natural disaster. For instance, fear of a timber famine in the late 19th century due to destruction of forests ushered in professional forest management, while the northern spotted owl (*Strix occidentalis caurina*) controversy of the 1990s shifted the focus of management to a more inclusive ecosystem approach (Dombeck et al. 2003).

In 2008, the USFS identified climate change as one of the three greatest conservation challenges of present-day forest management, and made it a priority issue. However, the only legislatively mandated course of action pertaining to climate change within the USFS is the 1990 Food Protection Act, which amended the 1974 Resources Planning Act (RPA). The Food Protection Act requires the USFS to assess the effects of climate change on renewable resources in forests and grasslands, as well as to identify any opportunities to mitigate carbon dioxide buildup along rural/urban gradients through forest management. These assessments have yielded an analysis of issues resulting directly from a changing climate and, perhaps more important, have evaluated both the ecological and socioeconomic responses of these systems (Joyce 2001). The Obama administration has made climate change a national priority. In 2009 both the US secretary of agriculture, Tom Vilsack, and the Forest Service chief, Tom Tidwell, made it clear that under the threat of ecosystem alterations due to climate change, restoring the resiliency of US forests is a first-priority issue to be addressed in coming years.

The idea of restoring resilience to ecosystems is cited as the most commonly suggested adaptive option within the context of climate change (Joyce et. al 2008, Millar et. al 2007). Ecosystems are regularly impacted by a number of biotic and abiotic influences that do not significantly alter the overall state of the ecosystem, but occasionally a disturbance is so great that an ecosystem's integrity is challenged and can be changed from one state to another. Thus, the resilience of an ecosystem is the magnitude of disturbance that system can sustain before it is fundamentally altered to a different state. In the context of climate change mitigation, the fundamental idea is to promote the resilience of an ecosystem so that it can withstand any of the uncertain impacts associated with a changing climate. The most common methods of achieving increased resilience include surplus seed-banking (Millar et. al 2007), intensive management during early years of species assemblage establishment (Millar et. al 2007, Spittlehouse and Stewart 2003), maintaining connectedness between large tracts of natural area (Joyce et. al 2008, Cross et al. this volume), removal of invasive species (Millar et. al 2007), and preserving and monitoring keystone species and functional groups (Noss 2001).

Promoting resilience has many drawbacks, especially considering the uncertain effects of climate change. First, resilience does not resist change, but instead absorbs the impact of a disturbance and then returns the system to a state similar to what existed before the disturbance. Climate change has the potential to prevent the return to a previous state. Further, managing for resilience may require intensive management and the use of funds, time, and energy for a short-term solution to a long-term problem. The idea of managing systems for increased resilience is probably best suited for short-term projects, endangered species management, and areas of low climate variability. Therefore, while land managers, policy makers, and scientists try to hash out the best response to climate change issues, management for resilience may be one of the best solutions available for the short term, though it should also be incorporated into any long-term solution. The recommendations we present below draw on short- and long-term solutions to the climate problem. In the absence of agency-wide science-based policy, the implementation of climate change strategies for the 155 national forests will certainly be highly variable. If the United States is to become a world leader in this field, an agency-wide and/or national climate policy focused on reducing greenhouse gas emissions must be developed. Now that the USFS has made climate change a priority issue, it must assess how forest management can be made more responsive to the impact on forest systems. First and foremost, the USFS should create climate change task forces at the national level as well as within each region or eco-region of the Forest Service. These task forces would be responsible for developing regional strategies that coincide with a national strategy to address and minimize the potential impact of climate change on forest systems. This will require extensive cooperation and understanding among policy makers, climate change scientists, and forest managers—as well as among members of various partner agencies, including the US Fish and Wildlife Service, the US Geological Survey, state agencies, nongovernmental organizations, and academic institutions—regarding impacts, threats, management options, and research needs. Only with this collaboration will the task forces be affective.

As suggested in the Climate Change Science Program document, "Adaptation Options for Climate-Sensitive Ecosystems and Resources," one of the first actions of the task force should be to carry out a science-based audit of all forest management plans (Joyce et. al 2008). This audit would be designed to achieve the following goals:

1. Gather information about the current level of preparedness and adaptive capacity of forests.

2. Identify areas in need of greater overall attention.
3. Evaluate current forest management plans and identify areas for potential improvement that relate directly to climate change issues.

An audit would have many practical implications. It should focus on current management direction and on-the-ground practices, including identifying and supporting high-priority administrative procedures. The task forces would identify specific areas where changes are needed within forest plans or project plans, and would also aid in helping managers decide the best course of action.

Finally, information about climate change effects and the tools with which to address them should be disseminated across all levels of management. A handbook or manual providing general information and management options should be provided to forest managers. Gaps in the understanding of risks, effects, and potential managerial responses to climate change are vast and must be filled as quickly as possible. Managers need to be able to understand the threats their forests face and how to address them. Policy makers need to be able to set realistic and workable policies, and scientists need to keep all groups up to date on the latest findings and research. Only after the threats and potential effects are understood can forest managers begin to assess whether changes are needed in the system.

Forest Climate Policy

Forests across the United States play an important role in the sequestration of carbon dioxide. It is estimated that they offset approximately 10% of America's carbon emissions, or approximately 780 million metric tons of carbon dioxide equivalent, each year and have the potential to sequester much more through reforestation of degraded lands and better management of lands already in forest (Kelly et. al 2008). The inclusion of carbon offsets, such as those potentially provided by forest systems in a national policy to reduce greenhouse gas emissions, could be a flexible, cost-effective way to contribute to emission reduction targets while providing numerous benefits to humans and wildlife.

To date, the United States has avoided legally binding agreements for reducing greenhouse gas emissions, opting instead for voluntary measures. The only national or international climate agreement ratified by the United States is the United Nations Framework Convention on Climate Change (UNFCCC). The Kyoto Protocol, which came directly from the need to establish legally binding agreements to reduce greenhouse gas emissions under the UNFCCC, focuses primarily on industry and energy-related emission reductions. Provisions for reducing emissions from agriculture, forestry, and other land use sources are included in the protocol, falling under the Clean Development Mechanism and Joint Implementation. But ultimately, these provisions are extremely limited in scope and reward only reforestation and afforestation projects in developing countries, where the vast majority of emissions associated with land use activities occur. The provisions also exclude efforts to reduce deforestation and forest degradation, which account for approximately 20% of global greenhouse gas emissions.

Over the past several years, the United States has made efforts to pass a domestic climate change policy, though many of those efforts have fallen short. For instance, the Leberman-Warner America's Climate Security Act of 2007 aimed to establish a cap-and-trade scheme to reduce greenhouse gas emissions. The legislation was approved in the Senate Environment and Public Works Committee, but was voted down on the Senate floor. Under this legislation, entities required to reduce greenhouse gas emissions could satisfy up to 15% of their obligations through domestic AFOLU offsets. Additionally, 5% of the funds collected through a cap-and-trade scheme would be made available on an annual basis to support AFOLU measures to reduce emissions, with another 2.5% of those funds being made available internationally for projects that address emissions from deforestation and forest degradation (Kelly et al. 2008).

A more recent piece of legislation, the Waxman-Markey American Clean Energy and Security Act (HR 2998), was approved by the House of Representatives in June 2009, though its Senate counterpart, the Kerry-Boxer Clean Energy Jobs and American Power Act (S 1733), died in the Senate in July 2010. The Waxman-Markey climate bill was centered around a cap-and-trade system with emission reduction targets of 17% below 2005 levels by 2020 and 83% below 2005 levels by 2050. Under the bill, offsets, such as those provided through the forestry sector, would have accounted for two billion tons of total emission reductions, half of which would have been mandated to be of domestic origin, and the rest could have come from international projects. While the Waxman-Markey climate bill was an all-inclusive clean energy and climate bill, the Kerry-Boxer bill focused primarily on reducing greenhouse gases.

Should either HR 2998 or S 1779 have become United States policy, both fell far short of what scientists recommend to avoid the worst effects of a changing climate. In light of new scientific findings, many climate scientists around the world now support the reduction of global atmospheric CO_2 levels to 350 ppm to avoid the most serious effects of climate change (e.g., Hansen 2008, Yale Environment 2009). This is down from the 450 ppm benchmark that scientists previously thought would be sufficient to avoid catastrophic consequences. To achieve the 450 ppm goal, the IPCC estimates that Annex 1 countries including the United States need to reduce greenhouse gas emissions by 25% to 40% below 1990 levels by 2020. Of the two bills outlined above, the one with the more strict emission reduction targets—the Kerry-Boxer bill, with a target of 20% emission reductions by 2020—will only reduce US emissions 7% below 1990 levels. Considering that the United States accounts for approximately one-fifth of all greenhouse gas emissions, it is apparent that the legislation that moved through the House and Senate was not enough to avoid the most severe effects of climate change by even the most conservative estimates. To get on the right path, the United States must create a domestic climate policy that meets the requirements mandated by the best available science, and must engage with the world community while doing so. Since emissions from deforestation and other land use activities account for approximately one-fifth of all CO_2 emissions worldwide, it is evident that any successful climate policy will tap into the potential of this sector to mitigate GHG emissions.

A Domestic Forest Carbon Offset Program

Reducing emissions through reforestation and afforestation projects, and even stopping deforestation and forest degradation, will not get at the true driver of the climate change problem: the combustion of fossil fuels. But if used carefully and responsibly, the AFOLU sector can help developed nations meet emissions targets or produce stricter ones, and it can also help developing countries meet their own climate targets and cope with climate effects while reducing the amount of CO_2 in the atmosphere. Reducing CO_2 emissions from the forest sector will also provide numerous other benefits such as clean air and water, protection of the largest terrestrial storehouse of biodiversity, and security for numerous cultures that rely on forest products. In this way we will protect the largest sinks of terrestrial carbon, as well as ecosystems that substantially influence local and regional climate patterns. Therefore, any US national and/or international climate policy must include incentives for protecting forests. The following is a recommended framework on how best to incorporate and ensure that the forest sector is used to the fullest within any climate policy at the national level.

A major component of any climate policy must be offsets provided through the forestry sector, which notably coincide with the provisions outlined within the climate bills described above. While the incorporation of forest offsets into a national climate policy may seem straightforward, it is actually highly complex. A number of issues must be considered and addressed for such a crediting scheme to work nationally and internationally. Because of the uncertainties that remain at the international level with regard to the logistics of a carbon offset program, we focus primarily on an offset scheme centered around the US national forest system, and provide some insight into how it could be incorporated into an all-inclusive national climate change policy.

First, a financing scheme to support a Domestic Forest Carbon Offset Program (DFCOP) would be established. This could be done in a number of ways: (1) the government could directly subsidize the entire program; (2) the program could be made to subsidize itself through a cap-and-trade carbon scheme, in which companies would directly fund the offset program and associated projects; or (3) in a hybrid approach, where the government could auction emission credits under a cap-and-trade scheme and use all or part of those revenues to pay for the program.

Regardless of the funding system used, putting a price on carbon emissions would allow for standing forests to compete economically with cleared forests. Those who choose to protect forests would be compensated either by the entity required to make emission reductions or by the government. Entities required to make emission cuts under a national climate policy could then supplement their emission reductions through such offsets in a way that is beneficial to biodiversity as well as to people who enjoy forests for the goods they provide and the uses they serve. Because of the enormous private ownership of existing US forests and land that has the potential to be afforested or reforested, the public also has the ability to participate in a national climate policy and a very lucrative carbon offset program that would benefit everyone from corporations to the individual landowner.

In the United States, the bulk of the offsets from the forestry sector would likely come from afforestation and reforestation projects as well as from forest conservation projects and improved forest management. As one of the largest forestry research institutions in the world, the USFS must play a key role in determining how forests under a DFCOP should be managed to maximize their carbon sequestering potential, as well as in providing expert guidance to people and organizations wishing to create offset projects of their own. Aside from managing 77 million ha of federal lands, the USFS assists states and other organizations in managing an additional 202 million ha of rural and urban forests. These two figures together add up to nearly 90% of all forested land in the country. In addition, the USFS research and development branch uses 83 experimental forests which have provided nearly 100 years of forest data. Finally, the USFS manages forests under a multiple-use approach that requires the agency to sustain healthy ecosystems while addressing the need for forest goods and services (USFS 2008). To meet the specifications required under a DFCOP, the USFS must continue capacity building related to managing forest systems in a changing climate, and must disseminate this information to other natural resource agencies and the public.

A system of monitoring, verifying, and reporting emission reductions needs to be developed at the national level. These actions would most likely be led by the Environmental Protection Agency in partnership with the USFS research and development branch. The details of some logistical items, such as the amount of carbon sequestered by a specific forest type, the region of the country where an offset project is located, and the species of trees used for sequestration, would need to be calculated, as these components are variable and sequestration rates vary from one location to another, differing according to tree species, age, and management protocols.

Another provision must be established to ensure an offset's permanence. Emissions are permanently reduced when an entity is required to change technology or when energy efficiency projects are undertaken, but planting a forest carries with it the risk that sequestered carbon could be emitted back into the atmosphere whenever the forest is destroyed. Permanent land easements on DFCOP projects would likely need to be established and run in the time frame of 100 to 150 years. After the easements are up, the project would need to be evaluated on the basis of climatic change, state-of-the-art science, and international climate policies. Projects could then be resubmitted for easement extensions.

Finally, the issue of leakage must be addressed. Leakage, or emission displacement, occurs when a reduction of emissions in one area leads to an increase of emissions in another. Basically, it occurs when the scale of the solution is smaller than the scale of the problem. For instance, if the USFS were to implement a DFCOP project without collaboration with private landowners, the amount of deforestation would simply shift from public forests to private forests with the net amount of deforestation likely remaining the same. Therefore, the benefits gained through a DFCOP project could easily be offset through emissions elsewhere.

There is no easy solution to the problem of leakage. If a DFCOP project were implemented in the United States, a company could offset the benefits gained by deforesting outside of the country. But there are options for dealing with leakage. Monitoring, primarily the understanding of existing forest stands, is probably the most effective method of responding to leakage issues. Monitoring of the project area, and of forests within the entire United States, will be critical to determining the amount of leakage occurring domestically. Also, as more people are allowed to participate in a DFCOP scheme, the likelihood of offset emissions from deforesting elsewhere is decreased. The monitoring of participants to be sure they are being rewarded for net emission reductions will play a critical role in reducing the likelihood of leakage.

The United States has seen very little change in the amount of its forest land over the past several decades (Heinz Center 2002). To safeguard against abusive deforestation ahead of any climate policy rewarding afforestation and reforestation efforts, a baseline for the determination of an eligible DFCOP forest must be created. This would involve the setting of an arbitrary date from which afforestation and reforestation activities could be measured. Because forest expanse in the United States has remained relatively stable over the past several decades, we suggest staying consistent with the Kyoto rule for the establishment of Clean Development Mechanism projects.[1]

Another important aspect of any DFCOP policy will be to determine what types of forest stands can be created through DFCOP projects. We recommend that all forest stands be composed largely of native vegetation types that mesh well with those already within the ecosystem. It follows that plantations and monocultures should be excluded from such a scheme. The primary purpose of DFCOP projects is to offset greenhouse gas emissions, but ways to maximize the additional benefits, such as for fish and wildlife habitat, watershed function, and biodiversity protection, also need to be considered.

Internationally, a program similar to DFCOP, known as Reducing Emissions from Deforestation and forest Degradation (REDD), has been flagged as the solution to incorporating forest offsets on a global scale. REDD is built around creating emissions offset credits from projects that avoid deforestation and forest degradation. These credits could be sold on the international carbon market as a way for companies to meet their reduction targets. It is not our intention to discuss REDD in detail here, as many of the logistics required to get such a program off the ground internationally still need to be worked out, but its goals are very similar to those we describe for a domestic forest carbon offset program. The opportunity for entities within the United States to purchase forest offsets internationally should be a part of any domestic climate policy that uses

---

1. For details regarding Clean Development Mechanism rules, see unfccc.int/kyoto _protocol/mechanisms/items/1673.php.

forests to offset emissions. This will help the United States to establish international leadership in climate change negotiations, as well as to reduce emissions from an entirely different sector in deforestation and forest degradation. It should be noted that although no binding agreement was reached at the Conference of Parties (COP) 15 in Copenhagen requiring attendant countries to reduce greenhouse gas emissions, the countries, including the United States, offered to give $30 billion in climate aid to developing nations between 2010 and 2012, and up to $100 billion by 2020. A portion of this money will almost certainly be used to prevent deforestation, thus exemplifying what an important aspect REDD and other forestry-related offset programs will be to any international climate agreement.

Some argue that the establishment of a REDD-ike mitigation program would be counterproductive to the grand scheme of reducing global temperature. Bala et. al (2007) suggest that the albedo, or solar radiation reflected by darker land areas as a result of increased forest area in temperate regions, will actually raise global temperatures through the retention of solar radiation. This is an understandable concern, but these authors believe that when it is taken into consideration with the other potential effects of warming, the benefits of promoting reforestation and afforestation outweigh the negatives. For instance, the melting of glaciers in northern latitudes, which has been well documented, leads to exposed areas of earth, which can decrease the albedo in those regions. The expanse affected by melting permafrost and glaciers is far greater than the land area affected by the addition of forest mitigation projects. In addition, the CO_2 released from the soil by wide-scale melting has the potential to go unchecked in the absence of mitigation measures.

While forests have an important role to play, the bulk of emission reductions must occur in the technology and energy sector if we are to avoid the most serious effects of climate change. What we have presented here is a framework for how the forest sector can be used to augment those emission reductions while safeguarding the biodiversity and human cultures that rely on these magnificent systems. There is much work still to be accomplished, and a number of technical aspects yet to be worked out, including determination of the amount of offsets allowed domestically, the percentage of offsets an entity can use, and how those offset credits would be measured on an open market against credits that come from the technology and energy sector. Ultimately, we hope a policy agreement will be achieved and implemented that results in significant emission reductions and real progress toward reversing the climate trend both nationally and internationally.

Table 18.1. Benefits associated with forest management options to mitigate climate change

Management activity	Climate change mitigation benefits	Wildlife benefits
Promote landscape connectivity	Potential for reforestation/ afforestation. Sequestration opportunities	Allows for species to move freely, promote gene flow and sustain processes such as pollination and dispersal.
Establish long-term seed banks	Reforestation/afforestation benefits. Sequestration opportunities	Provide a genetic library of species in the case of extinction or extirpation. Make available the option to re-establish or establish new populations.
Take early action against exotic and/or invasive species	Promotes resiliency of forests. Allows for maximum carbon sequestration opportunities	Ability to assess the impact potential of these species and develop treatment plans accordingly.
Increase monitoring efforts	Provides baseline data for current and future management schemes	Monitoring will provide necessary baseline data on the current trends of ecosystems from which managers will be able to compare future assessments to evaluate if current management techniques need to be changed.
Manage to restore watershed function	Keeps water within the system. Ecosystems can better deal with stressors associated with drought conditions	Allows for increased water retention and aquifer recharge, reduces run-off and soil erosion; improves riparian and aquatic habitats.
Manage to protect pristine unaltered habitats	Keeps carbon that is currently locked up within forest systems in those systems	These lands minimally or unaltered by humans are refugia for rare and endangered species and sources provide seed sources for expansion of species as conditions change.
Rethink what is done to significantly altered landscapes	Provides the option to evaluate the best possible opportunity to mitigate the impacts of climate change on the landscape	Restoration is critical and must be carried out where possible, however, in some instances rather than trying to re-establish and/or enhance habitats that will no longer support such systems, it may be practical to look at potential systems and assemblages that the land and climate will support while providing necessary ecosystem services
Manage for uncertainty	Allows for flexibility within management plans.	Allows for flexibility within management plans. Promotes the use of the best available science as management can change if need be according to new scientific findings.

Summary and Recommendations

Comprehensive policies must address the needs of both forest and human communities while providing global climate change solutions. Land use sectors, including forests, can play a key role in reducing the effects of climate change, and therefore US and/or international climate policy must include incentives for protecting forests. The USFS should take the lead in developing forest policies and forest management strategies to respond to climate change. Table 18.1 provides several examples of forest management goals and actions that can enhance adaptation to climate change, thus helping human and ecological communities respond and adapt to its mounting effects.

On a rapidly changing planet, with threats to biodiversity coming from many directions, an understanding of the importance of forest ecosystems will be essential. Protection of existing forests and the reestablishment of native forests will be major steps toward dealing with biodiversity loss and providing the means for carbon to be sequestered and ecosystem services to preserved, reestablished, or improved.

LITERATURE CITED

Bala, G., K. Caldeira, M. Wickett, T. J. Phillips, D. B. Lobell, C. Delire, and A. Mirin. 2007. Combined climate and carbon-cycle effects of large-scale deforestation. *Proceedings of the National Academy of Sciences, USA* 104:6550–55.

Dombeck, M. P. 2009. Testimony: Observations on the possible movement of the Forest Service to the Department of the Interior. House Committee on Appropriations: Subcommittee on Interior, Environment and Related Agencies. Accessed October 7, 2010, at www.uwsp.edu/cnr/gem/Dombeck/MDSpeeches/2009/Mike_dombeck_02_24_09.pdf.

Dombeck, M. P., C. A. Wood, and J. E. Williams.2003. *From Conquest to Conservation: Our Public Land Legacy*. Washington: Island Press.

Hansen, J., M. Sato, P. Kharecha, D. Beerling, R. Berner, V. Masson-Delmotte, M. Pagani, M. Raymo, D. L. Royer, and J. C. Zachos. 2008. Target atmospheric CO_2: Where should humanity aim? *Open Atmospheric Science Journal* 2:217–31.

Heinz Center. 2008. The state of the nation's ecosystems 2008: Focus on wildlife. H. John Heinz Center for Science, Economics and the Environment, Washington. Accessed at www.heinzctr.org/ecosystems/2008report/pdf_files/Wildlife_Fact_Sheet.pdf.

Houghton, R. A. 2005. Tropical deforestation as a source of greenhouse gas emissions. In *Tropical Deforestation and Climate Change*, edited by P. Moutinho and S. Schwartzman, 13–21. Amazon Institute for Environmental Research, Instituto de Pesquisa Ambiental da Amazônia, and Environmental Defense. Washington.

IPCC. 2007. Climate change 2007: Synthesis report. Contribution of working groups I, II, and III to the fourth assessment report of the Intergovernmental Panel on Climate Change. Edited by R. K. Pachauri and A. Reisinger. Geneva: IPCC.

Joyce, L. A, G. M. Blatte, J. S. Littell, S. G. McNulty, C. I. Millar, S. C. Moser,

R. N. Neilson, K. A. O'Halloran, and D. L. Peterson. 2008. National forests. In Adaptation options for climate-sensitive ecosystems and resources. US Climate Change Science Program. Accessed October 10, 2010, at www.climatescience.gov/Library/sap/sap4-4/final-report-Ch3_forests.htm.

Joyce, L., J. Aber, S. McNulty, V. Dale, A. Hansen, L. Irland, R. Neilson, and K. Skog. 2001. Potential variability and change for the forests of the United States. In *Climate Change Impacts on the United States: The Potential Consequences of Climate Variability and Change. National Assessment Synthesis Team Report for the US Global Change Research Program*, 489–522. Cambridge: Cambridge University Press.

Kelly, C., S. Woodhouse-Murdock, J. McKnight, and R. Skeele. 2008. Using forests and farms to combat climate change: How emerging policies in the United States promote land conservation and restoration. In *Climate Change and Forests: Emerging Policy and Market Opportunities*, edited by C. Streck, R. O'Sullivan, T. Janson-Smith, and R. Tarasofsky, 275–88. Washington: Brookings Institution Press.

Millar, C. I., N. L. Stephenson, and S. L. Stephens. 2007. Climate change and forests of the future: Managing in the face of uncertainty. *Ecological Applications* 17:2145–51.

Noss, R. F. 2001. Beyond Kyoto: Forest management in a time of rapid climate change. *Conservation Biology* 15:578–90.

Spittlehouse, D. L., and R. B. Stewart. 2003. Adaptation to climate change in forest management. *BC Journal of Ecosystems and Management* 4:2–11.

Streck, C., R. O'Sullivan, T. Janson-Smith, and R. Tarasofsky. 2008. Climate change and forestry: An introduction. In *Climate Change and Forests: Emerging Policy and Market Opportunities*, edited by C. Streck, R. O'Sullivan, T. Janson-Smith, and R. Tarasofsky, 3–10. Washington: Brookings Institution Press.

UNFCCC. 2010. United Nations framework convention on climate change: The Kyoto protocol. Accessed at unfccc.int/kyoto_protocol/items/2830.php.

USFS. 2001. US forest facts and historical trends. US Department of Agriculture Forest Service. Accessed October 10, 2010, at fia.fs.fed.us/library/briefings-summaries-overviews/docs/ForestFactsMetric.pdf.

USFS. 2007. The U.S. Forest Service: An overview. US Department of Agriculture Forest Service. Accessed October 10, 2010, at *www.fs.fed.us/documents/USFS_An_Overview_0106MJS.pdf*.

US Forest Service. 2007. Forest land ownership in the conterminous United States. Edited by M. D. Nelson and G. C. Liknes. Accessed October 10, 2010, at nrs.fs.fed.us/pubs/maps/map497_pg76.pdf.

Yale Environment 360. 2009. Pachauri supports goal limiting atmospheric CO2 to 350 PPM. Accessed December 1, 2010, at e360.yale.edu/content/digest.msp?id=2024.

World Bank. 2010. Protecting global forest values. Accessed October 10, 2010, at go.worldbank.org/42GAQVIU30.

Concluding Remarks

19

What to Expect and How to Plan for Wildlife Conservation in the Face of Climate Change

Daniel F. Doak, Jedediah F. Brodie, and Eric Post

The effects of climate change on even a single species will be complicated, and hence difficult to predict and plan for. At times the direct effects will be the strongest on a single species (Galliard et al. this volume), while in other cases the indirect alteration of ecological interaction will have the largest effects on population health and persistence (Paull and Johnson this volume). Not surprisingly, predicting effects on entire communities of species will present even greater challenges. Nonetheless, we are optimistic that over the next decade, we will be better able to anticipate how climate change will impact ecological systems, and also to determine how best to promote resilience of the most endangered populations in the face of these threats.

As many chapters in this volume emphasize, useful planning for conservation that includes responses to climate change effects must move beyond the overly simplistic understanding that has largely characterized the field in the past. This is especially important in efforts to promote resilience or resistance to the combined effects of climate change and other direct and indirect human impacts through management (Boyce et al. this volume, Cross et al. this volume, Owen-Smith and Ogutu this volume), and also in attempts to understand how and when natural processes will confer such resilience (Doak and Morris 2010, Austin et al. this volume).

Efforts in these directions are obviously increasing, but there are still daunting challenges: the history of correctly anticipating and then ameliorating complex impacts even on single populations is not encouraging. Nonetheless, the specter of climate change is focusing far more attention on the need to make such predictions and to build workable management plans that best use scientific input to hedge against uncertain outcomes. While the chapters collected here clearly do not cover all such approaches or activities, they do span an impressive range of threats, concerns, and future directions of conservation science for a world that will increasingly be altered by climate change.

To end this volume, we concentrate on four general lessons in climate change ecology that we feel are particularly important in setting the stage for future research and management, and which bridge multiple aspects of the work presented here. The themes of promoting resilience and of looking more deeply into the ecological details of climate change are tied together the work covered above, and in the four ideas that we now review.

Nonlinear Effects of Gradual Climate Changes

The ultimate concern about climate effects is that they may cause local, regional, or global extirpation of species and, hence, dramatic changes in communities, potentially triggering yet more extinction. However, what we can actually measure for any given population is generally only one or a handful of response variables, which we can usually link only roughly to climate effects. As a result, we cannot adequately predict the shape of population responses to gradual shifts in demographic rates and/or climate, particularly because we lack the data to quantitatively estimate nonlinearities in populations' dynamical responses. This limitation makes it critical for us to remember, and to try to educate the public about, the inherent nonlinearity of population dynamics to gradual shifts in demographic rates and/or climate. This nonlinearity manifests itself at multiple levels and will tend to make population stability and viability quite *insensitive* to climate change over large ranges of values, though highly sensitive past certain threshold points.

For example, if the only impact of warmer temperatures and increased aridity on a relatively long-lived species is reduced first-year survival (to which population growth will have low sensitivity), we expect little change in population growth with low to moderate climate change, and then increasing effects with further change (figure 19.1A). However, this quite straightforward nonlinearity hides two other complications of predicting climate effects: even a gradual change in growth *rate* will result in a step function in population viability, with annual maximal growth rates greater than 1 leading to a stable or growing population, and values of less than 1 leading to rapid declines (figure 19.1B). This highly nonlinear effect on the metric we care about—population extirpation—can easily be lost in the analysis of more proximate effects of climate shift (Harley and Paine 2009). An added complication is that the increasing stochasticity in the weather will result in further decreases in population growth rates and increase extinction risk, which is even more sensitive to variability in the environment than is average stochastic growth rate (Boyce et al 2006, Tews et al. this volume, Ray et al. this volume).

This viewpoint is especially important to bear in mind because of the

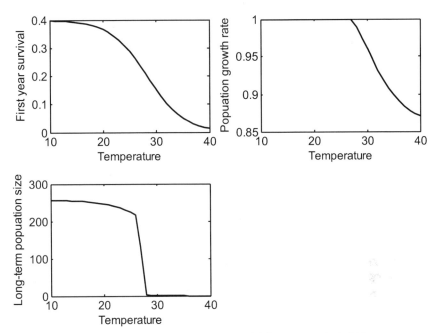

Figure 19.1. Effects of changing temperatures on (a) first-year survival; (b) long-term population growth rate; (c) population size after 200 years. Results are from a modification of a density-dependent model for stage-structure population dynamics of the Iberian lynx. See Morris and Doak 2002, chapter 8, for details.

limited data on future climate change and its relationship with individual vital rates and population growth rates. Climate models generally can predict trends only in large-scale climate measures, and we rarely have the data to link the effects of these changes to more than one or two vital rates for a population (although improvements in climate downscaling are rapidly changing this situation). These limitations tend to result in predictions of gradual change in population processes, rather than of the more likely step-function responses in population viability that will result from them. However, our current inability to make these fully integrated predictions should not prevent us from recognizing that they are what we should expect to see on the ground (Doak and Morris 2010).

Nonadditivity of Climate Change Effects

Like any potential threat to population viability, climate change is often written about as a separate and distinct impact on population dynamics and extinction risk. However, even more than for many other threats to populaiotn health, climate impacts are likely to have strong nonadditive effects on population growth rates and extinction risk. This nonadditivity is part and parcel of climate change impacts, both because of the non-linear responses of population viability just discussed, and also because of the basic math of population dynamics. For example, if both climate effects and harvest influence the juvenile survival of a species, the survival of any one individual for a year is the product (not the sum) of the probabilities of surviving both threats. This interactive effect on survival means that we will see strong interactions of these two mortality sources on population growth. As discussed in multiple chapters in this volume (e.g. Boyce et al. this volume, Cross et al. this volume, Fordham et al. this volume, Manne this volume, Owen-Smith and Ogutu this volume, Paull and Johnson this volume), these interactive effects mean that efforts to alleviate threats from climate change may in many cases be most effectively carried out by ameliorating other more manageable and more localized threats to the population (see also Linares and Doak 2010).

On the one hand, these interactions mean that analysis of climate effects cannot easily be discussed and dissected outside of the mire of other impacts that threaten many wildlife species. However, this recognition also means that planning for resilience in the face of climate change is also basically just doing conservation, and that is good. As we have noted in our introductory chapter, there is real danger in focusing too much on threats from climate change and in ignoring the many other clear and immediate problems that face wildlife species. To the extent that it helps us leverage more understanding and support for broad-scale conservation initiatives, we can use concern over climate change to improve overall conservation in the face of multiple threats.

Strong but Indirect Climate Effects

One of the most hopeful facets of the challenge that climate change poses for terrestrial vertebrate conservation is that for many species, climate itself is unlikely to be a major factor in driving distributional changes. Rather, the manner in which climate influences key food or habitat-forming species on which terrestrial vertebrates rely will drive these species' response. For example, there is ongoing concern that the grizzly bear (*Ursus arctos*) population in the Yellowstone ecosystem is unstable and could decline precipitously due to climate change (Mattson and Reid 1991, Koteen 2002). Before European settlement of North America, however, grizzlies lived far south of Yellowstone (Mattson and Merrill 2002), as well as far north; they have also occurred in both wetter and drier conditions, and it is very unlikely that under any future climate change scenario but the most catastrophic the Yellowstone region would be unsuitable for bears due to climate per se.

However, this absence of a direct climate effect on grizzlies should not dictate a lack of concern. A critical food source for bears in Yellowstone is whitebark pine (*Pinus albicaulis*), which is in widespread and rapid decline due to an introduced fungal disease and climate-linked bark beetle outbreaks (Koteen 2002, McKinney et al. 2009). Loss of this critical resource has direct nutritional costs for bears and also changes their habitat use, making them more susceptible to direct human-caused mortality (Mattson et al. 1992, Pease and Mattson 1999). Thus, climate effects are likely to be severe for this grizzly population, but through a very indirect set of effects and interactions.

What makes us hopeful about the prediction of climate effects in this system is that these species interactions (pines and bark beetles, grizzlies and rich food resources, humans and grizzlies) are ones for which we have considerably more and better information than we typically do about the direct behavioral and physiological adaptations of a wild species to climate itself. On the other hand, the multiple ways in which climate change can alter species interactions may result in complex sets of impact on wildlife species that are themselves quite difficult to fully anticipate. These indirect, ecologically mediated climate effects are also likely to generate time lags in response that in some cases will work for threatened species and in other cases against them. In the case of the Yellowstone bears, it is quite conceivable that with enough warming, alternative food plants could move to the region and provide new food sources. But the time lag between the movement of the species and the growth to maturity of any likely replacement food source (e.g., Gambel's oak, whose acorns are an important food source for vertebrates, which occurs just south of the Yellowstone ecosystem and is likely to migrate north with climate change, or Bur oak, which also may migrate into the Yellowstone region) and the rapid die-off of whitebark pines, means that bears could face decades without either food source.

Species Interactions as Climate Magnifiers or Buffers

The inherent sensitivity of population viability to climate changes will in many cases be modified by ecological interactions leading to the ecological mediation of climate impact. In particular, some species interactions are highly sensitive to incremental changes in temperature or precipitation, thus magnifying the ecological effects of very small climate changes. Among the best understood of such effects are interactions between conifers and bark beetles, with massive outbreaks of beetles over much of

North America almost certainly due to modest amounts of warming, tipping the balance in favor of the beetles over the trees (Raffa et al. 2008, Matsuoka et al. 2006).

For ecologists, it is easiest to think of "ecological magnifiers" of climate change, but physical processes will sometimes work similarly to, or in concert with, ecological effects. Sea ice retreat means that polar bears cannot access their prey at critical times of year, thus making them acutely sensitive to small changes in ice patterns (Regehr et al 2010). Mass balance of glaciers is highly sensitive to small changes in both precipitation (including the time of year it falls) and temperature, reacting with rapid advances or retreats. In turn, these shifts can dramatically change hydrological regimes in entire watersheds and whole mountain ranges, as has been well publicized (Arendt et al 2002, Woodword et al 2010). Similarly, small changes in climate may dramatically alter fire regimes, with large results for entire ecological communities (Hemp 2005, Westerling et al. 2006).

In most cases we will probably not correctly anticipate such tipping points or "ecological magnifiers" of climate change; but once they begin, we may be able to predict their spread and eventual extent well. The general lesson we draw from these effects is, again, that we should expect punctuated responses to climate change, not gradual ecological changes in response to gradual changes in climate.

In contrast to these magnifying effects, there is some hope that species interactions can buffer ecosystems against climate change, thus leading to greater ecological resistance to the direct or indirect effects of climate shift. As Wilmers et al. (this volume) discuss, there are some data to suggest that keystone predators may have this buffering role in some communities. Post and Pederson (2008) also show buffering effects of native herbivores on plant community responses to warming. Similarly, studies of artificially assembled communities have resulted in many claims that species diversity will yield higher productivity, as well as higher stability, in community and ecosystem functioning (e.g. Tilman 1996, Ives and Carpenter 2007). While these are intriguing ideas, most studies of these diversity effects have failed to demonstrate substantial effects of diversity itself (Cardinale et al. 2006). Indeed, in some cases strongly interacting species are likely to magnify and propagate anthropogenic perturbations rather than buffer community-wide effects against them (e.g., Springer et al. 2003, Matthews et al. this volume), or to have little effect either way (Harmon et al. 2009). The question of where and how to manage communities so that they are more resistant and resilient in the face of climate perturbation is still very much in its infancy, but could provide important insights into future wildlife conservation strategies.

Conclusions

In sum, ecologists and conservation biologists will increasingly have to merge approaches to management that don't simply emphasize climate shifts versus other threats and management tools, but consider those problems together. Perhaps it will be most difficult to predict when and why certain species or communities will be stable in the face of ongoing changes, and when or where they will respond rapidly. The recent exercise by Loarie et al. (2010) in defining how fast local climates will move across the landscape is, in our view, an excellent first step in understanding the underlying patterns of climate change to which ecological systems must respond. However, we agree with these authors that the ways in which populations and species will move and respond will not be like a simple mapping of their current ranges onto shifting climate envelopes. Instead, we anticipate that saltatory shifts in ranges will result, and that management efforts will have to be highly targeted to the places and times at which thresholds are reached for multiple species or for strongly interacting species whose loss will magnify other responses.

As we and other authors in this volume have emphasized, identifying these thresholds and tipping points is the other great challenge of climate change conservation. It will involve digging deeper into the ecological interactions that define the problems and constrain or enable alternate management solutions. We will be forced to do conservation in a shifting world under conditions different from those to which we have been accustomed in the past. One benefit that we already see is the drawing in of more and more academic ecologists. The boundaries between management-oriented scientists and those interested in basic problems in ecology have eroded over the last two decades, but the seriousness and difficulty of climate change impact will, we hope, further draw together ecologists with very different tools and perspectives to jointly tackling these problems. Doing so will be not only intellectually fruitful but clearly necessary to avoid grievous mistakes and best use our time and energy to save biodiversity in the face of mounting threats, both globally and locally.

LITERATURE CITED

Arendt, A. A., K. A. Echelmeyer, W. D. Harrison, C. S. Lingle, and V. B. Valentine. 2002. Rapid wastage of Alaska glaciers and their contribution to rising sea level. *Science* 297:382–86.

Boyce, M. S., C. V. Haridas, C. T. Lee, C. L. Boggs, E. M. Bruna, T. Coulson, D. F. Doak, J. M. Drake, J. Gaillard, C. C. Horvitz, S. Kalisz, B. E. Kendall, T. Knight, E. S. Menges, W. F. Morris, C. A. Pfister, and S. D. Tuljapurkar. 2006. Demography in an increasingly variable world. *Trends in Ecology and Evolution* 21:141–48.

Cardinale, B. J., D. S. Srivastava, J. E. Duffy, J. P. Wright, A. L. Downing, M. Sankaran, and C. Jouseau. 2006. Effects of biodiversity on the functioning of trophic groups and ecosystems. *Nature* 433:989–92.

Doak, D. F., and W. F. Morris. 2010. Demographic compensation and tipping points in climate-induced range shifts. *Nature* 467:959–62.

Harley, C. D. G., and R. T. Paine. 2009. Contingencies and compounded rare perturbations dictate sudden distributional shifts during periods of gradual climate change. *Proceedings of the National Academy of Sciences* 106:11172–76.

Harmon, J. P., N. A. Moran, and A. R. Ives. 2009. Species response to environmental change: Impacts of food web interactions and evolution. *Science* 323:1347–50.

Hemp, A. 2005. Climate change-driven forest fires marginalize the impact of ice cap wasting on Kilimanjaro. *Global Change Biology* 11:1013–23.

Ives, A. R., and Carpenter, S. R. 2007. Stability and Diversity of Ecosystems. *Science* 317:58–62.

Koteen, L. 2002. Climate change, whitebark pine, and grizzly bears in the Greater Yellowstone Ecosystem. In *Wildlife Responses to Climate Change*, edited by S. H. Schneider and T. L. Root, 343–411. Washington: Island Press.

Linares, C., and D. F. Doak. 2010. Modeling the interacting effects of multiple, disparate disturbances on the persistence of a threatened gorgonian coral. *Marine Ecology Progress Series* 402:59–68.

Loarie, S. R., P. B. Duffy, H. Hamilton, G. P. Asner, C. B. Field, and D. D. Ackerly. 2009. The velocity of climate change. *Nature* 462:1052–1111.

Matsuoka, S. M., E. H. Holsten, M. E. Shephard, R. A. Werner, and R. E. Burnside. 2006. Spruce beetles and forest ecosystems of south-central Alaska: Preface. *Forest Ecology and Management* 227:193–94.

Mattson, D. J., B. M. Blanchard, and R. R. Knight. 1992. Yellowstone grizzly bear mortality, human habituation, and whitebark pine seed crops. *Journal of Wildlife Management* 56:432–42.

Mattson D. J., and T. Merrill. 2002. Extirpations of grizzly bears in the contiguous United States, 1850–2000. *Conservation Biology* 16:1123–36.

Mattson, D. J., and M. M. Reid. 1991. Conservation of the Yellowstone grizzly bear. *Conservation Biology* 5:364–72.

McKinney, S. T., C. E. Fiedler, and D. Tomback. 2009. Invasive pathogen threatens bird-pine mutualism: Implications for sustaining a high-elevation ecosystem. *Ecological Applications* 19:597–607.

Moritz, C., J. L. Patton, C. J. Conroy, J. L. Parra, G. C. White, and S. R. Beissinger. 2008. Impact of a century of climate change on small-mammal communities in Yosemite National Park, USA. *Science* 322:261–64.

Morris, W. F., and D. F. Doak. 2002. *Quantitative Conservation Biology: The Theory and Practice of Population Viability Analysis*. Sunderland, MA: Sinauer Associates.

Pease, C. M., and D. J. Mattson. 1999. Demography of the Yellowstone grizzly bears. *Ecology* 80:957–75.

Post, E., and C. Pedersen. 2008. Opposing plant community responses to warming with and without herbivores. *Proceedings of the National Academy of Sciences, USA* 105:12353–58.

Raffa, K. F., B. H. Aukema, B. J. Bentz, A. L. Carroll, J. A. Hicke, M. J. G. Turner, and W. H. Romme. 2008. Cross-scale drivers of natural disturbances prone to anthropogenic amplification: the dynamics of bark beetle eruptions. *Bioscience* 58:501–17.

Springer, A. M., J. A. Estes, G. B. van Vliet, T. M. Williams, D. F. Doak, E. M. Danner, K. A. Forney, and B. Pfister. 2003. Sequential megafaunal collapse in the North Pacific Ocean: An ongoing legacy of industrial whaling? *Proceedings of the National Academy of Sciences, USA* 100:12223–28.

Tilman, D. 1996. Biodiversity: population versus ecosystem stability. *Ecology* 77:350–63.

Tingley, M. W., W. B. Monahan, S. R. Beissinger, and C. Moritz. 2009. Birds track their Grinnellian niche through a century of climate change. *Proceedings of the National Academy of Sciences, USA* 106:19637–43.

Westerling, A. L., H. G. Hidalgo, D. R. Cayan, and T. W. Swetnam. 2006. Warming and earlier spring increases western U.S. forest wildfire activity. *Science* 313:940–43.

Woodward, G., D. M. Perkins, L. E. Brown. 2010. Climate change and freshwater ecosystems: Impacts across multiple levels of organization. *Philosophical Transactions of the Royal Society B* 365:2093–2106.

Xu, J. C, R. E. Grumbine, A. Shrestha , M. Eriksson, X. F. Yang, Y. Wang, and A. Wilkes. 2009. The melting Himalayas: Cascading effects of climate change on water, biodiversity, and livelihoods. *Conservation Biology* 23:520–30.

Index

AC (assisted colonization), 9, 51, 238, 347–64; candidate species, 352, 356, 360, 362–63

adaptation: local, 41–42, 50, 115, 358–60; policy, 374

afforestation, 378, 380–81

Africa: eastern, 155, 157, 310; southern, 21, 155, 166, 168, 172–73, 177–78

African buffalo (*Syncerus caffer*), 157, 162, 164–65

amphibians, 58–61, 63, 65–66, 78, 110, 114, 226, 231; demographic patterns, 62

aquatic ecosystems, 23, 58–59, 79

aquatic species, 58, 62, 350, 366

Arctic, Canadian High, 214, 276–77

Arctic, coastal plain, 207

Arctic, western Alaska, 206–7, 215

arctic fox (*Alopex lagopus*), 207–8, 215–16

Arctic National Wildlife Refuge, 206, 218, 220

Arctic Ocean, 213–14, 219

asp viper (*Vipera aspis*), 187–89, 191–92, 194

assisted migration. *See* AC (assisted colonization)

Athi-Kaputiei Plains, 159–61, 168

Australia, 86–103, 232–33, 274

Bahamas, 237, 239

Bathurst Island complex, 277, 284

bioclimatic models. *See* climatic niche models

biodiversity: conservation of, 49, 102, 166–67; protection, 4–5, 380

biophysical models, 91, 119

birds, 140, 143, 193, 196, 208, 213, 217, 227, 229, 232, 234, 274, 308, 351, 357–61

BLM (US Bureau of Land Management), 206

Canada, 20, 216, 298–99, 370–71

caribou (*Rangifer tarandus*), 3, 219, 271–86, 293, 298, 300; Svalbard reindeer, 275, 300–301

carnivores, 9, 169, 188, 330–42, 360

carrion, 337–39

climate: drier, 184, 250; extreme weather events, 21, 52, 61, 63–65, 120, 219, 271–86, 293; heat waves, 20, 52, 271, 273–74; legislation (*see* climate policy); local, 18; optimal, 315–16; tolerances, 110, 370; variability, 120, 153–54, 166, 186, 271, 274, 300

climate models, 18, 20, 29, 97, 111, 251, 300, 335, 389

climate policy, 369, 371, 376–77, 380; international, 377, 379, 383; US, 372, 374, 377–78, 380; US Kerry-Boxer bill, 376–77; US Lieberman-Warner bill, 376; US Waxman-Markey bill, 376–77

climatic niche models, 2, 41, 48, 90–91, 97, 103, 116, 119, 130, 186, 195–96, 260

clutch size, 69–70, 237

coastal marsh species, 311

cold-adapted species, 196, 255–56, 259

connectivity, 307–24; areas, 310–13, 318–19; conservation, 50–51, 318, 321; conservation efforts, 309–10,

herbivores, 153, 159, 166, 330–32, 335, 340; large-bodied, 153, 160, 165, 171, 333, 335, 338–39; populations, 157, 159–60, 164, 170–71, 333

host populations, 51, 122, 350

human population, global, 17

human society, 1, 369–70

hydrology, 311, 314

hydropatterns, 59–63, 65, 78–80

hydroperiod, 59–60, 68, 71–73, 77–79

index, climate change vulnerability, 129–46

Indian Ocean, 155

indicator species, 263

insular populations, 227, 231, 236

interacting species. *See* species interactions

invasions, 6, 45, 86–103, 179–80, 192, 354

invasive species, 86–103, 180, 182–83, 194, 226, 309, 350, 352, 354, 363, 373; abundance of, 61, 103; cane toads, 45, 87, 89, 91–92; invasiveness, 89, 348–50, 354–55; plants, 354; rabbits, 86–103

IPCC (Intergovernmental Panel on Climate Change), 16–17, 21, 25, 38, 66, 316

islands, 215, 226–39, 334; oceanic, 226–27

IUCN (International Union for the Conservation of Nature), 237

IUCN Red List, 132, 242, 342

kelp forests, 331

Kenya, 155, 159–61, 168

keystone species, 89, 332–33, 349

Kruger National Park, 155–59, 161, 163–64, 166–68, 170

legislation. *See* climate policy, 360, 376–77

life-history: stages, 65–67, 335; traits, 60–61, 131, 189–91, 209, 363

lions (*Panthera leo*), 157, 164, 166, 168–69

lizards, 179–97; common (*Zootoca vivipara*), 187–89, 191–96

malaria: distribution, 110; risk, 110

mammals, 8, 46, 67, 140, 143, 146, 171, 179, 188, 193–94, 196, 227–28, 230, 234–35, 237, 247, 301, 307, 330, 351, 357, 359

management, 28, 52, 64, 74, 78, 145, 167, 169, 196, 235, 261, 341–42, 373–75, 387–88, 393; actions, 3, 66–67, 74, 77–78, 146, 261; plans, 31, 118, 122, 167, 382, 387; scenarios, 71–72, 74, 76

Masai Mara Reserve, 159, 161–62, 164, 168

matrix models, 6, 66–69, 77, 79, 191; simulations, 71–74, 76–77, 95, 140, 279–82

meadow viper (*Vipera ursinii*), 187–89, 195

mechanistic models, 91, 116, 119, 260

Mediterranean: basin, 21, 184–85, 195–96; species, 185–86

meta-analysis, 48, 116, 307

metapopulation dynamics, 87, 92, 96, 98, 103, 249, 309, 356; modeling, 98, 100; processes, 94, 96

migration, 1, 51, 145, 160–62, 194, 213, 217–18, 220, 236, 312, 315, 330, 347, 360; birds, 208–9, 219; wildebeest (*Connochaetes taurinus*), 157

mitigation, 4, 245

models: bioclimatic (*see* climatic niche models); climate-envelope (*see* climatic niche models); spatial, 92–93, 103, 320

moose (*Alces alces*), 334

mutations, 39, 41–42

Nairobi National Park, 160–63, 168

Namibia, 161, 163, 168

natural selection. *See* selection, natural

nesting, 197, 208–9, 212–13, 215–16, 218, 230, 232, 237–38; birds, 207–8, 215–17, 219

nest predators, 207–8, 216

Nevada, 140–46

niche models. *See* climatic niche models

SOI (southern oscillation index), 233, 235–36
soils, 28–29, 97, 219, 229, 370, 381
source populations, 50, 72, 358
South America, 110, 112, 370
specialists, 46, 341
species assemblages, 89, 316, 321, 370
species distributions, 48–49, 90–91, 93, 95, 307, 314, 320; historical reconstructions of, 180
species diversity, 121, 308–9, 336, 392
species interactions, 3–4, 39, 86, 90, 129, 314–15, 342, 348, 391–93
species invasions. *See* invasive species
species vulnerability, 129, 218
summer, 23, 28–29, 68–69, 153–54, 255, 312, 340

TEK (traditional ecological knowledge), 279
temperature: air, 23, 59–60; ambient, 255–56; annual, 260; body, 91, 179, 184, 263; daily, 155, 258; global, 3, 18–19, 38, 307, 381; increased, 20, 27–29, 77, 114, 145, 187–88, 196, 233, 388; projected, 19; sea surface, 155, 232–33, 273; subsurface, 254, 256; summer, 62, 246, 254–56; water, 28, 60, 62, 88, 340; winter, 20, 77
Teshekpuk Lake, 213, 219
TMIs (trait-mediated interactions), 342
translocation, 48, 51, 72, 167, 229, 262, 322, 348, 350–53, 355–57, 359–61

tropics, 59, 62–63, 212, 307, 371
tuataras, 229, 236–37
tundra ecosystems, 214–15, 222

UNFCCC (United Nations Framework Convention on Climate Change), 5, 375–76
ungulates, 7, 154, 157, 160, 162–64, 169, 337
United States of America, 9, 21, 129, 206, 299, 370–72, 374–81
US Endangered Species Act, 298, 361
USFS (US Forest Service), 371–74, 378–79, 383

vector species, 110–12, 117–18

wetland ecosystems, 27–28, 66, 145, 189, 207, 211, 214–15, 219–20
wildebeest (*Connochaetes taurinus*), 157–58, 160–66, 170
winter, 22, 63, 65, 153–54, 184, 187, 189, 192, 246, 275, 277, 334–35, 337; extreme, 274, 279, 284, 301; severity, 334, 337
wolves (*Canis lupus*), 298, 324, 330–42

Yellowstone National Park, 4, 76, 334, 337, 339, 341, 351, 390–91

zebras, 157–58, 160–62, 164, 166